AERIAL ROBOTICS IN AGRICULTURE

Parafoils, Blimps, Aerostats, and Kites

AERIAL ROBOTICS IN AGRICULTURE

Parafoils, Blimps, Aerostats, and Kites

By

K. R. Krishna

First edition published 2021

Apple Academic Press Inc.
1265 Goldenrod Circle, NE,
Palm Bay, FL 32905 USA

4164 Lakeshore Road, Burlington,
ON, L7L 1A4 Canada

CRC Press
6000 Broken Sound Parkway NW,
Suite 300, Boca Raton, FL 33487-2742 USA

4 Park Square, Milton Park,
Abingdon, Oxon OX14 4RN

First issued in paperback 2023

Library and Archives Canada Cataloguing in Publication

Title: Aerial robotics in agriculture : parafoils, blimps, aerostats, and kites / by K.R. Krishna.

Names: Krishna, K. R. (Kowligi R.), author.

Description: First edition. | Includes bibliographical references and index.

Identifiers: Canadiana (print) 20200371037 | Canadiana (ebook) 20200371169 | ISBN 9781771889261 (hardcover) | ISBN 9781003054863 (PDF)

Subjects: LCSH: Agricultural innovations. | LCSH: Drone aircraft. | LCSH: Robotics. | LCSH: Aerial photography in agriculture. | LCSH: Agriculture—Remote sensing. | LCSH: Crops.

Classification: LCC S494.5.I5 K75 2021 | DDC 338.1/6—dc23

Library of Congress Cataloging-in-Publication Data

....

CIP data on file with US Library of Congress

....

ISBN: 978-1-77188-926-1 (hbk)
ISBN: 978-1-77463-762-3 (pbk)
ISBN: 978-1-00305-486-3 (ebk)

DOI: 10.1201/9781003054863

About the Author

K. R. Krishna, PhD
Formerly Visiting Professor and Research Scholar at the Soil and Water Science Department at the University of Florida, Gainesville, USA; Independent Researcher and Author

K. R. Krishna, PhD, is an agricultural scientist. He has authored several books on international agriculture, encompassing topics in agroecosystems, field crops, soil fertility, crop management, precision farming, and soil microbiology. His more recent books deal with agricultural robotics, drones, and satellite guidance to improve soil fertility and crop productivity. He is a member of the International Society of Precision Farming, American Society of Agronomy, Soil Science Society of America, Ecological Society of America, Indian Society of Agronomy, and Soil Science Society of India.

Contents

Abbreviations

AAPS	automated pesticide sprayer
APADS	advanced precision airborne delivery system
AURORA	autonomous unmanned monitoring robotic airship
BAT	buoyant airborne turbine
Cd	drag coefficient
CEP	circular error probable
CIMMYT	International Maize and Wheat Improvement Center
CPU	central processing unit
CWSI	crop water stress index
DEMs	digital elevation models
DLIS	desert locust information services
DSM	digital surface model
EO	electro-optical
FAA	Federal Aviation Agency
FAO	Food and Agricultural Organization
GCF	ground cover fraction
GCS	ground control station
GHG	greenhouse gases
GPS	global positioning system
HABs	high altitude balloons
HADR	Humanitarian Assistance and Disaster Response
HAHO	high altitude high opening
HALO	high altitude low opening
ICE	image composite editor
IR	infra-red
ISS	International Space Station
JLENS	Joint Land Attack Cruise Missile Defense Elevated Netted Sensor System
JPADS	Joint Precision Airdrop System
KAP	kite aerial photography
LAI	leaf area index
LALO	low altitude low opening
LARS	low altitude remote sensing

LEMV	long endurance multi-intelligence
Lo-Po	low porosity
LTA	lighter-than-air
LTA	lighter-than-atmosphere
LVADS	low velocity aerial delivery systems
NASA	National Aeronautics and Space Agency
NDVI	normalized difference vegetation index
NH_3	ammonia
PA	precision agriculture
PADS	precision airdrop system
PAH	poly-aromatic hydrocarbons
PCDD	polychlorinated dibenzodioxins
PCDF	polychlorinated dibenzofurans
PM	particulate matter
PPC	powered parafoils
PTDS	persistent-threat-detection-system
SfM	structure from motion
SRDW	semi-rigid deployable wing
STMPAS	small tactical multi-purpose aerostat system
SWAMI	shortwave aerostat-mounted imagers
TARS	tethered aerial radar system
TAS	tethered aerospace system
UAV	unmanned aerial vehicle
UCM	universal camera mount
UHAB	ultra high-altitude balloon
VTOL	vertical take-off and landing
ZIPs	zip line cargo drones

Acknowledgments

During the preparation of this book, *Aerial Robotics in Agriculture: Parafoils, Blimps, Aerostats and Kite*s, I was offered reports, unpublished material, and research papers by several researchers from different institutions worldwide. Many of them were encouraging and offered permissions to reproduce the photographs and figures in this book. Several companies dealing with aerial robotics, parafoils, aerostats, and blimps have generously offered permissions to use photographs of their products. I wish to thank them profusely. The following are the administrators, scientists, professors, and their institutional affiliations.

Chapter 1

- Alvei Wall, Founder President, Infinity Power Parachutes Inc., East Lafayette Road, Sturgis, Michigan, USA.

Chapter 2

- Alvei Wall, Founder President, Infinity Power Parachutes Inc., East Lafayette Rd., Sturgis, Michigan, USA;
- Chief Technical Officer, Atair Aerospace Inc., 499, Van Brunt Street Suite 3B, Brooklyn, New York NY 11231, USA;
- Gene Engelgau, Chief Executive Officer, Fruitychutes Inc., Monte Sereno, California, USA;
- John Edling, Chief Executive Officer, Tetracam Inc., Chatsworth, California, USA;
- Dr. Kamron Blevins, President, Northwing Inc., Gala venue, Chelan, Washington State, USA;
- Maurice Ortiz, Aeromao Inc., Mississauga, Canada;
- National Aeronautics and Space Agency (NASA), Washington D.C., USA;
- Till Gottbrath, Marketing and Public Relations, Nova Performance Paragliders, Terfens, Austria;
- Director, SkyRunner Inc., 2929 Baird Rd., Building #50 Shreveport, Louisiana, LA 71118 USA;

- President, Buckeye Industries Inc., Mount Zion Rd., Cassopolis, Michigan, USA.

Chapter 3

- Dr. Alexander Mijatovic – Structure and Aerodynamics, AeroDrum Ltd, Vojislava Ilica 99a, 11000 Belgrade, Serbia;
- Dr. Charlie Steffan, Managing Director, SkyDoc Inc., Saskatoon, Saskatchewan, Canada;
- Dr. Clyde R Beaver, Creative Service Manager, International Centre for Maize and Wheat (CIMMYT), El Baton, Mexico;
- Daniel Geery, Director, Hyperblimp Inc., Salt Lake City, Utah, USA;
- Edward Markey, Vice President, Corporate Communications, Good Year Rubber and Tire Company, Akron, Ohio, USA;
- Cynthia Hess, Good Year Tire and Rubber Company Inc., 200, Innovation Way, Akron, Ohio, USA;
- Dr. Daniel Nelson, Director, Egan Airships, Seattle, Washington State, USA;
- Erica Tierney, Media Usage, Lockheed Martin Inc., Maryland, USA;
- Irene Guerrero, Aeroscraft Inc., Montebello, California, USA;
- United States Department of Homeland Security, Yuma, Arizona, USA (https://www.cbp.gov/frontline/frontline-november-aerostats);
- Dr. Kevin Hess, Chief Technical Officer and Founder, Drone Aviation Corporation, Central Parkway, Jacksonville, Florida, USA.

Chapter 4

- Dr. Alexander Mijatovic – Structure and Aerodynamics, Aero Drum Ltd., Vojislava Ilica 99a, 11000 Belgrade, Serbia;
- Dr. Charlie Steffan, Managing Director, SkyDoc Inc., Saskatoon, Saskatchewan, Canada;
- Irene Guerrero, Aeroscraft Corporation, Montebello, California, USA;
- M. E. Rogers Carolina, Unmanned Vehicles Inc., Raleigh, North Carolina, USA;
- Sandy Allsopp, Allsopp Helikites, Fordingbridge, Hampshire, United Kingdom;
- Dr. Taly Kosberg, Senior Vice President, SkyStar Inc., RT, Yavne, Israel.

Chapter 5

- • Chesapeake Bay Environmental Centre, CBEC, Discovery Lane, Grasonville, Maryland MD 21638, USA;
- Dyna-kites Corporation, Chicopee, Lancaster, Massachusetts, MA 01021, USA;
- Dr. Jim Christianson, Skydog Kites LLC., 220, Westchester Rd., Colchester, CT 06415, USA;
- Picture Pretty Kites Inc., Omaha, Nebraska, USA;
- National Aeronautics and Space Agency (NASA), Washington, D.C.;
- Robert Sutton, President, Sutton Agricultural Enterprises Inc, Harkins Rd. Salinas, USA.

I wish to thank Dr. Uma Krishna, Sharath Kowligi, Roopashree Kowligi, and offer my best wishes to Tara Kowligi.

Preface

Agriculture has involved the mending of soils, crops, and water resources with an aim to maximize grain/fruit yields. We have also kept a constant watch of forest soils, tree stands, understorey vegetation, the canopy, and applied amendments. However, the basics of agriculture and forestry may soon involve more of aerial vehicles-robotics, spectral analysis, and digital data collection relevant to higher grain/wood productivity. A new concept termed the 'Agriculture Sky' needs due consideration. The Agriculture Sky may soon become a region of intense activity involving aerial vehicles. It seems inevitable. Several types of aerial vehicles have been adopted to monitor the crops. The autonomous fixed-winged and copter drone aircraft have already made a mark in the agricultural sky. They seem to spread rapidly into agricultural farms in different continents. A few models of helicopters and multi-copters are now excellent sprayer drones too. There are still several types of aerial vehicles that are apt to be introduced into the agricultural sky. The parachutes, parafoils, blimps, aerostats, and even kites have been evaluated for their utility in the sky above large expanses of field crops. They need due consideration for regular use in the farming sector. This suggestion is based on the fact that such aerial robots, in general, reduce farm drudgery, make agronomic procedures easier, simpler, thoroughly electronically controlled, less time consuming, and, most importantly, economically efficient.

Our inquisitiveness to observe agricultural crops is great. It drives us to analyze the soil and its contents, crops and their genetic nature, the physiological response of crops to environmental changes, and few other factors. No doubt, there is a strong need to decipher the causes for fluctuations in crop growth and productivity. We have bestowed a great fraction of our time, intellectual ability, and resources on understanding and analyzing the below-ground portion, the crop's canopy, cropping systems, the effect of amendments, disease/pests, and weeds within agroecosystems. *However, we have glaringly neglected or bestowed the least interest on studying crop's response through aerial observation.* The satellite and aircraft methods have their lacunae. Hand-held canopy/leaf color meters are tedious, costly, and human errors could creep in. It is not advisable if the farms possess large landholdings. Aerial robots that transit rapidly at low altitudes over crop's canopy seem to offer intended details and best results, i.e., high-resolution images plus digital data.

During recent years, we have bestowed interest to study the agrarian fields, using the skyline above the crop's canopy. We can conduct a spectral analysis of crops using sensors. The recent trend is to use low flying unmanned aerial vehicles (UAVs), to obtain spectral data of crops. Several types of UAVs have been utilized to capture crop spectral data. Aerial vehicles such as small fixed-winged aircraft, copters, parafoils, blimps, balloons, aerostats, helikites, kites, etc. have been adopted. Each type of UAV has its advantages and lacunae. The farmers are still weighing out the options regarding UAVs. In due course, UAVs that are best suited for a specific or a cluster of tasks will emerge as more popular. Of course, a UAV may fade away as new improvised models arrive in the market.

In due course, the 'agricultural skies,' a concept mentioned in this book may take severe incursions and invasions, from a large posse of aerial vehicles. Such a phenomenon is not new to agriculture. Several types of implements and ground vehicles have periodically invaded and faded away from farmland.

The basic reasons to switch to aerial vehicles with sensors, to obtain crop's data and utilize computer chips with spectral data is very clear. Usually, such digital data is used to guide semi-autonomous or entirely robotic ground vehicles such as planters, sprayers, and fertilizers inoculators or helicopter/quad-copter aerial sprayers. The advantages are quite attractive to farms with large landholdings. They say, in the future, a versatile robotic tractor plus a versatile drone aircraft or a tethered aerostat combination should be the primary goal of a large farm. They can help farmers to accomplish most of the agronomic tasks easily, quickly, and yet be economically efficient.

The autonomous aerial vehicle is a major trend, right now, in agrarian regions of developed nations. The suggestion here is to try, test, and utilize the array of robotic aerial vehicles that have served military goals with flying colors, also in agrarian regions. Aerial robots are expected to perform peaceful tasks above agrarian regions with equal, if not, better results. This book and a couple of others on Small aircraft compiled recently by me are meant mainly to enthuse the farming companies, farmers, and agricultural researchers to adopt aerial robots. Particularly, to analyze the natural resources, vegetation, and crop. As stated earlier, this step eventually reduces farm drudgery to a great extent. It reduces inaccuracies that otherwise creep in due to human fatigue and other reasons. It hastens crop's data collection and processing. It provides an authentic basis for farmers' actions regarding inputs, agronomic procedures, and timing of harvest.

There is a very strong need to disseminate detailed knowledge about robotic aerial vehicles and their usage in the farming sector. The computer-aided spread of details about small aerial vehicles via the internet and websites is popular. Training courses and demonstrations too are becoming common in North America. While there are a couple of books on UAV aircraft, there are none so far covering the present status of knowledge about parafoils, blimps, aerostats and kites and their role in agrarian regions. Hence, this book is timely, apt, and useful to those interested in aerial robotics, agricultural farming, and general information about global technological development.

This book encompasses detailed information accrued, so far, on aerial vehicles such as parachutes, parafoils, blimps, free-floating aerostats, tethered aerostats, and kites. The emphasis is on their implications for global agriculture. At this juncture, we may note that efforts to adopt the above aerial vehicles in farming are still in the rudimentary stage. Published literature is indeed sparse. Hence to introduce the topic to agricultural experts and students, discussion in each chapter includes a good measure of introductory and general information about parafoils, blimps, aerostats, and kites. Yet the emphasis of this entire book is on the adoption of the above aerial vehicles in agrarian regions. Its immediate effect and extrapolations to larger agro-ecosystems have also been emphasized.

This book titled *Aerial Robotics in Agriculture: Parafoils, Blimps, Aerostats, and Kites*, has five chapters. The first chapter is introductory in nature and provides a general overview of the concerned aerial vehicles and their role in farming. The second chapter offers details about parachutes and parafoils. Initially, it offers information on types of parachutes, their components, and a few technical aspects. It highlights the applications of parafoils, particularly the autonomous ones that are of great utility in aerial observations of farms adopting precision farming methods. The third chapter on blimps informs readers about the history, their components, and uses in different aspects of human endeavor. A large portion, beyond one half of the chapter, deals with blimps and their role in agrarian regions. Aerostats, which are perhaps the most interesting and lowest-cost methods of the future, are dealt with in Chapter 4. The details include historical aspects, parts, and functioning of blimps, types of blimps, and their utility to humans, in general. Again, over a one-half portion of the chapter deals with the role of tethered aerostats in farming. Examples compiled include the adoption of aerostats to investigate various agricultural and ecological aspects. The role of aerostat in providing a platform for high altitude wind power generation has been delineated in detail. This aspect deals with gadgets that add to an electric power source to farmers. The last chapter deals with kites in agrarian regions. Kites are perhaps the earliest of the aerial vehicles evaluated for feasibility in aerial photography. Thanks to Frenchman Arthur Batut (1888). At present kites are being explored for their utility in the farming sector. Commensurately, the role of kites in obtaining spectral data about crops, wind power generation, bird-scaring, and identification of regions affected by diseases/pests, soil erosion, floods or drought form an important section of Chapter 5.

This book on parafoils, blimps, aerostats, and kites is perhaps most timely, because, all over the world, the interest in the use of aerial vehicles during farming is now conspicuous. The aerial robots are spreading rapidly into different agrarian regions. Agricultural researchers, farm companies, and farmers have to be conversant with the latest techniques and reap commensurate benefits. Students in educational institutions too need such treatises with the latest and detailed knowledge about aerial robots. Further, this book offers relevant information about recent developments in farming procedures and technological advances related to certain types of agricultural aerial vehicles to the general public.

This volume is a companion book to several other books published by Apple Academic Press by the same author: *Push Button Agriculture: Robotics, Drones and Satellite-Guided Soil and Crop Management; Agricultural Drones: A Peaceful Pursuit;* and *Unmanned Aerial Vehicle Systems in Crop Production: A Compendium.*

—*Krishna Kowligi, PhD*
Bangalore, India

CHAPTER 1

Introduction

ABSTRACT

Topics such as parachutes, parafoils, blimps, aerostats, and kites and their role in agriculture that forms the main theme of this treatise are introduced in this chapter. It begins with a few introductory remarks about aerial vehicles. Historical aspects of parachutes, blimps, and aerostats such as their demonstration to public in Paris and other European centers are mentioned. Blimps and aerostats were utilized for aerial photography, travel and cargo transport during early 20th century. Kite aerial photography began around 1888. Parafoils that are commonly referred as 'Ram-air Parachutes' are relatively recent. They were designed in 1970s and utilized frequently for recreation. During recent years, several models of parafoils, particularly semi-autonomous and entirely robotic versions are being evaluated for use above crop land. Blimps and aerostats (both tethered and untethered ones) are finding utility in agrarian regions and several other aspects related to military and civilian tasks. High altitude wind power generation using tethered aerostats is noteworthy. Helikites are a recent invention made in 1993, in Great Britain. They are being sought to conduct several different functions in a farm like aerial photography, surveillance, and bird scaring. The economic aspects of these aerial vehicles such as yearly turnover, expected markets in future years and their efficiency in conducting military and civilian tasks has been mentioned.

1.1 INTRODUCTION

Aerial vehicles have fascinated humans since several centuries. Human ingenuity and skills have led to the development of several different types of aerial vehicles. Earliest of the aerial vehicles invented is perhaps the tethered kites of different characteristics. During medieval period designs for aerial vehicles were already in place. The zeal to conquer airspace and develop vehicles that are of great utility began perhaps, during 18th and 19th century. Aerial vehicles such as parachutes, blimps, aerostats, ram-air parachutes (i.e., sparless parafoils), tethered balloons, airplanes, helicopters, robotic aerial vehicles such as small drone aircrafts, and satellites are all the result of human engineering skills. Since several decades these aerial vehicles have served us in variety of ways. Some of these aspects include military, recreation, aerial survey and vigilance, transport and supply of cargo, monitoring human activity, terrain, natural resources, vegetation, etc. Within the context of this book we are concerned more precisely with a few selected aerial vehicles such as parachutes, tethered parafoils, autonomous parafoils, blimps that are autonomous or semi-autonomous (driven by a pilot), aerostats that are tethered and kites. In particular, role of these unmanned aerial vehicles in agricultural regions have been highlighted. The four major types of aerial vehicles discussed in greater detail in this volume are the parafoils, blimps, aerostats and kites. These topics are discussed in detail in Chapters 2–5.

We may note that at present, unmanned aerial vehicles such as small aircrafts (e.g., fixed-winged, hybrids, helicopters, multi-copters) that are commonly referred as 'drones' are gaining popularity in the agrarian regions.

The drone aircrafts utilize the agricultural sky above the crop's canopy effectively and aid in spectral analysis of the crop and even in applying (spraying) inputs on the soil surface or crop's canopy. Indeed, several companies are engaged in manufacturing small drone aircrafts of utility in the farming belts of the world. There are treatises and manuals to help us in operating drones, Plus, there are private and governmental agencies that specialize and train personnel in operating drone aircrafts. However, it must be made clear that drone aircrafts are *not* within the purview of this volume.

Parachutes were in use in China during the reign of Han dynasty around 90 BC (Needham, 1965; Historylines, 2016). During medieval period Leonard Da Vinci (1483–1485) described for the first time the concept of parachutes meant for careful descent from a height (Bellis, 2017; Historylines, 2016; Kerman, 2012). Parachutes were demonstrated in France for the first time during the last quarter of 18th century by Sebastian Lenarmond (Bellis, 2017; Kerman, 2012). Parachutes were used predominately for safe landing and diving from airplane during a military operation. Paratroopers who dived from airplanes using parachutes were extensively used during World War II. Parachutes (circular types) originally used in 19th century depended on the resistance to air during descent. Whereas Sparless parafoils that possess air cells and are stiffened by air pressure is a relatively recent discovery made in 1970, by Domina Jalbert. They are also called as 'Ram-air Parachutes.' Parafoils could be totally autonomous. Guided parafoils are in operation in different continents. They allow us to surveillance the farms directly from low altitudes above the farmland. Autonomous parafoils fitted with two-stroke fuel engines and sensors are now a sought-after technology in agriculture (Pudelko, 2012; Thamm, 2011; John, 2011). They are recent introductions into agricultural farms. They offer sharp aerial images of land, water resources and crops. They are yet to make a strong mark in farm belts. Parafoils with 'spray bar' and fluid pesticide tanks are being manufactured. They are being experimentally tested as a method forest/disease control in agricultural farms (Buckeye Industries, 2018). Actually, during the past 10–15 years there has been a trend to try and utilize the autonomous (robotic) parafoils that fly at low altitude over the crop's canopy. The purpose is to conduct spectral analysis of terrain, its topography, natural vegetation, soil type, its fertility variations, crops and their growth pattern, nutrient status (leaf N) and grain ripening.

Blimp, also called 'Dirigible' in French language was first designed by Henri Giffard in 1852. Powered Blimp often called as 'airship 'was demonstrated by a Brazilian named Alberto Santos Dumont in 1898, in Paris France. Blimps did attract enthusiasts interested in aerial vehicles rather rapidly and intensely too. Blimps also called 'Zeppelins' dominated the sky during the early 20th century until 1930s (Bellis, 2017; Sharp, 2012; Colvin et al., 2017; Crouch, 2017; Valiulis, 2014; Lawless, 2018; Lord and Kolesnik, 1982). Blimps in fact were an important mode of air transport. Even trans-Atlantic voyages were made frequently using blimps. Blimps meant for travel became popular in 1920s. This period is called the 'Golden Age of Blimps (airships).' The usage of blimps/zeppelins subsided due to the disasters faced by European nations. Blimps filled with lighter-than-air hydrogen gas were prone to disasters. However, during recent years, the helium (lighter-than-air) gas filled blimps are utilized to conduct aerial imagery of high resolution. Helium is not inflammable. Blimps are also used to monitor public activity, urban traffic, buildings, agricultural farms, crops from vantage points in the sky. They are also utilized to transport goods. This resurgence in adoption of blimps for various activities including in agriculture began in 1990s (Botting, 1980; Wiesenberger, 2017).

Tethered aerostats were devised in 20th century. They are useful in long term surveillance of military and civilian installations, urban traffic, farms, etc. In agricultural farms, they could be utilized to conduct aerial surveillance of farm activity, vehicles, progress of agronomic procedures, irrigation equipment and crops. Aerostats are used to obtain detailed spectral maps of field, its terrain and crops. Aerostats are also used to identify variations in soil fertility, moisture status, crop growth, nutritional requirements, grain formation trends, etc. Of course, appropriate sensors and apt computer programs are mandatory to utilize aerostats in agriculture. Autonomous blimps and tethered aerostats are among the serious candidates for use above agricultural experimental stations. These aerial vehicles are useful in surveillance of experimental crops and in the collection of phenomics data about the crops (CIMMYT, 2014; Haire, 2004; EuroAirships, 2018a).

Kites were perhaps the earliest successful tethered aerial vehicles that were adopted by humans. They say, kites were in vogue in 232 BC in China and other regions in South Asia. Kites were utilized as a recreational aspect. They were also used for military purposes, mainly to scare the enemy. Chinese called them 'Incendiary kites.' Kites were popular during the mid-19th century. This period is termed the 'golden age of kite.' Kites were the first aerial platforms used to obtain aerial photography of ground surface features. It was demonstrated in Paris in 1888 by Arthur Batut, using diamond kites. Kites are relatively low technology aerial vehicles. They are tethered and are thoroughly controllable using tethers. They could be attached with sensors that provide crucial data about agricultural field, soil type, its fertility variations, moisture status, the crop and its growth pattern, leaf chlorophyll and N status, grain formation trends, if any, etc. Helikite is a recent invention with potential to transform the agricultural skyline world over. Helikites and Kites are low cost methods that may dominate the farm world, particularly in regions where it is difficult for farmers to muster capital and high technology items (Allsopp Helikites Ltd., 2019a,b; ICRISAT, .2001). Helikites were devised and developed in United Kingdom in 1994 (Allsopp et al., 2013; Helikite Hotspots Ltd., 2019). Reports from North American and European agrarian belts suggest that helikites and kites seem to offer excellent protection to crops against bird damage. They could gain acceptance because they are low cost technology and environmentally safe, since no harmful pesticide is employed.

The unmanned aerial vehicles (aerial robots) discussed in this volume, namely, parafoils, blimps, aerostats and kites have not attracted attention or imaginations of agricultural researchers to any extent. Reports are relatively feeble, sporadic and not pursued consistently. Therefore, this treatise aims at compiling information about role of these aerial vehicles that are useful to farming situations.

The above stated aerial vehicles such as parachutes, parafoils, blimps, aerostats, kites, and even others such as airplanes, drone air crafts were all first adopted by military establishments of different nations with great success. The UAV aircrafts (drones), for example, are among the most dreaded aerial military reconnaissance and offence vehicles. However, during past decade, these aerial robots have been modified to suit peaceful pursuits related to urban and agricultural settings (see Krishna, 2018, 2019). Autonomous parafoils, blimps, and aerostats were all initially, successfully utilized by military and border patrolling departments of different nations. However, at present, there is a spurt in usage of these UAVs in farms. Role of autonomous aerial vehicles in analyzing the agricultural terrain from vantage heights and locations needs greater attention.

Now, let us consider sensors. Satellites, airplanes, drone aircrafts (fixed-winged, helicopters, multi-copters), parafoils, blimps, aerostats, helikites, kites, etc. are some of the well-known aerial platforms utilized to place various sensors. Sensors form the central piece of UAV technology. The array of sensors offers the ground station (iPad) with spectral data and images of agricultural farm, installations, vehicles and crops. Such data helps farmers to evaluate the crops and decide about timing and quantity of inputs. We should note that aerial platforms such as helicopters, multi-copters and in the present context, the 'sprayer parafoils' and microlights also aid in aerial spray of crops with fluid and granular formulations.

At this juncture, we should to note that, small UAV aircrafts with fixed-wings, fused wings (i.e., with fuselage, e.g., Bramor, eBee, UX5, Delta-M), vertical lift-off-fixed-winged hybrids (Quantix, Alti-Transition, Terra hawk) helicopters (e.g., RMAX, Autocopter), multi-copters (e.g., MJ-Agras, Hercules, m-3000, Phantom) have gained airspace above agrarian regions (Krishna, 2019). They are being briskly disseminated into farmland. They can be deployed rapidly and in short intervals. Data relayed could be processed using computer programs such as Agisoft's Photoscan, Pix 4D Mapper, 'Inphos' or Map Knitter.

During past decade, several types of aerial platforms have been examined for their utility in agriculture. Parafoils, Blimps, Aerostats and Kites that have definite role in the conduct of agricultural procedures form the main focus of this book. Quietly, but surely, a race is on among different types and models of aerial platforms. We have to wait and watch which one dominates the 'agricultural sky' where, when and for how long? Obviously, we can forecast that a few types of UAVs and models may peak in certain agrarian locations and only for a while. They may fade as new models with greater sophistication and ease of operation arrive on the scene.

1.2 AERIAL VEHICLES (AUTONOMOUS, SEMI-AUTONOMOUS, AND TETHERED AERIAL VEHICLES)

Autonomous farm equipment such as unmanned driverless tractor, robotic combine harvesters (e.g., Kinze Autonomous Corn grain harvesting system), Autonomous GPS-guided seeders (e.g., Kinze autonomous seeders and planters; Vibro Crop Robotti; Prospero), fertilizer inoculators (e.g., Rowbot's robotic fertilizer applicator), pesticide sprayers (e.g., Ag Armor Pesticide Sprayer System), autonomous weeders (e.g., Wall-ye; Vitirover, Ag-Ant), etc. have attracted attention from agricultural researchers, farm related industries and most importantly the large grain producers and farmers (Zemlicka, 2019; Krishna, 2016). Large food grain producing companies in North and South America and European Plains have made rapid progress in adopting autonomous farm vehicles, particularly ground robots. The same is not true about aerial robots of relevance to farming. However, aerial robots are spreading rapidly when we consider farming belts of developed nations. Farmers have adopted small drone aircrafts (UAVs) (e.g., Sense Fly's *eBee*, Trimble's UX5, Precision Hawk) to obtain spectral data about the crop, by using various sensors (visual, infra-red, red-edge, LIDAR, etc.) (Krishna, 2016, 2018, 2019). It is true that fixed-winged UAVs have garnered more popularity among farmers. They are being sought by farm companies mainly to obtain aerial imagery of crops and digital data. Such data could help in ascertaining the variations in soil fertility, crop growth, disease/pest attacks and soil erosion. Fixed-winged drones have been utilized to make detailed aerial imagery using sensors. Crops such as rice (Bennet, 2013; Giles and Billing, 2014, Bendig et al., 2013), maize (Raymond Hunt et al., 2014; Mortimer, 2013; Jager, 2014; Agape Palilo, 2015) wheat (Torres-Sanchez et al., 2014,2015), soybean (United Soybean Board, 2014; Ruen, 2012), groundnut (The Economic Times, 2015), sunflower (Aguera et al., 2014), strawberry (Reider et al., 2014), citrus (Garcia-Ruiz et al., 2013), grapes (Primecerio et al., 2012; Turner et al., 2012; Lamb et al., 2001, 2013), coffee (Herwitz, 2002; Herwitz et al., 2013) and pastures (KSU, 2013; Ahamed et al., 2011) have all been assessed, using sensors placed on fixed-winged drones (Krishna, 2018). The UAV helicopters are among the most successful aerial robots used both for swift detailed imagery from low altitudes above the crop's canopy and to spray the crop, with granular and fluid formulation at a rapid pace (RMAX, 2015). Sprayer drones, namely helicopters and multi-copters are becoming a sought-after technology to supply fluid fertilizers, pesticides, herbicides and even apply (spray) life-saving irrigation. These sprayer drones are manufactured in different sizes. They are easy to operate. They are robotic and can utilize digital data supplied in a computer chip. Sprayer drones particularly multirotor types are expected to be frequently utilized during the in-season applications of agricultural inputs (see Krishna, 2019). So, they are apt to be used, in precision farming belts. The sprayer drones are highly efficient in applying agricultural inputs at a very rapid pace, consuming low energy (fuel) and monetary inputs. The urge to use sprayer drones is exceeding expectations in many states of USA. There are regular training courses held for farmers who wish to adopt aerial applications of in-season foliar fertilizer dosages, pesticides, fungicides, bactericides, weedicides, etc. For aerial imagery of crops, both medium and small UAV aircrafts have been utilized. Larger sized drones too have served farm agencies, by offering high resolution images through aerial photography. They also provide sharp images of agricultural terrain and crops when flown at higher altitudes (see Krishna, 2019).

There are now several treatises written periodically in short intervals, about the drone aircrafts that suit the military pursuits. Details about their range, endurance, reconnaissance and destructive abilities have been delineated via books. Also, through extensive reporting, both in classified and in the open access modes. Books on UAV aircrafts meant exclusively for farming situations are really feeble (see Krishna, 2018, 2019). However, recent surge in adopting drone aircrafts in agriculture has meant infusion of funds to popularize drones. Reports suggest that, in addition to UAV aircrafts of different capabilities (e.g., aerial imagery plus spraying), search for several other types of aerial vehicles is rapid. Aeronautical engineers and farm experts are trying to evaluate a range of other aerial vehicles in farming. Most frequently examined aerial vehicles are the Parafoils (both autonomous and guided), autonomous blimps, tethered aerostats and kites. Discussions about these aerial vehicles, in fact, form the center piece of this book. Each of these aerial vehicles has advantages and lacunae. Advantages to farmers seem to outweigh the difficulties. Plus, they are clearly less complicated and low-cost methods. Hence, a special volume, like this has been utilized to highlight the advances made about the use of these 4–5 types of aerial vehicles in the farming regions of the world.

1.2.1 PARACHUTES AND PARAFOILS

A parachute is an aerial vehicle that is commonly used to decelerate the descent of an object or a person from an altitude. Parachutes are used extensively by paratroopers who jump from military airplanes. They use it to descend slowly on the earth's surface. A parachute is capable of slowing both vertical (e.g., cargo, paratroopers) or horizontal movement of a vehicle (e.g., airplanes, space shuttle, cars, etc.). In practical situations, parachutes collect the air below the canopy and provide a drag force to decelerate the descent of a payload. Parachutes could be circular or rectangular (Parafoils) in shape. Parafoils were first developed and used for recreation in 1970. They are also called 'Ram-air Parafoils.' 'Ram-air parafoils' have cells that get filled due to air pressure. The inflated cells provide the rigidity to ram-air parafoils. Ram-air foils are easily steerable. This is unlike circular parachutes that are prone to wobble and loose direction during horizontal transit. There are also autonomous parafoils that float in the atmosphere based on pre-programmed flight path (e.g., Pixy ABS, SUSI 62, Hawkeye) (NASA STAFF, 2017; Knacke, 1991; Pudelko et al., 2012; Thamm et al., 2013). At present, based on shape we can recognize a few different types of parachutes/parafoils. They are Cruciform parachutes, Ring sail parachutes, Rogallo-winged para-chutes, Annular parachutes, and Ram air parachutes (also called parafoils). We can identify several types of parachute or parafoil based on design, size, their special features, source of power, electronic components and robotic features, purpose, etc. A few examples of major classes of parachutes/para-foils are: (a) circular or round parachutes. (b) Parafoils are rectangular. (c) Tethered parafoils, for example, 'Tethered Eye' produced by AeroVironment Inc., California, USA (AeroVironment Inc., 2018) is a tethered parafoil. The tethered parafoils are attached to ground through a string or rope or tether. They are suited to conduct aerial surveillance and photography of a location for long duration. (d) Powered parafoils, as the name suggests are provided with engine and propeller to provide thrust and movement in the air (Plate 1.1). Usually, a small 2-stroke engine attached to a propeller is utilized to provide trust. (e) The latest trend and of particular interest to farmers, farm companies and agricultural research centers is the 'autonomous powered parafoils.' They are usually small, light in weight and provided with fuel (petrol) or electric engine. These parafoils are fitted with electronics and computer processors that aid in autonomous flight of the parafoil. Usually, computer programs that offer pre-programmed flight path and way points to the parafoils such as 'Pixhawk' are used. The robotic parafoils could relay digital data and images directly to the ground computers. Otherwise, digital data could be stored in chips or in computer processor available on the parafoil. Currently, such robotic parafoils are being utilized to conduct aerial survey of farmland, mark the 'management blocks' for use during precision farming and in judging the crops in detail (e.g., SUSI 62; Pixy ABS) (Thamm et al., 2013; Pudelko et al., 2012). (f) Parafoils could be of different sizes. Some are small in size. They are called 'nano-parafoils.' Large parachutes have been designed for military and space science use. They are called 'Giga-parachutes' (Airborne Systems, 2018). They carry pay loads of 42,000 lbs and drop it from 15,000 ft. above ground surface (e.g., Giga Fly Ram-air parafoil). (g) Now, there are cargo parafoils that lift and transport cargo from one location to another. They are frequently in use by military establishments. They are utilized to drop large quantities of payload into military camps (e.g., G-12 Cargo Parachute) (Mills Manufacturing Inc., 2018a, b; Niccum et al., 1965). In addition, there are precision guided or pre-programmed autonomous cargo parafoils (e.g., ONYX System) (Atair Aerospace Inc., 2018). (i) A few very large parachutes are being tested for use as vehicles to transport space station cargo on to earth. No doubt, recreational parachutes/parafoils are popular in several nations. They are used to transit over cities, national parks and other locations of historical relevance. (j) There are also 'Drone-Parachutes.' Such parachutes are placed securely in the fuselage of small aircraft drones that fly rapidly across agricultural fields. The drone parachute helps the drone aircraft to avoid crash landing. Instead, as the small parachute opens, the drone lands safely without damage to fuselage or sensors (AeroMapper EV2; Delta-M UAV System) (Para Zero Safe Air Inc., 2018; Moore, 2018; AVASYS LLC., 2016).

There are parachutes that are exclusively meant to collect weather related parameters such as atmospheric temperature, moisture, relative humidity, dust particles, gaseous components, etc. (Pellone et al., 2017; Jensen, 2000; Huang et al., 2013; Woolsten, 2014). These are in vogue since several decades. They could be called 'Weather Parachutes.' 'Recreational parachutes and parafoils are perhaps among popular aerial vehicles used by humans. They are usually light and used to float in the sky or transit over natural parks, mountainous tracts, etc. (Buckeye Industries, 1997; Balena, 2005; SkyRunner, 2018).

Based on function, we can identify yet another type of parafoil. The new parafoils could be called the 'Agricultural Parafoils.' Of course, they are meant primarily for use above farms, crop land, pastures, dairy, cattle ranches, etc. The agricultural parafoils are provided with set of sensors such as the red, green, blue visual range cameras, multi-spectral sensor, near-infra-red and infra-red band width cameras as well as LIDAR sensor. They provide useful data to ground station about the crops. Within this class of parafoils, we can group the recently evaluated 'Sprayer parafoils.' Their function is to spray foliar nutrients or pesticide/weedicide based on the digital data provided to the computer processor on the parafoil. These are autonomous sprayer parafoils (Buckeye Industries Inc., 2018).

PLATE 1.1   (Top). A Powered Parafoil taking-off from an Agricultural field; (Bottom). A Powered Parafoil navigating in the sky.

Source: Alie Wall, Infinity Powerchutes LLC, Sturgis, Michigan, USA.

1.2.2   BLIMPS

Blimps are non-rigid airships while Zeppelins possess rigid framework made of light weight wood or aluminum. These airships were initially filled with a lighter-than-air gas such as hydrogen. However, since the disasters experienced in 1930 in New Jersey, USA with Hindenburg, in Russia with several zeppelins and in other European nations, hydrogen as lifting-gas was abandoned. Instead, a lighter-than-air gas such as helium was utilized in blimps. Helium offers strength, rigidness and shape to the airship through the internal pressure that it causes (Gerke et al., 2013; Carrivick et al., 2013)

Since the resurgence of interest in blimps/zeppelins in 1990s, several nations have started their production. Highly sophisticated instrumentation and robotics are being incorporated into the blimps meant for military cargo transport, travel, regular surveillance and spectral analysis of terrain and natural resources. A few examples are companies such as Goodyear at Akron, in Ohio, USA (Goodyear Inc., 2018); Egan Airships in Seattle, USA (Egan Airships Inc., 2018); Lockheed Martin in Maryland, USA (Lockheed Martin Inc., 2018); Augur-RosAero Systems in Moscow, Russia (RosAero Systems, 2019; Gertcyk and Lamb, 2016); Hybrid airships in Bradfordshire in United Kingdom (Airlander 10, 2018); Aerodrum Ltd. Belgrade, Serbia (AeroDrum Inc., 2018), etc.

Blimps/zeppelins could be entirely autonomous. For such blmps, flight path, speed and descent are pre-programmed using appropriate computer programs such as Pixhawk, etc. They could be semi-autonomous. Such airships are piloted by a crew that has all the necessary navigation instruments in the gondola. Usually, 6–15 crew members plus passengers are accommodated on the blimps meant for travel or recreational flights (EuroAirships 2018b.

Blimps/zeppelins vary in size and electronic sophistication based on the purpose. Blimps used for aerial survey of agricultural terrain and crops are relatively smaller. There are large versions of blimps such as Airlander-10, Lockheed Martin's Hybrid Airships. They are also called 'Giga-Blimps.' These are meant for long distance cargo transport. They are said to carry huge payloads into military camps, disaster affected zones or storage locations. Small and medium sized blimps such as the Goodyear blimps (e.g., Wingfoot One) are apt for advertisement, aerial survey of ground features, surveillance, small cargo transport, urban policing, etc. In future, they may be among the most useful blimps in monitoring large agrarian zones. This is because, they are robotic, their area of operation could be pre-programmed and results, i.e., aerial imagery and digital data could be transmitted instantaneously to ground station. In addition, these are long endurance aerial vehicles. So, they could be kept floating for entire crop season. They could travel a good distance covering large agricultural patches. Certain types of hybrid blimps are a combination of airplane and blimp. They show better forward thrust in the air and utilize lighter-than air gas helium for lift. They are being examined for use above agrarian regions to monitor crop growth, disease/pest affliction if any, disasters such as floods, large scale erosion, etc. Such blimps are being referred as 'Plimps' (Egan Airships Inc., 2018).

Blimps are versatile aerial vehicles, although only to a certain extent because of their size. Their slow movement restricts their use in instances when rapid transit and small size is preferred. Blimps too found their utility, first of all, in military establishments. Later, they were adopted in many other aspects of human endeavor. In military, currently blimps are being employed in persistent threat detection systems. Recently, they were used effectively in Afghanistan, to surveillance and act as early warning systems in the air. They are being used as sentinels above missile storage sites. Blimps have found excellent use in border patrol functions across long borders between USA and Mexico (US Department of Homeland Security-Science and Technology Directorate, 2018; also see Euro Airships 2018c). In fact, USA maintains a series of blimps above its southern borders to surveillance illegal migration, smuggling and disasters if any. Blimps are used by many city municipalities for policing the urban traffic and other public activities. Blimps could be useful in space science expedition. They have been examined as possible robotic vehicles to hover in the atmosphere above planetary bodies (Wall, 2010; Elfes et al., 1998, 2000, 2003; Heun et al., 2018; Jet Propulsion Lab., 2000). There are too many civilian related uses for blimps/zeppelins. Firstly, they are used to monitor and conduct aerial survey of geographic locations, mark their topography, study the natural resources including soil, water resources, vegetation and crops. As stated earlier, they are good aerial surveillance vehicles. Particularly, to conduct policing above urban loca-tions, monitor traffic and proceedings in a public event. They can be kept floating at a place above the mines and monitor ore movement, etc. Blimps could become useful cargo transport vehicles in future. They say, blimps are being seriously considered as candidates for transport of food supplies to different retail outlets. Blimps could get revived as possible long-distance air travel systems. Blimps have several functions in agricultural zones. They may also be utilized to make observations related to ecological aspects above mountainous sites, hazardous sites such as volcanoes, craters, riverine belts, etc. In future, blimps could be preferred in greater proportion because of their low carbon footprint (Wiesenberger, 2017). Recent thrust by the International Agricultural Research Centers and Universities to adopt blimps to monitor crops in the experimental stations, out fields, and in the large expanses (or agroecosystems) is noteworthy. For example, wheat and maize germplasm and advanced breeder's lines are being intensely monitored and phenomics data of each genotype has been noted using sensors on tethered blimps (CIMMYT, 2012, 2014; Tattaris, 2015). Sooner, we may find blimps, both autonomous and tethered types floating constantly over the crops. They would be able to offer crucial data about crop's growth rate, leaf area, leaf-N status, and crop's water status. Blimps have already been evaluated for use in agricultural experimental stations and crop production farms in USA. Particularly, to note down data about drought tolerance of crops (e.g., cotton) and screen them for future use in the outfield (Grace, 2004; Haire, 2004).

1.2.3   AEROSTATS, HELIKITES, AND KITES

The tethered balloons which are the forerunners of modern aerostats were in vogue in mid-19th century. Its utility in aerial surveillance was first demonstrated by Thaddeus Lowe in 1861. Tethered aerostats were used during American civil war by the Unionists, to track the positions of troopers. The information gathered from vantage heights were relayed using 'Morse code.' The telegraphic wires were strapped to aerostat's tethers. For the past several decades, improvised versions of such aerostats have been utilized by military establishments.

Firstly, an aerostat is lighter-than-air vehicle that could be free -floating or tethered. The buoyancy is usually created by lighter-than-air gas such as helium. Tethered aerostats could be fastened to the ground and allowed to stay floating in the atmosphere for longer duration. A tethered aerostat requires no extra energy. Its long endurance allows it to be used during surveillance of ground activities for lengthier periods of 30 days and above, at a stretch (Howard, 2007). In farming situations, perhaps, a low-cost tethered aerostat could be lofted over crop fields and held for the entire crop season. So that, the sensors on such aerostats can relay images of the crop's growth and nutrient status. The crop maturity trend could be monitored continuously, from seeding till senescence. At present, aerostats of different sizes and capabilities are available for use by military, civilian and agricultural agencies. Low altitude aerostats are lofted at 10 m -50 m above the ground surface or crop's canopy. A high-altitude aerostat, for example, that is utilized in high altitude power generation could reach up to 5 km altitude (Altaeros Inc., 2018; Altaeros Energies Inc., 2019). The payloads carried by aerostats too varies from mere 3 kg to 50 kg depending on the purpose.

In the near future, it is expected that aerostats could be a common feature above an agricultural field. Tethered aerostats could be relaying vital information about agricultural terrain, natural vegetation, crops and their status to ground station computers. They are already popular as telecommunication towers in the rural regions of North America, Europe and Australia. Aerostats are being introduced into rural Africa. Also, above important public places, buildings, industries, mines, etc. In all, tethered aerostats could eventually be a popular aerial vehicle above farms and other installations in Africa. Tethered aerostats may dominate the 'Agricultural Sky.' However, it is very clear that farmers have not yet exploited the full potential of tethered aerostats. Recent reports suggest that, tethered aerostats with a full complement of sensors were deployed over agricultural experimental stations, to collect data about performance of wheat and maize germplasm (CIMMYT, 2014). Low altitude tethered aerostats were tested for utility in relaying information about drought tolerant cotton varieties grown in Georgia, USA (Haire, 2004). Tethered aerostats lofted at early seedling stage could be an effective measure taken to control bird pests. This is an environmentally safe procedure since it does not involve application of harmful chemicals, to suppress birds (Bird Control Systems, 2018; Bajoria et al., 2017; Gaskite Helikites Ltd., 2019).

High altitude wind energy generation is a concept that envisages use of tethered aerostats effectively. Tethered aerostats attached with wind turbines are lofted and wind power generated is relayed to the ground station via wirings strapped to tether (Altaeros Energies Inc., 2018; Altaeros Inc., 2019). The energy derived, it seems satisfies the needs of a few homes and farms. The other important function of tethered aerostats is their role in acting as mobile towers in telecommunication. Tethered aerostats are being introduced into rural belts of Australia (Livson, 2016), USA (Cadogan, 2018), France (Airstar Aerospace SAS, 2010), Brazil (Altave Inc., 2019a,b; Vasconcelos, 2019), South Africa (DefenseWeb, 2017), and India (Hindustan Times, 2018), etc. Tethered aerostats are quick to install. They could be moved from one location to other on pick-up vans. They are relatively a low-cost technology. They could replace a high-cost permanent concrete/metal telecommunication tower in many places.

Aerostats have wide applicability in military. In military and rescue missions, they are adopted for surveillance and reconnaissance, for border patrol and urban policing, and in detecting disasters and reporting them quickly to control stations. They are apt for use in variety of civilian tasks too. Firstly, aerostats are useful in aerial photography and survey of natural resources and vegetation. Aerostats and balloons are useful in collecting weather data. During recent years, they are also adopted to detect and quantify greenhouse-gas emissions from urban and agrarian regions. Now, considering the central theme of this book, we have to note that, aerostats and helikites can play a vital role in aerial photography of land resources, soil types, water resources, crops and their growth pattern. Tethered Aerostats and helikites are cost efficient. They can exhibit relatively longer endurance for floating above the crop land. They are expected to flourish in the agrarian sky. As stated earlier, high altitude tethered aerostats are currently sought-after platforms for wind power generation. They may add a certain quantity of electric power to farms (Altaeros Energies Inc., 2019). Aerostat towers may well become commonly utilized mobile telecommunication towers. They are easy to install and even move from one place to another (Cadogan, 2018; Livson, 2016; Glass, 2018). Aerostats are being touted as possible relay towers on other planetary bodies. Aerostats have found use in archaeological studies.

Helikites may soon be the conspicuous aerial vehicle above the crop fields. They may spread into different agrarian regions. Their versatility and ease of operation makes them a popular aerial robot in farm belts. Farming regions are expected to be dotted with tethered helikites lofted into the sky above crop's canopy. No doubt, they may occupy the 'agricultural sky' in a big way. Helikites are hybrid combination of tethered helium balloon (aerostat) and a kite. Helikites are among the most recently invented aerial vehicles. They were designed and developed by Sandra Allsopp of Fordingbridge, United Kingdom, in 1992. In a helikite, lighter-than-air helium balloon and kite, both are glued careful to each other. Helikites utilize the upward thrust caused by the helium balloon and simultaneously derive thrust created by the blowing wind due to kite portion of the aerial vehicle. Helikites of different sizes are being manufactured and utilized (Allsopp et al., 2013; Allsopp Helikites Ltd., 2019 a; Droneco, 2019). The helikites are relatively more stable in the air. They do not collapse like aerostats or parafoils in the air, if wind is less than threshold. Helikites have longer endurance. Therefore, helikites attached with sensors can relay images for longer duration about cropped fields. Helikites are versatile and withstand vagaries of weather and geographic locations. It seems British Antarctic Survey groups have utilized helikites to obtain aerial images of ice sheets. At present, helikites are being manufactured and sold to clients in all major agrarian regions. This is because they are relatively low-cost aerial vehicles, they can be held stable and guided or shifted from one place another with great ease. They can surveillance the agricultural terrain, fields, crops and farm vehicles on a long-term basis, say, a few weeks (Valour Consultancy, 2019; Altave Inc., 2019a,b). The helium balloon has to be leak proof. Helikites are a good bet to survey, monitor and enumerate botanical species and their diversity in natural vegetation zones and reserve areas. Helikites sold in North America are being effectively utilized to conduct aerial surveillance of military camps (Allsopp Helikites Ltd., 2019b,c) urban locations, installations, and public events. Meteorological helikites are already in use (Allsopp Helikite Ltd., 2019, d, f). Helikites are adopted by policing agencies and telecommunication departments (Helikite Hotspots Ltd., 2019). Helikites have provided detailed imagery of archaeological sites, on a long-term basis (Allsopp Helikites Ltd., 2019 e). It is interesting to note that, helikites are most popular bird scaring aerial vehicles. Helikites with bird scaring noises (taped) helps in reducing the bird related damage to crops (Bird Control Systems, 2018; Perigrine Ltd. 2018). More important is the fact that helikite is relatively a low technology item. It is much less costly than a drone aircraft. Definitely its cost is negligible if we considered the large robotic blimps or aerostats. Most of the models of helikite now available to farming companies are smaller. They are easily transportable and storable until next use.

Now, let us consider kites. Globally, we have now innumerable types of kites being manufactured and sold. Perhaps over a thousand types of kites grouped into different classes are available. However, those of concern to kite aerial photography, particularly in agrarian regions are only few (Wikipedia, 2017, 2019a,b). The kite types (classes) meant for use above agricultural farms for aerial photography are: Eddy Diamond kites, Delta kites, Rokkaku kites, Box kites, Train kites, Sparless parafoil kites (2 lines or 4 lines) and Flowform kites.

Kites have been utilized to conduct geomorphological surveys. It includes extensive aerial photography of natural resources and topography of large districts/state (Aber et al., 2010; Lorenze and Scheidt, 2014; Kite Aerial Photography, 2017). Kite aerial photography offers visual and infrared images. It also provides data to develop Digital Surface Models (DSM). High resolution LIDAR photography is also possible. For example, in Poland they have mapped a few provinces using Sutton's Flowform kites with facility for kite aerial photography (Aber and Galazka, 2000). Kite aerial photography has helped in geomorphological assessment of coastal zone and its vegetation in relation to construction of a new harbor in southern Srilanka (Madurapperuma et al., 2019; Madurapperum and Dyllesse, 2018). Kites with sensors have offered excellent data and imagery of ice sheets, their surface features and glacial movement if any (Boike and Yashikawa, 2003). Such studies show that kites could be adopted even in difficult terrain with erratic wind conditions. Kite aerial photography has helped ecologists in studying the vegetation, its spatial and temporal changes and even botanical diversity of species. They were able to relate the vegetational changes to water resource such as a canal and changes in its flow, human activity and cattle grazing (Fastie, 2012). Kites have been used to monitor and obtain aerial images of archaeological sites (Klemas, 2015).

Kites are in use for recreation, traditional purposes, aerial surveillance and photography using cameras attached to Kitelines. Kites known as power kites are utilized to drag and draw ground vehicles, trolleys, boats, even ships that have been used for trans-Atlantic journey. Kites lofted to different altitudes help us to generate power using small turbines.

'Agricultural Kites' are yet to appear in agrarian regions in a big way, although their potential seems immense. This is attributable to reasons such as low-cost involved in kite-aided aerial photography and surveillance of cropping regions/fields. Kites are such a low technology item that it could be learnt and mastered rapidly by anyone with a little enthusiasm. There are a few instances where in kites have been evaluated in agricultural farms. For example, in the millet cultivation zone of Sahelian West Africa. Here, kites have offered sharp images of the terrain. The sand dunes, sandy soils, sparse vegetation and pearl millet crop has been photographed and studied. In particular, agricultural researchers have been able to demarcate the zones affected by disease, drought and water paucity. Similarly, in other regions, kite aerial photography has helped groundnut farmers to detect crops affected by disease/pest ICRISAT, 2001; Marzollf et al., 2014; Marzolff and Poesen, 2009). Helikites and kites are gaining acceptance by farmers as a low-cost but effective method to scare birds away from crops. Research reports indicate good control of bird pests over cereal and fruit crops. This method is environmentally safe since it does not adopt harmful chemicals to control bird species (Brittingham and Falker, 2010; Ontario Ministry of Agriculture, 2006; Fazlulhaque and Broom, 1985; Mofokeng and Shargie, 2016). Research reports indicate that kites with turbines offer energy to farmers (Arnolds, 2019; Mortenson, 2016; Owens, 2013; Coleman et al., 2017; Archer and Caldiera, 2009).

1.3 ECONOMIC ASPECTS OF USING PARAFOILS, BLIMPS, AEROSTATS, AND KITES

We should note that economic aspects such as capital, production, sales, and turnover of aerial vehicles such as parafoils, blimps, aerostats and kites are highly relevant. These aerial vehicles have to be efficient and profitable, to both the industries that manufacture them and to those who utilize them, such as military and civilian cargo transport agencies and recreational organizations. Their usage in agriculture may get manifested in a big way in the near future. They will be sought by farmers mainly to obtain aerial imagery of land resources and crops repeatedly in a season. At this juncture we have to realize that, the above five types of aerial vehicles discussed in the book have to compete for agricultural air space. These aerial vehicles also have to be efficient in providing aerial imagery and digital data. A few aerial vehicle types could also act as 'sprayer aerial vehicle.' Such sprayer aerial vehicles are useful in dispensing fluid fertilizers as in-season foliar spray. They could also apply pesticide, fungicide and pre- and post-emergent herbicide sprays.

Let us consider manufacturing and market data about the aerial vehicles. Parachutes and guided parafoils are predominantly used by the military for cargo transport and drop. There are also private cargo transport companies that adopt large parafoils. It is said that in 2019, the market size for parachutes/parafoils is 1.2 billion US$. Large parafoils procured in North America constitutes 47% of global market (Rohan, 2018; Sargar, 2018). Parafoils meant for surveillance, policing, power (drag) and aerial photography constitute only a small portion of parafoil market. Agricultural parafoils as a concept has to become conspicuous in future, for it to get listed under global parafoil market trends. Its usage in farming, particularly, robotic parafoils is still rudimentary.

Blimps were meant primarily for military, cargo transport, passenger travel and advertisement. Their usage in survey of natural resources, vegetation and above agricultural farms is relatively recent. Major market for blimps/zeppelins/tethered aerostats is in North America. It is followed by Europe, and Southeast Asia in that order. In 2019, the market size is forecasted at 174 million US$. The demand for blimps is expected increase to 304 million US$ by 2024 (Variant Market Research, 2018; Sambit, 2019). Blimps/aerostats of different size and capabilities are being introduced and evaluated in agrarian regions. Soon, market for them in farming belts may increase. There are a few market evaluations and forecasts about aerostats. In 2014, global market for aerostats was 3604.7 million US$. Another report states that global market for aerostats was about 6.94 billion US$, in 2017. It could increase to 29.12 billion US$ by the year 2026. Clearly, agricultural usage of aerostats is feeble right now. It is expected to be better, if adoption of aerostats in farming regions increases.

There is an underlying competition among several different types of aerial vehicles that are meant for aerial survey, regular surveillance, reconnaissance, aerial photography, spraying, etc. The type and model of aerial vehicle that suits the 'agricultural sky' best is in question. In the recent past agricultural drones, particularly the small drone aircrafts have gained popularity in farming belts. However, there are also autonomous parafoils, blimps, aerostats and even kites that compete for similar function in 'agricultural sky.' It is a matter of time, economic considerations, ease of operation and utility of spectral data obtained that decides the coemption among aerial vehicles.

The following four chapters in this volume deal with wide range of aspects about parachutes, parafoils, blimps, aerostats, helikites, balloons and kites. Their relevance to agricultural crop production and related activities are highlighted.

KEYWORDS

- **aerostats**
- **blimps**
- **digital surface models**
- **kites**
- **parafoils**
- **unmanned aerial vehicles**

REFERENCES

Aber, J. S., & Galazka, D., (2000). Potential of kite aerial photography for quaternary investigations in Poland. *Geological Quarterly, 44*, 33–38.

Aber, J. S., Marzolff, I., & Ries, J. B., (2010). *Small Format Aerial Photography: Principles, Techniques and Geo-science and Applications* (pp. 213–228). Elsevier, Amsterdam, The Netherlands.

Aero Drum Inc., (2018). *10 m Long Blimps* (p. 161). Aero Drum Inc., Belgrade, Serbia. https://www.rc-zeppelin.com/outdoor-rc-blimps.html (accessed on 30 July 2020).

AeroVironment Inc., (2018). *Tethered Eye* (pp. 1–6). https://www.avinc.com/resources/press_ release/u.s.-combattingterrorism-technical-support-office-evaluating-new-aeroviron. (accessed on 30 July 2020).

Agape, P., (2015). *Monitoring and Management of Maize Rust (Puccinia sorghi) by a Drone Prototype in Southern Highlands* (pp. 1–15)*.* Tanzania. http://www.academia.edu/8063999/ monitoring_and_managementofmaize-rustdisease (accessed on 30 July 2020).

Aguera, F., Carvajal, F., & Perez, M., (2014). Measuring sunflower nitrogen status from an unmanned Aerial vehicle-based system and an on the ground device. Conference on unmanned aerial vehicle in geomatics, Zurich, Switzerland. *International Archives of the Photogrammetry, Remote Sensing and Spatial Information Sciences, 38*, 1–22.

Ahamed, T., Tian, L., Zhang, Y., & Ting, K. C., (2011). A Review of Remote Sensing methods for Biomass feedstock production. *Biomass and Bioenergy, 35*,2455–2469.

Airborne Systems, (2018). *World's Largest Autonomously Guided Ram-Air Parachute: GigaFly* (pp. 1, 2). http://www.defense-aerospace.com/article-view/release/99894/ new-payload-record-for-gps_guided-parachute.html (accessed on 30 July 2020).

Airlander 10, (2018). *Airlander: Rethink the Skies* (pp. 1–12). Hybrid Air Vehicles, Bradfordshire, United Kingdom. https://www.hybridairvehicles.com/ (accessed on 30 July 2020).

Airstar Aerospace SAS, (2019). *Aerostats in Telecommunication Services and Mobile Broadband Services in Rural Area* (pp. 1–12)*.* https://cnim-air-space.com/en/(accessed on 30 July 2020).

Allsopp Helikites Ltd. (2019a). *Aerial Photography* (pp. 1–7). Allsopp Helikite Ltd, Hampshire, England, United Kingdom, http://www.allsopphelikites.com/index. php?mod=page&id_pag=33 (accessed on 30 July 2020).

Allsopp Helikites Ltd. (2019b). *The Skyhook Tactical Helikite Aerostat* (pp. 1–5). Allsopp Helikites Ltd., Hampshire, United Kingdom, http://www.allsopp. co.uk/index. php?mod=page&id_pag=24/ (accessed on 30 July 2020).

Allsopp Helikites Ltd. (2019c). *Skyhook Helikites* (pp. 1–5). Allsopp Helikites Ltd, Hampshire, United Kingdom. http://www.allsopp.co.uk/index.php?mod=page&id_pag=10 (accessed on 30 July 2020).

Allsopp Helikites Ltd., (2019d). *Helikite Case Studies: Oil Spill detection and Clean-Up* (p. 1). Arctic ocean, Norway 2012-Ongoing. http://www.allsopp. co.uk/index. php?mod=page&id_pag=55/ (accessed on 30 July 2020).

Allsopp Helikites Ltd. (2019e). *Helikite Case Studies: Aerial Photographic Survey of Ancient Armarna, Egypt* (pp. 1, 2). Archaeology Department, Cambridge University, United Kingdom. http://www.allsopp. co.uk/index.php?mod=page&id_pag=55/ (accessed on 30 July 2020).

Allsopp Helikites Ltd. (2019f). *Lifting Multiple Tethersondes for Meteorological Research* (p. 1). University of Millersville, USA. http://www.allsopp. co.uk/index.php?mod=page&id_ pag=55/ (accessed on 30 July 2020).

Allsopp, S., Reynaud, L., & Mohorcic, M., (2013). *Integrated Project: ABSOLUTE-Aerial Base Stations with Opportunistic Links for Unexpected and Temporary Events* (pp. 21–25). https://cordis.europa.eu/docs/projects/cnect/2/318632/080/deliverables/001-FP7ICT20118 318632ABSOLUTED23v10isa.pdf (accessed on 30 July 2020).

Altaeros Energies Inc., (2019). *Clean Energy* (pp. 1–3). Altaeros Energies, Somerville, Massachusetts, USA.

Altaeros Inc., (2018). *The Super Tower ST 20* (p. 1). Altaeros Inc., Somerville, Massachusetts, USA http://www.altaeros.com/technology.html2 (accessed on 30 July 2020).

Altave Inc., (2019a). *Altave Omni* (pp. 1–4). http://www.altave.com.br/en/produtos/altave-omni/ (accessed on 30 July 2020).

Altave Inc., (2019b). *Innovative Impact Technologia* (pp. 1–3). http://www.altave.com.br/en/tecnologia/ (accessed on 30 July 2020).

Archer, C., & Caldeira, K., (2009). Global assessment of high-altitude wind power. Energies, 2,307–319.

Arnolds, R., (2019). *Will Kite Power Fly High as the Next Renewable Energy Solution* (pp. 1–3). Earth.com. https://www.earth.com/news/kite-power-renewable-energy/ (accessed on 30 July 2020).

Atair Aerospace Inc., (2018). *Atair Aerospace Awarded 3.2 Million US $ Contract by US Army to Supply Onyx Precision-Guided Parafoil System* (pp. 1–18). https://www.prweb.com/releases/2006/09/prweb433958.htm (accessed on 30 July 2020).

AVASYS LLC, (2016). *Delta-M UAV Systems* (pp. 1–8). Autonomous Aerospace Systems-Geoservice. Krasnoyarsk, Siberia, Russia http://uav-siberia.com/en/content/delta-m-uav-system?gclid=CNKdmZ3–0NACFdi (accessed on 30 July 2020).

Bajoria, A., Mahto, N. K., Boppana, C. K., & Pant, R. S., (2017). Design of a tethered aerostat system for animal and bird hazard management. *First International Conference on Recent Advances in Aerospace Engineering (ICRAAE)* (pp. 1–12).

Balena, G., (2005). Powered parachute takes off from water. *Farm Show Magazine, 29*, 4.

Bellis, M., (2017). *History of Parachute* (pp. 1–4). Thought Company Inc., https://www.thoughtco.com/history-of-the-parachute-1992334 (accessed on 30 July 2020).

Bendig, J., Wilkomm, Tilly, N., Guyp, M. L., Bennetz, S., Quing, C., Miao, Y., et al. (2013). Very high-resolution Crop Surface Models (CSMs) from UAV-based images for Rice growth monitoring in Northeast China. *International Archives of the Photogrammetry, Remote Sensing and Spatial Information Sciences, 22*, 45–50.

Bennett, C., (2013). *Drones Begin Decent on US Agriculture* (pp. 1–3). Western Farm Press http://westernfarmpress.com/blog/drones-begin-descent-us-agriculture (September 6th, 2013)

Bird Control Systems Ltd., (2018). *Helikites: Bird Control Research* (pp. 1–7). https://www.birdcontrol.net/helikites-for-farms/ (accessed on 30 July 2020).

Boike, J., & Yashikawa, K., (2003). Mapping of periglacial geomorphology using kite/balloon aerial photography. *Permafrost and Periglacial Processes, 14*, 81–85. doi: 10.1002/ppp. 437.

Botting, D., (1980). *The Giant Airships* (p. 180). Time-Life Books, Alexandria, Virginia, USA.

Brittingham, M. C., & Falker, S. T., (2010). *Controlling Birds on Fruit Crops* (pp. 1–4). Pennsylvania State Agricultural Research and Cooperative Extension Service Pennsylvania, USA.

Buckeye Industries, (1997). *Introducing the Aquatically Enhanced Buckeye Explorer* (p. 14). http://buckeyedragonfly.com/contact.htm (accessed on 30 July 2020).

Buckeye Industries, (2018). *Summary of Aircrafts and Powered Parachutes* (pp. 1–3). https://en.wikipedia.org/wiki/Buckeye_Industries (accessed on 30 July 2020).

Cadogan, S., (2018). *Aerostats Emerge in US for Rural Broadband Delivery* (pp. 1, 2). https://www.irishexaminer.com/breakingnews/farming/aerostats-emerge-in-us-for-rural-broadband-delivery-838133.html (accessed on 30 July 2020).

Carrivick, J. L., Smith, M. W., & Quincey, D. J., (2013). Developments in budget remote sensing for the geosciences. *Geology Today, 29*(4). 138–143. https://doi.org/10.1111/gto.12015 (accessed on 30 July 2020).

CIMMYT, (2012). *Obregon Blimp Airborne and Eyeing Plots* (pp. 1–4). http://www.cimmyt. org/obregon-blimp-irborne-and-eyeing-plots/ (accessed on 30 July 2020).

CIMMYT, (2014). *An Aerial Remote Sensing Platform for High Throughput Phenotyping of Genetic Resources* (pp. 1–24). International center for Maize and Wheat, Mexico. www.slideshare.net/CIMMYT/an-aerial-remote-sensing-platform-for-high-throughput-phenotyping-of-genetic-resources/ (accessed on 30 July 2020).

Coleman, J., Ahmad, H., & Toal, D., (2017). Development of testing of a control system for the automatic flight of tethered parafoil. *Journal of Field Robotics, 34,*519–538.

Colvin, J. D., (2017). *History of Airship* (pp. 1–8). Airship Research labs. Houston, Texas, USA. http://airship-research-lab.com/History_of_Airships.html (accessed on 30 July 2020).

Crouch, T., (2017). *Blimps* (pp. 1–17). Smithsonian National Air and Space Museum. https:// airandspace.si.edu/stories/editorial/blimp (accessed on 30 July 2020).

Defense Web, (2017). *CSIR Develops Active Aerostat* (pp. 1–4). Council of Scientific and Industrial Research, South Africa. https://www.defenseweb.co.za/joint/science-a-defense-technology/csir-develops-active-aerostat/ (accessed on 30 July 2020).

Droneco, (2019). *Helikites Aerostat Capabilities* (pp. 1–3). http://thedroneco.com.au/helikite-aerostat/ (accessed on 30 July 2020).

Egan Airships Inc., (2018). *Plimp-Airship Revolutionizing Flight* (pp. 1–4). https://plimp. com/ (accessed on 30 July 2020).

Elfes, A., Bueno, S. S., Bergerman, M., & Ramos, J. J. G., (1998). A semi-autonomous robotic airship for environmental monitoring missions. In: *Proceedings of the 1998 IEEE International Conference on Robotics and Automation* (pp. 1, 2). Leuven, Belgium., https:// ieeexplore.ieee.org/abstract/document/680971 (accessed on 30 July 2020). doi: 10.1109/ ROBOT.1998.680971 (Abstract).

Elfes, A., Bueno, S. S., Bergerman, M., Paiva, E. C. D., Ramos, J. G., & Azinheira, J. R., (2003). Robotic airships for exploration of planetary bodies with an atmosphere: Autonomy challenges. *Autonomous Robots, 14,*147–164 https://link.springer.com/ article/10.1023/A:1022227602153 (accessed on 30 July 2020).

Elfes, A., Campos, M. F. M., Bergerman, M., Bueno, S. S., & Podnar, G. W., (2000). A robotic unmanned aerial vehicle for environmental research and monitoring. In: *Proceedings of the First Scientific Conference on the Large-Scale Biosphere-Atmosphere Experiment in Amazonia (LBA)* (pp. 12630–12645). Belém, Pará, Brazil, LBA Central Office, CPTEC/ INPE, Rod. Presidente Dutra, km 40,12630–000 Cachoeira Paulista, SP, Brazil.

EuroAirships, (2018a). Agriculture (p. 1). http://www.euroairship.eu/index.php/agriculture/ (accessed on 30 July 2020).

EuroAirships, (2018b). *Unparalleled Touristic Experience* (p. 1). http://www.euroairship.eu/ index.php/luxury-tourism/ (accessed on 30 July 2020).

EuroAirships, (2018c). *Extended Regional and Border Surveillance* (p. 1). http://www. euroairship.eu/index.php/military/ (accessed on 30 July 2020).

Fastie, C., (2012). Swamp NDVI (pp. 1–3). http://fastie.net/salisbury-swamp/ (accessed on 30 July 2020).

Fazlulhaque, A., & Broom, D. M., (1985). Experiments comparing the use of kites and Gas Bangers to protect crops from Wood pigeon damage. *Agro Ecosystems and Environment, 12,*219–228 doi: 10.1016/0167–8809(85)90113–6.

Garcia-Ruiz, F., Sankaran, S., Maje, J. M., Lee, W. S., Rasmussen, J., & Ehsani, R., (2013). Comparison of two imaging platforms for identification of huanglongbing-infected citrus orchard. *Computers and Electronics in Agriculture, 91,*106–115.

Gaskite Helikite Ltd., (2019). *Gaskite Aerostat Bird Control* (pp. 1–3). Gaskite Helikite Ltd. Summerland, Canada, https://www.environmental-expert.com/companies/gaskite-aerostats-82821 (accessed on 30 July 2020).

Gerke, M., Masar, I., Borgolte, U., & Rohrig, C., (2013). Farmland Monitoring by Sensor Networks and Airships. *Fourth IFAC Conference on Modeling and Control in Agriculture, Horticulture and Post-harvest Industry* (pp. 321–326). Finland. doi: 10.3182/20130828–2-SF-3019.00024.

Gertcyk, O., & Lamb, D., (2016). *Hi-Tech Airships Heralded as the Future of Transport in the Region* (p. 17). The Siberian News. https://siberiantimes.com/business/others/features/f0126-hi-tech-airships-heralded-as-the-future-of-transport-in-the-region// (accessed on 30 July 2020).

Giles, D., & Billing, R., (2014). Deployment and performance of an unmanned aerial vehicle for spraying of specialty crops. *International Conference on Agricultural Engineering* (pp. 1–7). http://www.eurageng.eu (accessed on 30 July 2020).

Glass, B., (2018). In: Soaring 'Super Towers' Aim to Bring Mobile Broadband to Rural Areas (pp. 1–7). (by Steadler, T. 2018). ITU News https://news.itu.int/soaring-supertowers-aim-to-bring-mobile-broadband-to-rural-areas/ (accessed on 30 July 2020).

Goodyear Inc., (2018). *Wingfoot One* (pp. 1–5). https://www.goodyearblimp.com/behind-the-scenes/current-blimps.html (accessed on 30 July 2020).

Grace, F., (2004). *Farmers Buddy up to Blimps* (pp. 1–5). The University of South Georgia, USA. https://www.cbsnews.com/news/farmers-buddy-up-to-blimps/ (accessed on 30 July 2020).

Haire, B., (2004). *Blimp Helps Fine-Tune Irrigation* (pp. 1–6). University of Georgia Extension Service, Georgia, USA. http://grains.caes.uga.edu/news/story.html?storyid=2045&story=Infrared-cotton/ (accessed on 30 July 2020).

Helikite Hotspots Ltd., (2019). Airborne Wi-Fi (pp. 1–3). https://www.helikite-hotspot.com/about/ (accessed on 30 July 2020).

Herwitz, S. R., (2002). *Coffee Harvest Optimization Using Pathfinder-Plus (solar powered aircraft)*. National Aeronautics and Space Agency, USA, http://www.nasa.gov/missions/research/FS-2002–9-01ARC.html (accessed on 30 July 2020).

Herwitz, S. R., Johnson, L. F., Dunagan, S. E., Higggins, R. G., Sullivan, D. V., Zheng, J., Lobitz, B. M., et al. (2013). Imaging from an unmanned aerial vehicle: Agricultural surveillance and decision support. *Electronics in Agriculture, 44*, 49–61.

Heun, M. K., Jones, J. A., & Neck, K., (2018). Solar/infrared aero bots for exploring several planets. JPL new technology report: Npo-20264, jet propulsion laboratory (JPL), Pasadena, CA, USA. *Autonomous Robots, 14,*147–164.

Hindustan Times, (2018). Uttarakhand Launches Aerostat Balloon Facility to Boost Internet Connectivity (p. 1). https://www.hindustantimes.com/topic/aerostat-balloon-facility (accessed on 30 July 2020).

Historylines (2016). *The Parachute Was Invented by Ancient Chinese-Not Leonardo Da Vinci* (pp. 1–4). Ancient History facts. http://www.messagetoeagle.com/the-parachute-was-invented-by-ancient-chinese-not-leonardo-da-vinci/ (accessed on 30 July 2020).

Howard, A. J. G., (2007). *Experimental Characterization and Simulation off a Tethered Aerostat with Controllable Tail Fins* (p. 124). Department of Mechanical Engineering, McGill University, Montreal, Canada, http://digitool.library.mcgill.ca/webclient/StreamGate?folder_id=0&dvs=1552406509939~264/ (accessed on 30 July 2020).

Huang, Y. B., Thomson, S. J., Hoffmann, W. C., Lan, Y. B., & Fritz, B. K., (2013). Development and prospect of unmanned aerial vehicle technologies for agricultural production management. *International Journal of Agriculture and Biological Engineering, 6*(3), 1–10.

ICRISAT, (2001). It's a Bird! It is a Plane. No, It's a super scientist. ICRISAT Newsletter SA Trends (13), 1, 2. http://www.icrisat.org/what-we-do/satrends/01dec/1.htm (accessed on 30 July 2020).

Jager, J., (2014). Flying in to increase crop yields and reduce losses. *Seed and Crop Services.* http://www.SGS.com (accessed on 30 July 2020).

Jensen, M. L., (2000). *Tethered Lifting System for Measurements in the Lower Atmosphere.* PhD dissertation, University of Colorado PhD dissertation, https://cires1.colorado.edu/science/groups/balsley/people/jensen.html (accessed on 30 July 2020).

Jet Propulsion Lab, (2000). *Planetary Aerovehicles: Balloons and Ballutes* (pp. 1–3). http://www.jpl.nasa.gov/adv tech/balloons/ summary.html./ (accessed on 30 July 2020).

John, P., (2011). *Tetracam's Hawkeye UAV: Flying Multispectral Camera Platform* (p. 1) https://www.youtube.com/watch?v=ERSAm7k6clI (accessed on 30 July 2020).

Kerman, B., (2012). *A Brief History of the Parachute* (pp. 1–17). http://www.popularmechanics.com/flight/g815/a-brief-history-of-the-parachute/ (accessed on 30 July 2020).

Kite Aerial Photography, (2017). *A Bit of History: Arthur Batut of Labraguire* (pp. 1, 2). France. http://kap.ced.berkeley.edu/background/history1.html (accessed on 30 July 2020).

Klemas, V. V., (2015). Coastal and environmental remote sensing from unmanned aerial vehicles: An overview. *Journal of Coastal Research, 31,* 1260–1267 https://doi.org/10.2112/JCOASTRES-D-15-00005.1 (accessed on 30 July 2020).

Knacke, T. W., (1991). *Parachute Recovery Systems Design Manual* (pp. 1–51). Recovery systems division. Naval weapons Center, China Lake, California, USA paper No NWC TP 6575.

Krishna, K. R., (2016). *Push Button Agriculture: Robotics, Drones and Satellite-guided Soil and Crop Management* (p. 446). Apple Academic Press Inc., Waretown, New Jersey, USA.

Krishna, K. R., (2018). *Agricultural Drones: A Peaceful Pursuit* (p. 394). Apple Academic Press Inc., Waretown, New Jersey, USA.

Krishna, K. R., (2019). Unmanned Aerial Vehicle Systems in Crop Production: A Compendium (p. 699). Apple Academic Press Inc., Palm Beach, Florida, USA.

KSU, (2013). *Small Unmanned Aircraft Systems for Crop and Grassland Monitoring* (pp. 1–4). Farms Company http://www.Agronomy.k-state.edu/document/eupdates/eupdates04513.pdf (accessed on 30 July 2020).

Lamb, D., Hall, A., & Louis, J., (2001). Airborne remote sensing of vines for canopy variability and productivity. *Australian Grape Grower and Winemaker, 5,* 89–94.

Lamb, D., Hall, A., & Louis, J., (2013). *Airborne/Space borne Remote Sensing for the Grape and Wine Industry* (pp. 1–5). Geospatial Information and Agriculture. http://regional.org.au/au/gia/18/600lamb.htm (accessed on 30 July 2020).

Lawless, J., (2018). *Giant Helium-Filled Airship Airlander Takes off for First Time* (pp. 1–6). https://phys.org/news/2016–08-giant-helium-filled-airship-airlander.html (accessed on 30 July 2020).

Livson, B., (2016). *Aerostats All Australia Mobile Coverage* (p. 65). Canberra, Australia, http://www.bal.com.au/AAA.pdf. (accessed on 30 July 2020).

Lockheed Martin Inc., (2018). *Hybrid Airship* (pp. 1–5). https://www.lockheedmartin.com/en-us/products/hybrid-airship.html (accessed on 30 July 2020).

Lord, V., & Kolesnik, E. M., (1982). *Airship SAGA: The History of Airships Seen Through the Eyes of the Men Who Designed, Built and Flew Them* (p. 189). Blandford Press, Poole, England.

Lorenz, R. D., & Scheidt, S. P., (2014). Compact and inexpensive kite apparatus for geomorphological field aerial photography with some remarks on apparatus. *Geographical Research Journal, 3, 4,* 1–8 https://doi.org/10.1016/j.grj.2014.06.001 (accessed on 30 July 2020).

Madurapperuma, B. D., Dellysse, J. E., Zahir, I. L. M., & Iyooba, A. L., (2019). Mapping shoreline vulnerabilities using kite aerial photographs at Oluvil Harbor in Ampara. In: *Proceedings of the 7th International Symposium SEUSL* (pp. 1, 2). E-Repository URL: http://ir.lib.seu.ac.lk/handle/123456789/3010 (Abstract) (accessed on 30 July 2020).

Madurapperuma, D. D., & Dyllesse, J. E., (2018). Coastal fringe habitat monitoring using kite aerial photography: A remote sensing-based case study. *Journal of Tropical Forestry and Environment, 8,* 25–35.

Marzolff, I., & Poesen, J., (2009). The potential of 3D gully monitoring with GIS using high-resolution aerial photography and a digital photogrammetry system. *Geomorphology, 111,* 48–60.

Marzolff, I., (2014). The sky is the limit? 20 years of small aerial photography taken from UAS for monitoring geomorphological processes. Proceedings of European geological union general assembly (. Vienna, Austria. *Geophysical Research Abstracts, 16,* 2014–7005.

Mills Manufacturing Inc., (2018a). *Cargo Parachutes* (p. 19). MillsManufacturing.com (accessed on 30 July 2020).

Mills Manufacturing Inc., (2018b). *G-12 Cargo Parachute Assembly* (pp. 1–3). http://www.millsmanufacturing.com/products/g-12-parachute/ (accessed on 30 July 2020).

Mofokeng, M. A., & Shargie, N. G., (2016). Bird damage and control strategies in grain sorghum production. *International Journal of Agricultural and Environmental Research, 2,* 264–269.

Moore, J., (2018). *Unmanned Aircrafts* (pp. 1–3). AOPA Institute for Air Safety. Aircraft Owners and Pilots Association. https://www.aopa.org/news-and-media/articles-by-author/jim-moore (accessed on 30 July 2020).

Mortenson, E., (2016). *Oregon Firm Developing Airborne Wind Energy System* (pp. 1–5). https://www.capitalpress.com/state/oregon/oregon-firm-developing-airborne-wind-energy-system/article_4e1227e1–4a47–5261–907a-76008e16d9fd.html (accessed on 30 July 2020).

Mortimer, G., (2013). *'SkyWalker' Aeronautical Technology to improve Maize yield in Zimbabwe* (pp. 1–5). DIY Drones. https://www.suasnews.com/2013/04/skywalker-aeronautical-technology-to-improve-maize-yields-in-zimbabwe/ (accessed on 30 July 2020).

NASA STAFF, (2017). *A Canopy of Confidence: Orion's Parachutes* (pp. 1–8). http://www.nasa.gov/exploration/systems/mpcv/canopy_of_confidence.html (accessed on 30 July 2020).

Needham, J., (1965). *Science and Civilization: Physics and Physical Technology* (p. 816). Part 2 Mechanical Engineering. Cambridge University Press, England.

Niccum, R. J., Haak, E. L., & Gutenkauf, S., (1965). *Drag and Stability of Cross Type Parachutes* (p. 111). University of Minnesota, Minneapolis, Technical Report No FD TDR-64–155 http://www.dtic.mil/docs/citations/AD0460890 (accessed on 30 July 2020).

Ontario Ministry of Agriculture, (2006). *Food and Rural Affairs: Guide to Fruit Production: Bird Control*, (p. 78). Publication 360.

Owens, N., (2013). *Delft Professor Puts Kites High on List for Renewable Energy*. Phys.org. https://phys.org/news/2013-07-delft-professor-kites-high-renewable.html (March 10th, 2019)

ParaZero SafeAir (2018). *Para Zero Drone Safety Systems* (pp. 1–7). https://www.bhphotovideo.com/c/product/1416005-REG/parazero_pz_sa_2_parazeros_smart_autonomous_recovery.html (accessed on 30 July 2020).

Pellone, L., Salvatore, A., Favaloro, N., & Concilio, A., (2017). SMA-based system for environmental sensors released from an unmanned aerial vehicle. Italian aerospace research center-CIRA. *Aerospace, 4*, 1–22. doi: 10.3390/aerospace4010004.

Perigrine Ltd. (2018). *Light weight Helikites (36 inch)* (pp. 1–3). https://www.peregrinehawkkites.com/lightweight-helikite-instructions/ (accessed on 30 July 2020).

Primecerio, J., Filippo, D., Gennro, S., Fiorillo, J., Generio, L., Matese, A., & Vaccani, F. P., (2012). A flexible Unmanned Aerial Vehicle for Precision farming. *Precision Agriculture, 13,*517–523.

Pudelko, R., Stuczynski, T., & Borzecka-Walker, (2012). The suitability of an unmanned Arial vehicle (UAV) for the evaluation of experimental fields and crops. *Zemdirbyste=Agriculture, 99,*431–436 UDK 631.5.001.4:629.734 (accessed on 30 July 2020).

Raymond-Hunt, E., Hively, W. D., Fujikawa, S. J., Linden, D. S., Daughtry, S. S. T., & McCarty, G. W., (2014). Acquisition of NIR-Green-Blue digital photograph from unmanned aircraft for crop monitoring. *Remote Sensing, 2,*290–305.

Reider, R., Pavan, W., Carre, M. J. M., Fernandes, J. M. C., & Pinho, M. S., (2014). A virtual reality system to monitor and control disease in Strawberry with drones: A project. *7th international Congress on Environmental modeling and Software* (pp. 1–8). International Environmental Modeling and Software Society. San Diego, California, USA.

RMAX, (2015). *RMAX Specifications* (pp. 1–4). Yamaha Motor Company, Japan. http://https://www.yamaha-motor.com.au/products/sky/aerial-systems/rmax (accessed on 30 July 2020).

Rohan, M. C., (2019). *Military Parachutes Market Worth 1.21 Billion US$ by 2020* (pp. 1–27). Markets and Markets Research, Pune, India. https://www.marketsandmarkets.com/PressReleases/parachute.asp (accessed on 30 July 2020).

RosAero Systems, (2019). *ATLANT 30-Hybrid Aircraft* (pp. 1–12). Ros Aero Systems, Moscow, Russia, http://www.aerall.org/projet_RosAreosystems-Atlant30.htm (accessed on 30 July 2020).

Ruen, J., (2012). Put crop scouting on autopilot. Unmanned Aerial Vehicles Association, Corn and Soybean Digest: Exclusive Insight. Accession No 84562598 http://connection.ebscohoost.com/c/articles/84562/.html pp. 1–13 (accessed on 30 July 2020).

Sambit, K., (2019). *Airships Market Report, Include Product Scope, Overview, Opportunities and Risk, Driving Force Analysis with Global Forecast* (pp. 1–14). https://industrynewstrend.com/1376/airships-market-report-includes-product-scope-overview-opportunities-and-risk-driving-force-analysis-with-global-forecast/ (accessed on 30 July 2020).

Sargar, S., (2018). *Global Parachute Market to Experience Significant Growth During Period 2016–2025* (pp. 1–55). https://www.fiormarkets.com/report/global-parachute-market-research-report-2019-346236.html#tableofcontent (accessed on 30 July 2020).

Sharp, T., (2012). *The First Powered Airship: The Greatest Moments in Flight* (pp. 1, 2). https://www.space.com/16623-first-powered-airship.html (accessed on 30 July 2020).

SkyRunner, (2018). *Introducing SkyRunner: The World's First Flying Off-Road Vehicle GO-ANYWHERE* (pp. 1–3). https://www.prnewswire.com/news-releases/introducing-skyrunner-the-worlds-first-flying-off-road-vehicle-go-anywhere-300407206.html (accessed on 30 July 2020).

Tattaris, M., (2015). *Obregon Blimp Airborne and Eyeing Plots* (pp. 1–3). CIMMYT, International Maize and Wheat Center, El Baton, Mexico.

Thamm, H. P., (2011). SUSI 62 A robust and safe parachute UAV with long flight time and good payload. *International Archives of the Photogrammetry, Remote Sensing and Spatial Information Sciences, 38*, 19–24.

Thamm, H. P., Menz, G., Becker, M., Kuria, D. N., Misana, S., & Kohn, D., (2013). The use of UAS for assessing agricultural systems in AN Wetland in Tanzania, in the wet season, for sustainable agriculture and providing ground truth for Terra Sar X data. *ISPRS International Archives of the Photogrammetry, Remote Sensing and Spatial Information Sciences*, pp. 401–406. XL-1/w2: doi: 10.5194/isprsarchives-XL-1-W2–401–2013.

The Economic Times, (2015). *Drones to Help Rajasthan, Gujarat Farmers Detect Crop Diseases* (pp. 1–3). https://economictimes.indiatimes.com/news/economy/agriculture/drones-to-help-rajasthan-gujarat-farmers-detect-crop-diseases/articleshow/46293237.cms?from=mdr (accessed on 30 July 2020).

Torres-Sanchez, J., Lopez-Granados, & Pena, J. M., (2015). An automatic object-based method for optimal thresholding in UAV images. Application for vegetation detection in herbaceous crops. *Computers and Electronics in Agriculture, 114*, 43–52.

Torres-Sanchez, J., Pefia, J. M., DeCastro, L., & Lopez-Granados, F., (2014). Multispectral mapping of the vegetation fraction in early-season wheat fields using images from UAV. *Computers and Electronics and Agriculture, 103*,104–113.

Turner, D., Lucieer, A., & Watson, C., (2012). *Development of an Unmanned Aerial Vehicle (UAV) for Hyper Resolution Vineyard Mapping Based on Visible, Multi-Spectral and Thermal Imagery* (pp. 1–12). http://citeseerx.ist.psu.edu/viewdoc/summary?doi=10.1.1.368.2491 (accessed on 30 July 2020).

United Soybean Board, (2014). *Agriculture Gives UAVs a New Purpose* (p. 1). AG Professional. https://www.suasnews.com/2014/04/agriculture-gives-unmanned-aerial-vehicles-a-new-purpose/ (accessed on 30 July 2020).

US Department of Homeland Security-Science and Technology Directorate, (2018). *Tethered Aerostat System Application Note* (pp. 1–18). https://www.dhs.gov/sites/default/files/publications/TetheredAerostat_AppN_0913–508.pdf (accessed on 30 July 2020).

Valiulis, A. V., (2014). Dirigibles. *A History of Materials and Technologies Development* (p. 444). Vilnius: Technika.

Valour Consultancy, (2019). *Our Fine Tethered Friends* (pp. 1–3). https://www.valourconsultancy.com/fine-tethered-friends/ (accessed on 30 July 2020).

Variant Market Research, (2018). *Airships Market Overview* (pp. 1–4). https://www.variantmarketresearch.com/report-categories/defense-aerospace/airships-market/ (accessed on 30 July 2020).

Vasconcelos, Y., (2019). *Balloons for Internet Access* (Vol. 257, pp. 73–76). Research Project of the Foundation for Research Support of the State of São Paulo (FAPESP).

Wall, M. D., (2010). *Titan from Above: Blimp Survey Options to Study Saturn's Moon* (pp. 1–8). National Aeronautics and Space Agency. https://www.space.com/10544-titan-blimp-survey-options-study-saturn-moon.html (accessed on 30 July 2020).

Wiesenberger, (2017). *Flying High, 7 Post-Hindenburg Airships* (pp. 1–5). https://www.livescience.com/58988-post-hindenburg-airships.html (accessed on 30 July 2020).

Wikipedia, (2017). *Kite Types*. https://en.wikipedia.org/wiki/Kite_types (accessed on 30 July 2020).

Wikipedia, (2019a). *Kite Applications* (pp. 1–14). https://en.wikipedia.org/wiki/Kite_applications (accessed on 30 July 2020).

Wikipedia, (2019b). *Kites* (pp. 1–17). https://en.wikipedia.org/wiki/Kite. (accessed on 30 July 2020).

Woolston, V., (2014). *China successfully Tests Smog-Fighting Drones that Spray Chemicals to Capture Air Pollution* (pp. 1, 3). Mail online. https://www.dailymail.co.uk/sciencetech/article-2577347/China-successfully-tests-smog-fighting-drones-spray-chemicals-capture-air-pollution.html (accessed on 30 July 2020).

Zemlicka, J., (2019). *Autonomous technology in Agriculture* (pp. 1–3). Precision Farming Dealer. https://www.precisionfarmingdealer.com/articles/3961-new-autonomous-precision-fertilizer-applicator-to-be-unveiled (accessed on 30 July 2020).

CHAPTER 2

Parachutes and Parafoils in Agricultural Crop Production

ABSTRACT

Parachutes and parafoils are highly useful aerial vehicles. Parachutes are predominantly utilized for deceleration during descent of military cargo or personnel. While parafoils (also known as ram-air parachutes) with their ability for steering, guidance, and rapid thrust in the air are preferred during cargo lifting and transit, descent, aerial survey (multi-spectral imagery) and recreational flights. This chapter begins with the mention of several historical aspects about parachutes/parafoils. Designs for a slow descent parachute was initially provided by Leonardo Da Vinci during the medieval period. A functional parachute was first demonstrated in Paris, France, by Sebastian Lenarmond in 1783. Since then, this technology has experienced improvements. In comparison to parachutes, design and development of parafoil technology is recent. It began during the 1970s primarily for recreational purposes. Historical facts about parachutes/parafoils pertaining to military, civilian tasks and recreational aspects are discussed in greater detail. The adoption of parafoils, in particular, in agriculture is still in its initial stages. The trend to adopt parafoils in agrarian regions seems grow as time lapses. They are being evaluated because they have certain advantages in aerial surveying of crops, surveillance, and transit. Economic advantages related to parafoils may outweigh other aerial vehicles, and therefore, parafoils could be the most sought-after aerial robot above farms worldwide.

A brief section in this chapter deals with basics of parachutes and parafoils such as definitions, explanations, and parts of these aerial vehicle. There are now innumerable types of parachutes/parafoils designed, standardized, and regularly used by various clientele in military, civilian, and agricultural realms. Types of parachutes/parafoils discussed in greater detail in this chapter are: circular parachutes, parafoils (ram-air parafoils), tethered parafoils, powered parafoils (manned), trikes or microlights, autonomous (entirely robotic) parafoils, nano or very small parachutes, giga parachutes, drone parachutes, agricultural parafoils, cargo transport parachutes/parafoils, parafoils with electro-optical sensors and electro-chemical probes for assessment of weather parameters, and recreational parafoils.

Parachutes/parafoils have several uses. They are grouped as those pertaining to military and civilian uses, agricultural uses, utility in space science, in ecological monitoring, and in archaeological studies.

A major portion, over one half, of the chapter encompasses detailed discussions about the potential role of parafoils in agrarian regions. Topics include role of parafoils in aerial photography and assessment of natural resources including vegetation and agricultural crops. Parafoils utilized to judge crop growth using a set of visual, multispectral and infrared sensors plus Lidar have been highlighted. The parafoils with infra-red sensors aid in assessing a crop's water status. The infrared aerial imagery provides a map depicting variations in a crop's water stress index and influence of drought, if any. The spectral imagery obtained using parafoils helps in mapping the crop fields. Using standard spectral signatures available in the databanks, we can map disease/pest attacks in agricultural fields. Weeds and their spread too could be mapped. Such digital data is highly useful during precision farming. Digital data could also be utilized in autonomous farm vehicles (ground). Parachutes/parafoils are useful during collection of data pertaining to weather above crop fields. Parafoils may play the role of being sentinels above agricultural experimental stations. Plus, they can collect data about a large number of germplasm and elite genetic material of crops sown in experimental fields.

The economic and regulatory aspects pertaining to use of parachutes and parafoils, particularly in agricultural sector, has been discussed. It is said that the rules dealing with parafoils are less rigid. Therefore, parafoils may be preferred by the farming community to aerial vehicles where rules are many. This chapter also provides a few forecasts about rapid adoption of parafoils in agrarian regions.

2.1   HISTORICAL ASPECTS OF PARACHUTES AND PARAFOILS

The history of parachutes dates back to 200 B.C. In China, parachutes were used to conduct stunts in front of court audience (Patworks, 2014). A book dealing with Chinese ceremonies titled *Zhou Li* (*The Rites of Zhou*), dating 200 B.C. contains information on parachutes (Needham, 1965). It seems toy parachutes were in vogue during middle ages in European region (Mines, 2016). There are statements in the book *Si Ji* (*Chi Jhi*) written by Chinese historian *Si Ma* that, in 90 BC, Chinese Emperors of 'Han dynasty' did employ parachutes (Historylines, 2016; Needham, 1965). The Chinese archived literature suggests the existence of parachutes in the ancient period. A few other literature-based evidences suggest that, parachutes were in vogue in 12th century China. There are also historical accounts dating 1308 AD indicating that, parachute related activities and acrobatics occurred in the palace of the Yuan Emperor. Reports suggest that, in 1650 AD, parachute was being used in Siam (modern Thailand). It seems parasol parachutes were used in Siam in 1687 AD (Needham, 1965; Eric and Chun-Chi, 2016).

The concept of a parachute that helps in slow descent from altitudes was first propounded by Leonardo da Vinci (1452–1519), between 1483 AD and 1485 AD. He has drawn diagrams and sketched a parachute. The figure depicted a large four-sided cloth-covered framework. It resembled a pyramid in shape. Leonardo da Vinci is known as "Father of the Parachute" (Bellis, 2017; Kerman, 2012). We may note that, the term 'parachute' itself was coined a clear four centuries later. There were many experimental parachute descents made before the turn of the twentieth century. Fauste Verancic constructed a device, like the one depicted by Leonardo da Vinci's drawing. He jumped from a tower in Venice in 1617 AD.

The parachute was first demonstrated by a Frenchman, Sebastian Lenarmondin 1783. He successfully jumped and descended *slowly* from a tower. This event proved that a parachute can help in *slow decent* from high altitudes. He used a 14 ft wide parachute. It was touted as a method helpful to escape from buildings, if they were under fire. Later, Joseph and Etienne Montgolfiers designed and tested parachutes of different sizes, shape and size. They also developed balloons that lifted a payload, including humans. Paintings by Etiene Chevalier de Lorimier suggests use of parachutes in 1797.

Frenchman Jean Pierre Blanchard (1753–1809) was probably the first person, to use a parachute for a emergency. In 1785 AD, he dropped a dog in a basket, using a parachute. In 1797, Jacques-Andre Garnerin demonstrated parachute-aided safe jump from a hydrogen balloon, in London (England). He jumped from an altitude of 8000 ft. from the terrain below (Bellis, 2017; Kerman, 2012). During this period, parachute jumpers noticed that severe oscillations occurred during descent. Atmospheric (wind) conditions were affecting the descent process, severely. Such oscillations were effectively overcome by making holes (vents) in the canopy. In 1890 AD, a German lady, Kathe Paulus, developed expertise in using parachutes and jumping.

She was given the title of 'Professional Parachute Jumper.' The parachute harness and strap were devised by Thomas Baldwin in 1887 AD. Methods for safe folding and preservation of parachutes were standardized, in 1890 AD (See Australian Parachute Federation Ltd., 2017).

In 1901, Charles Broadwick designed a parachute that was used by aviators. It was nicknamed 'Pack of Aviators.' It was demonstrated for regular use in locations such as Raleigh in North Carolina, also in Los Angeles and San Diego in California. The United States of America Military formed its first parachute section in 1918. Later, in 1921, a private parachute making company was started in Ohio, named 'Irving Parachute Company' (NASA STAFF, 2017).

The first freefall parachute descent was made by the American, Leslie Irvin, on 28 April 1919. He jumped from a height of 1800 ft. (Mines, 2016). The paratroop training program in the United States of America was launched, unofficially, in 1928. Parachutists learnt to jump from a Martin bomber over Kelly Field, in Texas. Although limited experimentation continued, it was not until the spring of 1940 that the United States of America established an official paratroop training program. It was situated at Fort Benning, in Georgia State (see Australian Parachute Federation, 2017).

The development of modern parachutes deployed at high speeds and high altitudes started in the 1930s. Knacke and Madelung (1992) developed the ribbon parachute in Germany, for decelerating heavy and high-speed payloads (Meyer, 1985; Australian Parachute Federation Ltd., 2017). In 1933, the Germans investigated the feasibility of developing parachutes suitable for in-flight and landing deceleration of aircraft. The ribbon parachute was developed and successfully tested in 1937, as a landing brake for a Junker's W-34 aircraft. The ribbon parachute was adequately stable and opened reliably. It showed up only a low opening shock. Also, it did not interfere with the controllability of the aircraft. The Germans used mass airborne troops to support the invasion of Holland, Belgium and Luxembourg. They dropped 35,000 paratroops in these areas. After World War II, Knacke invented the ring slot parachute. It was used for moderate sub-sonic speeds (Bellis, 2017; Kerman, 2012). During World War II, Germans used ribbon parachutes as landing deceleration parachutes and retractable dive brakes.

Aerial delivery and cargo parachutes were gradually evolved from personnel parachutes. They were utilized to perform the specific cargo related job, for which they have been designed. By the end of World War II, five sizes and weights of parachutes were in use. The aerial delivery parachute, known as type G-1 had a 24 ft. canopy. It was meant for use with loads up to 300 pounds.

It was standardized in December of 1942. Cargo parachutes were being made larger and constructed of heavier material than the aerial-delivery parachutes. They were dropped from higher altitudes. We may note that as the size or weight of the canopy increases, then, greater is the opening time required. The Air Force Board considered 200 feet to be the best altitude, for dropping supplies and equipment from aircraft, using parachute. The variations in altitude were dependent upon the type of load. Dropping at such low altitudes eliminates drift and assures greater accuracy in landing supplies, in the desired drop zone. It also eliminates oscillation of the load, to a great extent.

The 'Rogallo wing' for parachutes was conceived in the late 1940s, by Francis and Gertrude Rogallos (Wikipedia, 2018a; Australian Parachute Federation Ltd., 2017; NASA, 1967). Francis Rogallo was an aeronautical engineer at the Langley Research Center for the NACA (NASA). Rogallo created the flexible wing and flew a prototype in 1948. He patented the design in 1951 (see NASA, 1967; Wikipedia, 2018a). Pioneer Aerospace and Irvin Industries obtained a license from NASA and manufactured these para-wings. The US Army's Golden Knight Precision Parachute Team were the first to adopt Rogallo-winged parachutes. They demonstrated their unique steerability. The Rogallo-wing made the sport of hang gliding one of the most widespread forms of flying throughout the world (Meyer, 1985; Australian Parachute Federation Ltd., 2017).

The 'Parafoils' or 'Ram-air Parachutes' was preferred. They eventually replaced 'Para-wings' by 1970s. Parafoil ram-air parachutes were invented around 1965, by Domina Jalbert. He was a professional kite maker. Regular parachute/parafoil aided sports and gliding became common in 1960s (Bellis, 2017; Kerman, 2012). The parafoil or ram-air parachute is a technology first developed by the US Army. It is a deformable aerofoil that maintains its profile, by trapping air between two rectangularly shaped membranes. The membranes are sewn together at the trailing edge and sides, but open at the leading edge (Meyer, 1985; Global Security. Org, 2018).

Austrian airforce was the first to use an emergency parachute. The 'Heineken parachute,' was developed under with the advice of a German professional parachutist, Kätchen Paulus (1868–1935). It was attached to the airplane. It opened automatically when the aviator jumped from the aircraft. The Guardian Angel parachute developed in England was regarded as the best of the emergency parachutes developed, during the First World War. It was used in large numbers by the Italian Air Force but not by the Royal Flying Corps in Britain. They thought its availability would tempt aviators to abandon the valuable aircraft (Mines, 2016).

Among Australians, Vincent Patrick Taylor (Captain Penfold) had expertise in parachuting. It was not till the era of sport parachuting that Australia really began to move into the international mainstream of parachuting. In the late 1930s, civilian parachuting was about to become popular in the Australian society. However, the advent of the Second World War brought an abrupt end to civilian parachuting. In 1930s and 1940s, parachuting in Australia remained largely a derivative from the parachuting activities of other countries (Australian Parachute Federation Ltd., 2017; Mines, 2016). Parachute jumps were made in Australia during this period. Beginning early in 1909, parachute descents were made every Sunday at Clontarf, usually, from about 3000 feet. Patrick Taylor also took the opportunity of his balloon rides to take aerial photographs. He was the first aviator to take photos from the air, in Australia. Patrick Taylor set up a factory in Sydney to manufacture balloons and parachutes made of Japanese silk cloth.

Reports state that in Australia, the first emergency descent from a balloon using a parachute happened on 14 April 1879. The first intentional parachute descent in Australia was done in December 1888, in Sydney by one Mr. J. T. Williams, a watchmaker (Mines, 2016). The first descent by a woman from a balloon was made at a racecourse in Newcastle, New South Wales, Australia, on 8 February 1890. Two sisters who carried out the first jump were Valerie and Gladys Van Tassell. They made a descent at the Newcastle Cricket Ground. Most of the parachuting in Australia in the early 1890s appears to have been done, by four women; Valerie and Gladys Van Tassell (Freitas) and Millie and Elsie Viola.

Mines (2016) has listed a series of names of parachutists of Australia along with dates on which they demonstrated intentional parachute jumping. They are:

- J. T. Williams, in 1888 at Sydney, from 4,000–5,000 feet, from a coal gas filled balloon;
- Val Van Tassell and Miss Gladys Van Tassell, in 1890, at Newcastle, NSW, from 1900 feet, from a hot air balloon;
- Fernandez, at Sydney in March 1891;
- Professor Price, at Perth, Western Australia, in 1891, from 2000 feet, from a hot air balloon;
- Millie Viola, at Perth, in 1891 from 3000 feet, from a hot air balloon;
- Zahn Rinaldo, in 1908, at Broken Hill, NSW, from 7300 feet, involving a triple cutaway;
- C. Sebphe, in 1910, at Perth, from 400 feet, from a hot air balloon;

- Captain Penfold, in 1914 from the North Sydney suspension bridge.

In Australia, the first public parachute descent from an airplane was made at the Epsom Racecourse, Mordialloc, in December 1919. It was part of an aeronautical display. The descent was made by Captain G. C. Wilson MC, AFC, DCM, of the Australian Flying Corps. Parachutes opened automatically, at a height of 1500 ft.

2.1.1 HISTORICAL FACTS AND EPISODES RELATED TO PARACHUTES IN MILITARY

Parachutes have taken slightly over a century to reach military establishments, since the first demonstration of parachute jumping in France. Reports indicate that during World War 1, early aircraft models were almost flying deathtraps. The dry wooden frames and fabric were highly flammable. Parachutes did exist, although rudimentary by current standards (Ross, 2014). Regular use of parachutes began during World War I. Parachutes were used to escape from aircrafts and blimps. It seems, American Military General Billy Mitchel first suggested its regular use, as early as in 1917, by forming a para-troopers regiment. Italian Military made the first combat use of parachutes in 1918. By 1920s, military establishments of many other European nations started to train and build 'Paratrooper Regiments.' Soviet (USSR) military had well trained Parachute troopers by 1933 (Australian Federation of Parachutes Ltd., 2017; Mines, 2016). They used parachutes to drop soldiers into forward positions. Germany and Italy had used paratroopers by 1936. They had well trained airborne regiments. However, it is said that, first well focused army attack using paratroopers was made by Germans, during World War II, in May 1940. They made large scale air drops of troopers into Holland. The British Military has been using parachutes since 1940s. American military used parachutes during World War II. One of the earliest major effort using airborne troops was made by Americans, in Holland. They used 35,000 parachutes to place their army in the southern Holland. Parachutes were used by French in Indochina. The British military used them during 1955, to place their army on the banks of Suez Canal (Meyer, 1985). At present, military set up in most countries possess specialized para-trooping regiments. However, they say, with the advent of helicopters, parachutes had an alternative. Their adoption has relatively subsided. Yet, the argument is that there are instances where parachutes are apt, to land troops in deep enemy territory. Also, large scale drops of paratroopers are best done using parachutes. Parachutes have found excellent use during covert operations by military. Primarily, parachutes are utilized in military to jump from an aircraft. Incidentally, there are at least three types of jumps from an aircraft. They are LALO (Low Altitude Low Opening), HAHO (High Altitude High Opening), and HALO (High Altitude Low Opening) (Meyer, 1985).

During World War I, most aviators were provided with parachutes. It was meant to escape in case of perilous situations. Still, the parachutes used during World War I were, by modern standards, makeshift contraptions. They had proven their worth. They formed the basis for series of experiments in design, which were initiated immediately after the war (Australian Parachute Federation Ltd., 2017). There was a great zeal to develop parachutes that were safe, better in design and easier to construct. In 1929, parachute unit at Wilbur Wright Field, Dayton, Ohio, started the development of a radically new type of parachute canopy, the triangular type. The triangular canopy was standardized and adapted (Australian Parachute Federation Ltd., 2017). In the mid-1930s, the Russians pioneered large-scale airborne and air-supply operations. The Italian army used airdrop procedures in its campaign against the Ethiopians, in 1950s.

The need for parachutes in military, it seems, was first proved by an Austrian pilot who ejected from a burning plane. He descended inside Russian territory. Germans demonstrated the use of parachutes in rescue mission in 1917. They said replacing a burnt airplane is not difficult but replacing a well-trained air force pilot is difficult. Training a pilot is a time-consuming task. We cannot afford loss of pilots. The Germans used mass airborne troops to support the invasion of Holland, Belgium and Luxembourg. They dropped upwards of 35,000 airborne troops on the Isle of Crete. After World War II, Knacke invented the ring slot parachute which was used for moderate sub-sonic speeds (Meyer, 1985). This parachute was used primarily for cargo delivery and aircraft deceleration.

No doubt, training military personnel in para-trooping became very important. General William Mitchell, Commander of the US Air Force organized parachute test and development program in the United States of America. The aim was to provide better parachutes for air force pilots. Later, a parachute facility was established at McCook Field, Dayton, Ohio, USA. It began functioning in the summer of 1918 (Australian Parachute Federation Ltd., 2017). Parachutes were utilized by US Air force and air borne divisions during World War II. For example, 101st Airborne Division of Louisiana, USA known as 'Screaming Eagles' airdropped paratroopers in 1942. It was done to thwart the advancing German army, into Belgium. It delayed the conquest of Belgium (Bennet et al., 2015).

The establishment of paratroop units and their training was initiated earlier in some European nations. The French army organized its first battalion of air troops in 1938. Both, Russia and Germany were training paratroopers long before this time. The first recorded jump, made by the national hero Mikhail Gromov, occurred in Russia, in 1927 (Gromov, 1939). A concept called "Vertical Invasion" was envisaged by employing parachutes and paratroops. Germany's paratroop program was underway as early as 1935.

During first half of 20th century, Russians and Germans took greater interest in air-borne transport of goods and military supplies. During the early part of Second World War, i.e., 1939 to 1941, it seems, Germans used large number of 'Falshirmjager' (para-troops) and paraglider borne infantry. It was done with great effectiveness. Such paratrooper land-ings aided Germans and resulted in fall of Norway, Denmark, Holland, Belgium and France (Matheson, 2017). However, after a few setbacks and heavy casualties during invasion of Crete, it seems, Adolf Hitler decided to disband para-trooper regiment. That is because the element of surprise created by first time use of parachutes was over. Paratroopers became less effective. During 1950s, development of large parachutes helped in trans-port of personnel and military supplies weighing 1–2 ton. Air dropping became a more common choice during combat (Kovacevic et al., 1998). During 1970s, advancement in parachute technology further improved the load that could be carried by them. Aircrafts transported up to 40 t cargo and parachutes in them could airdrop loads of 12–16-ton cargo at a time, safely at the military regiments.

Canada, who got involved in World War II, in 1939, formed the initial Canadian Airborne Division by 1941. These paratroops were commanded by Major H. D. Proctor of Ottawa. They were initially trained at the U.S. Airborne School at Fort Benning, Georgia, USA (Meyer, 1985; Australian Parachute Federation Ltd., 2017). The Canadian contingent was saddened by the loss of their Commanding Officer who was killed during his first jump at Benning. The parachute 'Drop Zone' at Canadian Forces Base Shilo, Manitoba was subsequently named 'Proctor Field' in his memory. The 1st Canadian Parachute Battalion went on to fame when they combined with the British Parachute Division. Together, they dropped into combat on 06 June 1944, in advance of the massive seaborne invasion of France. These paratroopers later jumped into Germany during the attack at the Rhine (Australian Parachute Federation Ltd., 2017).

"In the South East Asian region, the Air-Landing School at Willingdon Airport near New Delhi, in India, was moved in 1942, to Chakiala, near Rawalpindi (now in Pakistan). In Rawalpindi, No. 3 Parachute Training School was established by Ringway instructors. In early 1944, Major-General E. E. Down formed 44th Indian Airborne Division from the 50th Indian Parachute Brigade. On May 1st, 1945 an improvised battalion group, made up from 1st Indian, 2nd and 3rd Gurkha Parachute Battalions, was dropped on a successful operation ('Dracula'). It aimed to destroy Japanese gun positions at Elephant Point, at the mouth of the Rangoon River (see Mines, 2016).

In Australia, the earliest effort to train military personnel in parachute jumping began at Melbourne, in 1926. "The minimum height for this type of parachute jumping was 120ft. The main envelope of the parachute used was 24 ft. in diameter. The dummy weighed approximately 150lb. A few other demonstrations used parachutes of 26 ft. and then 28 ft. canopy. Australian artillery unit along with the American 53rd Parachute Infantry Regiment jumped with guns at Nadzab in the Markham Valley, during September 1943 (Mines, 2016). During the Second World War, a much larger group of Australians than at any comparable previous period was given training, in parachuting as part of their military or special forces assignments (Mines, 2016).

Mass production of parachutes was a problem faced by Military establishments of many nations involved in the World War II. It was solved in different ways. In USA, mass production of parachutes meant for world war II front lines was primarily done, by women. For example, in Utah, women with tailoring skills were employed to sew as many silk parachutes. Such companies were started in many places across USA. They were simple conical parachutes and meant to airdrop paratroopers (Borneman, 2006). At present, industrial automation and pre-programmed assembly lines allow us, to churn out as many parachutes of different sizes and payload capabilities. There are indeed several companies in each nation that offer specialized parachutes for various tasks. In due course, we may come across companies that specialize in agricultural parachutes/parafoils, exclusively.

After World War II, the use of parachutes continued as a mass military transportation media. In May 1954, the French used vast numbers of paratroops in a futile but heroic attempt, to relieve their besieged garrison at Dine Bien Phi in French Indochina. During the Korean conflict, the Americans, as part of the United Nations Force, carried out a mass airborne drop in South Korea.

During the 1956 Middle East crisis, British and French paratroops combined to form an airborne assault force against Egypt. It was formed to capture the vital Suez Canal. Originally, the British Parachute Regiment was formed in 1942. Their uniform included the distinctive 'Maroon beret.' Hence, they were nicknamed "the red devils" by their German adversaries (Lee, 2015). In 1964, the red devils formed a sporting team for sky diving with different formations. It was an army display team. An army display team was the brainchild of Lt Edward Gardener. The Golden Knights, the US Army's version of The Red Devils offered the initial training in skydiving and displays. So, this way, the era of recreational parachuting got initiated.

During Korean War, in 1953, problems in delivering large amounts of war material to ground troops in combat zones became conspicuous. Hence, methods to deliver air cargo safely to ground units were explored. The conventional airdrops and ground-level parachute extraction techniques were examined. The Parachute group at El Centro, California, USA, completed the C-124 Globemaster II Aerial Delivery System test program, in May 1953. The group successfully airdropped A-22 cargo containers, a M-29C cargo truck, and a 105mm howitzer from a C-123B Provider. They used techniques that later proved invaluable, in Vietnam (Global Security. Org., 2018). The United States Army conducted airborne attack using paratroopers. They used at a time over 800 paratroopers, to gain advantage over North Vietnamese forces. During early 1970s, they evaluated the performance of several models and different sized parafoils. They used wind tunnels and compared them for actual performance (Nicolaides and Tragarzz, 1971).

The free-fall type parachutes have been examined in a military offensive role. These are halo-type parachutes. They possess canopies that allow the parachutist to steer himself. It offers the trooper maneuverability. The paratrooper can make accurate landings. Such parachutes are usually small. They are used to drop guerrilla groups from great heights. They go undetected and fall thousands of feet. Then, they deploy their parachutes at the lowest feasible height above ground level (Australian Parachute Federation Ltd., 2017).

A hybrid system made of parachute and a parafoil was designed during last decade, starting in 2000 AD (e.g., Onyx Systems parafoil cargo airdrop system). It is an autonomous parachute/parafoil system fitted with GPS receiver and on-board computer. It also takes command from ground control center. During trials, it offered accurate landing and air drop of military and civilian cargo. It is currently in use with US military for accurate air drop. There are three versions namely one that carries 20 lb. cargo, the others 500 lb. and 2000 lb. pay load. The accuracy of landing is 100 m from the spot pre-programmed on computers (Onyx Systems, 2018).

2.1.2   PARACHUTES AND PARAFOILS IN CIVILIAN, RESCUE, AND RECREATIONAL PURSUITS: HISTORICAL ASPECTS

Parachutes have been utilized to accomplish an array of different civilian tasks. Most importantly, they are utilized in rescue acts during and after a disaster. Records show that, both, parachutists and supplies were dropped at disaster scenes in the US Great Plains, during the 1920s and 1930s, particularly when dust bowl conditions prevailed (Australian Parachute Federation Ltd, 2017). Clearly, parachutes/parafoils were not just meant to be utilized in the realm of military. Civilian purposes such as cargo transport, air drop and rescue missions from difficult terrain are also important. Air drop of essential supplies in flood or quake affected zones are other major activities with which the parachutes/parafoils are related. Parafoils have been utilized during recreational flights since 1970s. Since a few decades, 'Ram-air parachutes or Parafoils' have been consistently adopted, during recreational activities in the air. For example, such as sky diving, navigation and aerial acrobatics. Now, let us consider a few salient historical facts pertaining to above aspects. We begin with the use of parachutes in USA, in early 1900s.

2.1.2.1   CIVILIAN AND CARGO PARACHUTES/PARAFOILS

During past 2–3 decades, beginning in mid-1980s, powered parafoils have been utilized in variety of functions that require low flying and rapid take-off. Powered parafoils have been utilized for surveillance and aerial photography. Parafoils have been utilized by police and vigilance departments, in tracking down vehicles (Wikipedia, 2018b; Wikimilli, 2018; Powerchute Educational Foundation Inc., 2018; Ripon Police Department, 2012).

Parachute aided air drop of essential supplies and other types of cargo has been in vogue, since a few decades. Cargo parachutes were in use during 1930s and 40s. Most of these designs, it seems, are still in use. Of course with some minor modifications. They are being utilized in civilian cargo transport. Larger cargo parachute of 24 ft width was developed in 1960s. They are even now useful to lift and recover larger sized cargo (Mills Manufacturing Inc., 2018 a,b).

During mid-1990s, relatively large autonomous guided parafoils of light weight were tested for performance. They were evaluated for their ability to transport and air-drop larger cargo (Patel et al., 1997). Matheson (2017) states that, during 1990s, several new transport and delivery systems were explored by the United States of America. They examined high altitude transport and delivery or dropping, using parachutes/parafoils. The project was called 'Advanced Precision Airborne Delivery System (APADS).' It involved a gliding parafoil with autonomous GPS connectivity and navigation. It could drop pay loads from 25,000 ft altitude. It could easily deliver goods weighing 800 kg per drop. Later, several variants were tested for high altitude transport of cargo. The largest consignment tested was 19 ton from a distance of 40 km. It could carry and deliver the said pay load.

The semi-rigid deployable wing (SRDW) is a self-inflating glider. It has a rigid frame. It is able to transport cargo weighing 275 kg and drop it from high altitudes. There were also many guided-parafoil models examined, during 1990s. The USA army procured and adopted them for use. Particularly, to transport cargo up to 700 kg (Matheson, 2017). They could carry and drop the cargo from a distance of 20 km. No doubt, such parafoils with ability to transport cargo relatively quickly and at low altitudes are highly useful. They could have a role even in different agricultural zones. We may note that, Guided Parafoil Air Delivery System is a US Army development. It enables precision delivery of a payload of about 700 kg, from a drop distance of 20 km (Matheson, 2017; Braunberg, 1996).

During 2016 to 2018, very large autonomous parafoils (Giga-parafoils) capable of airdropping 42,000 lb (19,000 kg) from 15,000 ft. above ground have been designed and utilized. Historically, although very recent since it is in 2016–2018, the large or giga-parafoils of 195 ft. wingspan are a remarkable achievement (Airborne Systems, 2018; Benolol et al., 2018).

2.1.2.2 *RECREATIONAL AND SPORTS PARACHUTES*

After World War II, sport jumping using parachutes became a recreational activity. This sport, also called 'sky diving" started with round parachutes. Such parachutes ranged in size from 20 to 30 ft. in diameter. Parachutes later evolved into a steerable vehicle. For example, rectangular ram-air powered parachute (PPC) which is approximately 38 ft. wide was steerable. On October 1, 1964, Domina Jalbert applied for a patent for his "Multi-Cell Wing" named "Parafoil" (also known as a "ram-air" wing). It was a new parachute design (United States Department of Transportation, 2007).

In the early 1960s, Lowell Farrand flew a motorized version of parafoil called "The Irish Flyer." Farrand fitted an engine on a ram-air inflated parachute wing. This step initiated the evolution of the 'powered parafoil.' It has led to modern powered parachute canopies. They include rectangular, elliptical, semi-elliptical, and hybrid wings.

In USA, the use of parachutes for recreational jumping got initiated a bit later than couple of European nations. For example, France had 20 skydiving centers by the time the first "Jump Town' in USA had its sky-diving club. Similarly, Russia had over 100 parachute jumping schools and recreational clubs, by 1959 (United States Parachute Association, 2016).

In USA, recreational parachuting and parachute jumping were first initiated in 1959, at the Orange Municipal Airport in Massachusetts, USA. A recreational parachuting company was founded in 1959 by a group that included persons named Nate Pond, Lew Sobourn, Jacques-Andre Istel, and George Flynn. This recreational parachute company was supplied parachute units with 'low porosity (Lo-Po),' by a company named 'Pioneer Aerospace Corporation.' At present, the company is still in production and is called 'Zodiac Aerospace Inc..' The recreational parachute center at 'Jump Town' was provided with several parachutes out of excess supplies found in military units (United States Parachute Association, 2016). Recreational parachuting was also practiced in Australia. During 1960, The Wonderland City Venue near Sydney, Australia was the venue for parachutists descending from balloons floating at thousands of feet altitude. Powered parafoils were also demonstrated in Sydney, in 1960s.

2.1.2.3   METEOROLOGICAL AND SPACE SCIENCE PARACHUTES

The development of parachutes suitable for space science vehicles began earnestly in 1950s and 60s, in USA and USSR (Wright, 2017). Actually, parachutes suitable for high altitude weather balloons and space vehicle recovery were developed in 1950–60s, in USA and USSR. It involved some light-weight miniaturized versions of parafoils (Benton and Yakimenko, 2013). These parafoils brought home the payload of weather balloons, if and when they exploded at high altitudes.

The light weight guided parafoils or fully autonomous powered parafoils were examined and used successfully by space explorers, in 1960s and 1970s (Stein and Strahans, 2018). Such Autonomous Parafoils helped in recovery of personnel (astronauts). Since early 1980s, drag parachutes have been regularly and efficiently utilized, to slow down Space shuttle each time it landed after a ride into space. Rapid deceleration required was partly achieved, by using such round parachutes fixed at the rear end of shuttle vehicle (Plates 2.1 and 2.2).

During past two decades (2000 to, 2018), guided (remote controlled) and autonomous parafoils have been tested and utilized regularly, in space science. They have been successfully deployed at very high altitudes of over 55,000 ft., to aid slow and steady descent of rocket portions, modules and personnel (astronauts) (Dunker et al., 2015; Fields et al., 2014; McKinney and Lowry, 2009; Rumermann, 2009; Van Rossem, 1997).

PLATE 2.1   A drag parachute used on a space shuttle.

Note: A parachute slows down the shuttle on the runway after landing at the Kennedy Space Station. The parachute offers horizontal drag.

Source: Kennedy Space Center, National Aeronautics Space Agency, Florida, USA.

During recent past (2014–2018), NASA of USA has aimed to develop an inexpensive on-demand capability, to return samples from the International Space Station (ISS). The Small Payload Quick Return (SPQR) system is being developed by NASA's Ames Research Center. Here, parafoils were explored for regular use in such airdrop of small payload from high altitudes and space. In this regard, during 2014 to 2018, researchers at University of Idaho evaluated a very small parafoil (or nano-parafoil) that can be deployed, at heights near space or just below. The idea is to transport or air-drop small pay loads from International Space Station, frequently. The autonomous small parafoil is controlled using radio signals from ground control station or its' descent could be pre-programmed. Incidentally, Kologi (2015) has evaluated several aspects of design and development of small parafoils of great utility to space science projects.

During January 2017, NASA tested world's largest parafoil as part of a program related to International Space Station. The parafoil was tested in Arizona. The aim was to examine the use of large parafoils to recover large payloads from International Space Station. They would be used during periodic re-entry of payloads. Largest parafoil until now was flown in January 2018. The parafoil recently tested in Arizona has a span of 143 ft. and a total surface area of 7,500 ft^2. These characteristics make it the largest parafoil in the world (Hartsfield, 2018).

PLATE 2.2   Soyuz Landing: A slow descent using a round parachute.

Note: The round parachute uses vertical drag and slows down the descent of space craft's module and personnel, on a desert surface, in Kazakhstan (see Wright, 2017).

Source: SoyuzLanding, Soyuz page, NASA, Washington D.C. https://www.nasa.gov/mission_pages/station/structure/elements/soyuz/landing.html.

2.1.3 PARACHUTES AND PARAFOILS IN AGRICULTURE AND FORESTRY: HISTORICALLY A RECENT TREND

Parachutes were first utilized to survey forest stands and to locate forest fires, sometime around early 1930s, in California, USA. In 1935, they experimented with the idea of dropping large number of fire fighters, at the spots affected within forest stands (USDA, 2018). A few professional parachute jumpers demonstrated the utility of parachutes in air dropping of personnel, in forest areas. It was abandoned for a while, but later, it became a good option for the USDA Forest Service to use parachutes. Also, experiments using fire fighters continued in Washington D.C. area, until 1938 and later too. Simultaneously, USDA Forest Service tested the idea of use of chemicals to control forest fires. They used parachutes to drop material required for fire control, at different spots in the Western regional forests. In late 1930s, parachutes from companies such as Eagle Parachute Company and Derry Parachutes Company were used in fire-fighting exercise inside deep forests. Air dropping were successfully conducted. The USDA Forest Service gave several contracts to produce suitable parachutes, fire-fighting equipment and clothing. They also trained forest fire fighters in parachute jumping.

In USA, Parachutes had found their way into agrarian regions and forest stands with in just one to two decades, after it was being inducted into military (USDA, 2018). By 1940s, parachutists were used regularly at Chelan Forest Experimental Station, Washington D.C. area. The parachutists jumped from 2000 to 6000 ft. above forest stands or timber yards and landed safely (USDA, 2018).

Research reports during 2006 to 2013 clearly indicate use of autonomous parafoils, in agricultural farms. Autonomous parafoils were used to obtain detailed aerial photography of terrain, water resources and cropping systems. Such light parafoils were first adopted in German and Polish plains to detect growth variations, crop-N status, disease/pest attack on crops such as wheat, barley and lentil (Thamm and Judex, 2006; Thamm, 2011; Thamm et al., 2013). During the same period, parafoils were first used during precision farming, particularly to mark 'Management Blocks.

During this past decade (2005 to, 2015), parachutes and parafoils have been evaluated for use in aerial survey of agricultural terrain, spectral analysis of crops, detection of disease, drought affected patches or soil erosion, etc. Researchers in Germany and Poland have shown that parafoils are useful in assessing agricultural fields (Pudelko et al., 2012; Thamm and Judy, 2006,

Thamm, 2011). Parafoils have been deployed and utilized efficiently during formation of 'management blocks' in large crop fields (Thamm et al., 2013). Remote controlled powered parafoils were used to monitor crops in experimental stations in the Polish plains and other regions of Europe (Pudelko et al., 2012).

Recently (i.e., 2015 to, 2018), parachutes/parafoils have also been utilized to detect crop disease/pest attacked zones in a large farm. Further, it is interesting to note that parafoils with pesticide tank in the pay load have been effectively used, to spray plant protection chemical on to crops below (Akhramovich, 2017; Pudelko et al., 2012; Thamm et al., 2013).

Agricultural drones, particularly, small fixed-winged and copter models are gaining in acceptance by farmers in different agrarian regions. Such autonomous drones are prone to mishap due to mechanical and computer programming related errors. These small aircrafts are fitted with a 'parachute pod' that pops up in times of distress. Such parachutes are named 'Drone Parachutes' or 'Drone Rescue Parachutes.' They rescue defective and crashing drone aircrafts. The parachutes are small and about 40–50 g in weight. They help in slow and safe landing of drone air crafts. The drone parachutes are in vogue since 2000 AD, i.e., when small drones became more popular, in agricultural farms.

2.2   DESCRIPTION OF A PARACHUTE AND PARAFOIL

The word parachute means against the fall. A parachute is designed to slow the motion of an object. Normally, it is used by people, to slow down their descent to earth or to other celestial bodies. It is also used to slow down vehicles transiting at great speeds like aircrafts, space shuttle, high speed cars, etc. It can also be used to slow down the descent of objects into the atmosphere. The parachute is capable of deceleration in both vertical and horizontal direction. Parafoils are derivatives developed using technology *similar* to parachutes. However, parafoils are steerable. They are more efficient in terms of flight control and smooth landing.

Parachutes have been used extensively for airdrop of military personnel and cargos. Also, by civilians in recreational and sport jumping. The two most common types of parachutes used are round parachutes and rectangular shaped parafoils. Parachutes capture air below the canopy. It provides a drag force to decelerate a payload. The velocity of free fall reduces. Parafoils are inflated with air inside rectangular foil shaped cells. It provides a lift force as well as a drag force. Hence, parafoils can be steered to deliver payloads accurately on the ground. Based on this capability, parafoils are currently actively pursued by the military for precision aerial delivery of cargos (See Knacke, 1991; National Aeronautics and Space Agency, 2017; United States Department of Transport. Federal Aviation Administration, 2013).

2.2.1 PARTS OF A PARACHUTE AND PARAFOIL AND THEIR FUNCTIONS

The design and construction of a parachute and its components are based on the old idea that strength and durability is determined by weakest links. Every component, or link from the jumper to the canopy must carry its share of the maximum load that is applied, during the opening shock. The five major parts of a standard service parachute are the pilot chute, main canopy, suspension lines, harness, and pack. The pilot chute has the job of anchoring itself in the airstream, then, pulling the remaining packed components out of the parachute pack (Figures 2.1 and 2.2).

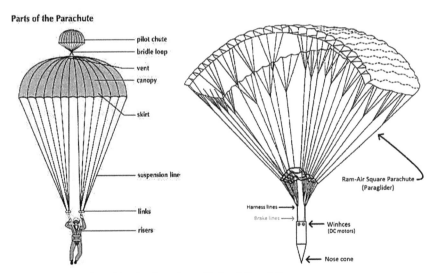

FIGURE 2.1 Left: A Circular Parachute. Right: A Parafoil

Note: A simple parachute/parafoil and its major parts: Pilot chute, bridle loop, vent, canopy, skirt, suspension lines, links and raisers.

Source: Gene Engelgou, Fruitychutes Inc, Monte Sereno, California, USA; National Aeronautics and Space Agency, Washington D.C., USA.

There are indeed several types and sizes of parachutes and parafoils used by different agencies. A few different factors influence the manufacture of a parafoil/parachute. Factors that influence are purpose, weather conditions, altitude for cruising and airdrop, size and nature of payload, descent rate required, material used for the parachute, and economics. So far, considering the frequency with which a particular sized parachute/parafoil is adopted, we can state that, most parachutes range between 17 ft. to 35 ft. in sides and 28 ft. in diameter. For example, a 28-ft. canopy contains approximately 796 ft² of nylon cloth, plus 2,400 yards of nylon thread. Sewing on a parachute varies from 8 to 10 stitches per inch. Parachutes and parafoils are manufactured adopting machine stitching, except for zigzag. The canopy is provided with vents. Such vents help in releasing small amounts of air through the canopy. Vents reduce the high internal pressure within the parachute, at the instant of opening. A few vents are a necessity. Otherwise, opening the canopy at high speed could result in the rupture and damage of canopy. The vents are invariably tubular nylon webbing with a tensile strength. It prevents tears from completely separating the canopy. There is a process where the nylon is rolled under heat and high pressure to lower the porosity of the nylon without the need for coatings. The result is a very strong light weight fabric with low permeability. As a consequence, if we get a hole or tear in the rip-top it does not spread (Fruitychutes LLC., 2018a,b; Lingard, 1995; Table 2.1).

TABLE 2.1   Specifications of a Typical Ram Air Parafoil

Canopy:
Model: Modified MC-4; Chord (m):1.27; Span (m): 3.22; Aspect Ratio: 2.54; Inlet height (m); 0.153
Airframe:
Power: 5.3KW BLDC; Weight: 1.2 kg; Actuator Type: Winch Type

Source: Kim and Song (2017).

A brief description of parts and their purpose is as follows:

- Pilot chute: It is a small parachute located on top of the main canopy. It opens the parachute;
- Bridle: It is used to connect pilot chute and main parachute;
- Vent: It allows the escape of air from main parachute. The vent avoids stray loss of air from outside the canopy. Which otherwise can cause destabilization and wobbling in the air.

- Canopy: This is the main part of the parachute and it is made of special type of nylon or silk or tarpaulin. As the parachute size goes up, the number of gores increases. This allows more connection points to the parachute and keeps the load capability per shroud line more constant. Small chutes have just 8 gores, while a 192 inch parachute has 18 (Fruitychutes LLC., 2018a,b).
- Skirt: It forms the lower part of the parachute's canopy. Its edges have to be sewn without any tear.
- Suspension lines: These are threads made of tough material and are attached to skirt of the canopy. They help in distributing the weight of payload (e.g., a person, large container, aluminum cage for sensors, etc.);
- Links: They are crucial parts that connect the suspension lines to risers;
- Risers: They connect the links to harness;
- Control lines: They are useful in steering and braking the parachute/ parafoil.
- Harness: It is used to tightly fasten the paratrooper. A container is used to hold all the parts of the parachute, when not in use. It is also held in a cotton backpack. The pack is fastened to the parachutist with a harness. It is specially constructed so that the forces of deceleration, gravity, and wind are transmitted to the wearer's body as safely and comfortably as possible. Modern parachutes are nearly always worn on the back and are rarely worn on the chest.

Regarding parafoils, commonly listed and essential parts of a powered parafoils are as follows: A parafoil canopy with air-foil pockets, suspension line, harness, motor attached to propeller, cage-quarter part, large exhaust donut, exhaust spring with nut and bolt, fuel tank, maintenance kit, Intake silencer, pull starter, re-drive assembly, throttle assembly, common helmet, air filter, kill switch, propeller cover, spark plug cap, soft j bar suspension, speed bar riser, trim riser and starter pawl (Poynter, 1991; Kurt, 2018, Lingard, 1995; Plate 2.1).

A clear knowledge about descent rate is most important when we adopt parachutes/parafoils, to air drop a cargo. The descent rate of a parachute depends on a payload (e.g., supplies, persons, rocket recovery module, etc.). The descent rate depends also on drag force that a parachute develops to counter the gravitational fall due to the weight of the payload. Now, the drag force itself is dependent on; (a) dynamic pressure created by the air striking the canopy of the parachute and keeping it inflated; (b) diameter of parachute decides the area upon which the air strikes the canopy; (c) drag coefficient (Cd) of the parachute also affects descent rate. The dynamic pressure generated on the canopy is a function of velocity of air and air density. It depends on altitude and temperature. The drag coefficient of any parachute is dependent on factors such as: shape of the canopy, surface area of the canopy, gliding characteristics, air flow density and pattern, permeability of material used for canopy, descent velocity and length of the shroud line.

The vertical descent rate (i.e., velocity of descent – Ve) of a parachute is calculated using a formula (see Potvin, 2018):

$$V_e = \sqrt{\{2\ W_t / S_o\ C_d \rho\}}$$

where, C_d is the parachute *drag coefficient* which is approximately 0.75 for a parachute without holes or slits cut in the fabric – same value in both Metric and English unit systems; ρ *(rho)* is the air density – near sea level, its value is given by 0.00237 sl/ft^3 (English units) and 1.225 kg/m^3 (Metric), and near 4000 ft. (or 1219 m) above sea level its value is 0.00211 sl/ft.^{3} (English units) and approx. 1.07 kg/m^3 (Metric); W is the weight of the parachute + load, in pounds (English) or Newtons (Metric); V is the vertical descent velocity, here expressed in ft/sec (English) or m/sec (Metric); S is the *total surface area* of the fabric used to build the parachute, plus the areas of the holes and vents cut in the fabric – if present, the units of S are in square feet (English) or square meters (Metric). This definition is such that when vents are cut in the fabric, *the value of S remains the same, but the value of CD becomes smaller*, where, W_t = total weight of body and parachute (lb.), S = canopy reference (surface) area (ft^2), and ρ = air density (slug/ft^3) (Potvin, 2018).

Often the company that produces the parachute also offers a user manual and computer based- parachute descent rate calculator. For example, in the present case, a parachute company called 'Fruitychutes LLC' offers a plotted graph. Using it, we can decipher the descent rate of a particular weight pay load attached to a round parachute (see Figure 2.2). We have to know the weight of cargo and a few characteristics of parachute model (e.g., Cd drag coefficient; diameter of parachute, weight of the payload, etc., to extrapolate and arrive at descent rate of parachute. The calibration graph is updated as often as possible with more points.

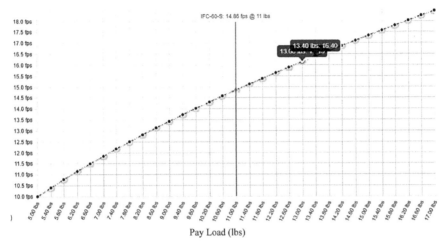

FIGURE 2.2   The Descent rate calculator for a parachute carrying a specific weight of pay load.

Source: Gene Engelgau, Fruitychutes LLC., Monte Serena, California, USA.

Note: The descent rate calculator graph has been plotted using weight of the payload (lbs) on X axis (horizontal) and descent rate ft. s^{-1} on Y axis (vertical). The above calibration graph is provided by a company Fruitychutes Inc., Monte Sereno, California, USA. For a given weight of the payload, usually, the computer program offers at least two different models and their descent rate.

2.3   TYPES OF PARACHUTES AND PARAFOILS WITH EXAMPLES

There are several models of parachutes designed, tested and utilized by different agencies. Parachutes are made of wide range of materials. They differ in size. The models are usually designed based on the purpose, such as height at which it should get deployed and pay load it must carry. The design also depends on many other functions such as gliding, transport or need to float above crops to obtain spectral data, regular surveillance of ground features, etc. Parachutes/parafoils are grouped or classified based on different criteria. A long list of parachutes, parafoils, deployable para-wings, steerable parafoils, rotating parachutes, aerial recovery parachutes is available (Wikipedia, 2018 a,b). Several designs have been tested for deceleration, for five decades, by agencies such as National Aeronautics and Space Agency of USA, Soviet Space Agency, etc. (National Aeronautics and Space Agency, 2017; NASA, 2018; Donaldson, 2009; Lee, 2005).

Broadly, parachutes are divided into two types: ascending and descending. Ascending refers mostly to paragliding and recreational parafoils.

- **Based on design**: The parachutes are classified into three groups based on the design (Sanchez, 2017):

  1. *Conventional Parachutes:* (a). Solid Textile Parachutes: – Flat Circular – Hemispherical – Conical – Guide surface – Biconical – Annular – Triconical – Cross. (b). Slotted Textile Parachutes: – Flat Ribbon – Ringsail – Conical Ribbon – Ring Slot – Ribbon (Hemisflo) – Disk-gap-band.
  2. *Rotational Parachutes:* e.g., Rotafoil, Vortex Ring, Sandia RFD.
  3. *Maneuverable or Gliding Parachutes:* e.g., TOJO, T&U slots, LeMOIGNE, Para wing, Ram-air/Parafoil Sail wing, Volplane.

- **Based on shape:** We can group parachutes into circular, conical, oval, triangular, rectangular and parafoil types.
- **Round-type Parachutes:** These are used in the military and as emergency rescue parachutes. The circular or round parachutes rely on drag to slow a descent, rather than having any lift. They are made of dome-shaped canopies. They are often referred to as 'jellyfish chutes. They are less frequently used by modern jumpers.
- **Cruciform Parachutes:** Cruciform parachutes are designed to provide a steady descent. They show relatively reduced oscillation. Such square-shaped parachutes have been modified. They can reduce descent speed by as much as 30%. Therefore, cruciform parachutes reduce landing injury.
- **Ringsail Parachutes:** This type of parachute was developed by modifying the ring slot one. It was designed by Edward Ewing in 1955. Initially, interest in this type of parachute was feeble. However, now a days, it is used as a method of escape system during landing at the end of space mission. It has a good opening reliability, damage tolerance and low shock (Buhler and Wailes, 1989; Delurgio, 1999; Ewing, 1972).
- **Rogallo-Winged Parachutes:** Rogallo wings are used to increase forward speed and decrease landing speed. Many trials and experimentations have been conducted with Rogallo-winged parachutes. These Rogallo wings are frequently used during sports flight and gliding. There were difficulties encountered during production of Rogallo wings. Plus, the introduction of ram-air parachutes or parafoils meant low demand for Rogallo wings.

- **Annular Parachutes:** Annular parachutes have a ring – or a series of concentric rings in the canopy. They have a lower drag factor than conventional round parachutes. The rear position of the vents can give the jumper considerable forward speed in descent (McWilliams, 2013).
- **Ram-air Parachutes:** Most of the modern parachute types deployed are ram-air, specifically, in sports jumping. The self-inflating airfoils, known as parafoils, give the jumper greater control of speed and direction. The air foils will spread the stress of deployment (a major problem on some older chutes). Two layers of fabric allow air to penetrate from vents in the front and form cells. The use of sail sliders allows the jumper to adjust the canopy during descent. Sail sliders maintain greater control of airspeed, descent speed and direction. Further, vents and brake lines can also allow a greater degree of control.

We use parachutes and parafoils to accomplish several different tasks. We can also classify parachutes based on purpose for which they are adopted. We can classify them into drag parachutes, air drop parachutes, gliders, recreational parafoils (sky diving and gliding), cargo transporters, aerial photography and survey parachutes/parafoils, sprayers, geological mineral and water resource prospecting parafoils, etc.

A few types of parafoils are controlled using remote control signals. Some of them are entirely autonomous. These types of parachutes are pre-programmed, regarding flight path and aerial photography. Specific computer software such as 'Pixhawk' or *'e-Motion'* are used. Parachutes derive power from different sources we can classify powered parachutes/parafoils based on power source. Powered parafoils and microlight gliders use petrol engines, electric engines or sometimes even solar powered engines. However, the source of power is commonly petrol/diesel. A two-stroke engine gets connected to a propeller. Parachutes without powered engine are usually gliders. They utilize wind current and float through the distance. They are usually guided by a pilot. Powered parachutes could be guided, using GPS and a remotely placed computer control (e.g., Pixy, SUSI, Hawkeye, etc., see Krishna, 2016, 2019).

2.3.1 CIRCULAR OR ROUND PARACHUTES

There are indeed innumerable research papers and treatises published about circular parachutes, their designs, functioning and applications. Several different operation and maintenance manuals have also been written and released periodically, during the past 4–5 decades. Circular or round parachutes are the earliest types of parachutes designed, developed and utilized, in large numbers. Circular parachutes vary regarding their size, length, breadth, canopy characteristics, pay load area, harness, packing method, etc. They were utilized to conduct airdrop from tall buildings, aircrafts and balloons. Circular parachutes are among the simplest to design and manufacture. They are mostly used by military establishments, to achieve careful deceleration of personnel and equipment from large cargo airplane, such as C-130. Dhobrokhodov et al. (2003) have reviewed our knowledge regarding aerodynamic decelerator system and related technology. They mostly deal with circular parachutes. It is said that circular parachutes have been the simplest and cheapest devices used for the delivery of materials, people, and vehicles. This is ever since their first recorded use by Jacques Garnerin, who jumped from a balloon over Paris in 1797. However, aerodynamics of circular parachutes is not easy to model. Significant research has been conducted for over 60 years, towards gaining absolute control of the air flow and canopy's descent rate. Yet, we have not mastered handling of parachute's descent rate. Parachute drifts and wobbling that does occur. and the drift in the atmosphere too has not been mastered. The payload carried by a circular parachute differs enormously. Usually, circular parachutes are selected after careful evaluation of payload area, type of payload, its weight, material and other characteristics (Fruitychutes LLC., 2018a,b; Figure 2.3).

2.3.2   PARAFOILS

A parafoil is a parachute-like device that can be steered, used especially as paragliders. Parafoils have good stability, excellent gliding performance, and maneuverability, if we compared them with round parachutes. Such characteristics have made them popular in the area of spacecraft recovery. At present, among aerial drones, autonomous parafoils, have become the focus of research pertaining to cargo transport and recovery technology.

Parafoil is an air drop device consisting of a non-rigid (textile) air foil with an aerodynamic cell structure. It is inflatable by wind (and used without engine or motor). It is designed to slow the descent and impact of cargo, equipment or personnel dropped from an airborne aircraft. Ram-air inflation forces the parafoil into a classic wing cross-section. Parafoils are most commonly constructed out of ripstop nylon (e.g., Parafoil ZD 141).

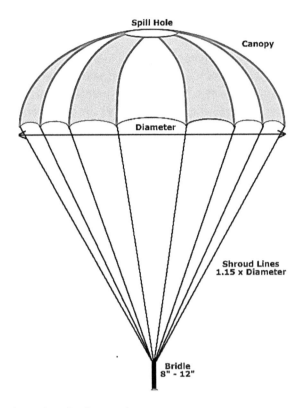

FIGURE 2.3 A round or circular parachute.

Source: Gene Engelgou, Fruitychutes LLC., Monte Sereno, California, USA.

A parafoil has a soft fully inflatable wing. Parafoils are familiar in the sports of hang-gliding and parachuting. This system is designed to retard the vertical velocity and provide a relatively soft touchdown. The lifting parafoil has three advantages over the conventional type:

1. being able to reduce the dispersions associated with trajectories, by using its maneuverability to glide to a predetermined point;
2. ability for manual control; minimized landing area impact dispersions and;
3. by flaring, to reduce the impact shock at touchdown.

The advantages of a conventional parachute are reduced weight and less complexity. If a parafoil is deployed at an altitude of approximately 6000 meters, typical performance characteristics provide a maneuvering circle of about 16 km in radius.

The parafoil or "square" parachute is popular in sport parachuting, but it has limited use in the military. They are used mainly during special operations and demonstration teams. Most military parachutes are round and have limited or no steering capability. Such features are important to large scale paratrooper operations. They say, it has undesirable effect to have several hundred paratroopers independently steering their parachutes, because of the risk of collision. During recent years, autonomous parafoils with pre-determined path and way points are finding acceptance. They are an excellent choice for low altitude aerial survey of agrarian regions. The visual and multi-spectral images from parafoils are used in precision farming (Thamm et al., 2013; Pudelko et al., 2012; Krishna, 2018, 2019; Lee, 2005).

Parafoil trajectory planning systems must be able to accurately guide the highly non-linear, under-actuated parafoil system. Say, from the drop zone to the pre-determined impact point. Parafoil planning systems are required to navigate highly complex terrain scenarios (Sugel, 2013). Flight control systems are important during agricultural surveys. In agricultural locations, perhaps, it is equally important to navigate under harsh weather patterns. The autonomous parafoil should transit over flat, undulated and hilly terrain covering pre-determined way points. This is to obtain high resolution images of crop fields. Sugel (2013) states that, in general, parafoil dynamics are highly non-linear, relying on aerodynamic drag effects for lateral control. Parafoils are subject to uncertain and variable wind environments. Such deviations must be compensated suitably. Otherwise, it results in large errors between predicted and actual transit path. High-risk military, humanitarian and even agricultural applications often have tight landing and aerial photography restrictions. They could be narrow valleys, crop fields, small patch adjacent to water body or a patch with other terrain related problems, or along the edge of a cliff.

A parafoil can carry greater payload compared to its weight than a fixed-wing or rotary aircraft. This is a clear advantage that makes powered parafoils useful for bulk air delivery. Further, they can now be GPS-guided to a target or location on land, accurately. So, parafoil system can be a good means for civil uses. Particularly, to carry food and medical supplies. No doubt, they are of immense importance in transport of military cargo. Parafoils are useful in locations afflicted by disasters. However, the parafoil system is very susceptible to wind disturbances, which makes it difficult to design a good autopilot for it (Kim and Song, 2017).

2.3.3 TETHERED PARAFOILS

Lingard (1995) defines a tethered, ram-air parafoil as an aerial vehicle whose geometry is determined, by the fabric panel elements and the system of rigging. The fabric usually forms an upper and lower airfoil surface. The fabric is subdivided into small span of ribs and cells. Tethered parafoils are made of tough fabric. The fabric withstands high tension. It is made of non-porous material. The ram-air has vents at one edge of cells. The parafoil attains its aerodynamic shape by ram-air pressurization through vents, in the leading edge of the foil. The pressure inside the cell can be made uniform by making cross-ports. It has two symmetrical "power" lines and two steering lines. The powerlines carry the force developed by the foil to the control pod, where they merge onto the main tether. The foil is controlled by adjusting the relative length of the steering lines. The steering lines are also called 'brake-lines.' Deflection of the brake lines will induce a coupled yaw and roll output. It helps in steering the parafoil to the desired direction. In the brake-line zero position, the control lines are very lightly loaded. To minimize airborne weight, extremely high strength-to-weight ratio polymers such as Dyneema braided rope (Bosman et al., 2013) are used, both in the rigging lines and main tether.

Tethered parafoils have certain clear advantages. They offer longer continuous flight in the air. That is as long as it is connected to ground station or a moving vehicle. They are much easier to handle than a totally electronically controlled autonomous parafoil. Reports from Brazil and USA suggest that, tethered parafoils have been utilized effectively, to conduct un-interrupted long-term surveillance of buildings, farms and crop land. For example, AeroVironment Inc., (California, USA) produces a tethered parafoil called 'Tethered Eye.' It is a kind of perpetually floating parafoil connected to ground station. It draws power from GCS to operate a set of high-resolution sensors. The gimbal carries visual, multi-spectral and infra-red cameras. This 'Tethered eye' – a parafoil has been useful in the surveillance of defense installations and military camps. Surveillance could be conducted continuously and for long stretch of time (Aero Vironement Inc., 2018). This tethered parafoil has 10 days endurance. Therefore, it keeps excellent continuous vigil of the happenings on ground. The locations and items monitored could be a farm, farm installations, crops, irrigation lines, urban buildings or city traffic, etc. Tethered parafoils costs much less relative to other UAVs, such as autonomous drone aircrafts (fixed-winged or copters), etc. Tethered parafoils are easy to deploy from ground. Reports state that tethered parafoils could be flown above the farm or field, in 10–15 minutes. Forecasts indicate that tethered parafoils and balloons could dominate the agricultural skyline, in future. They both are relatively low-cost instruments and re-usable several times, above a farm or a crop field. They could be customized accurately to just pick aerial images of a certain area or an individual farm. Therefore, farmers may prefer to fly them consistently above the fields. Say, for the whole season and monitor the crop and agronomic procedures.

Parafoils are usually tested in a wind tunnel for performance. Also, to collect data relevant to its flight pattern. Generally, tethering a parafoil helps in testing the parafoil better. Tethering technique has been adopted on parafoils of 18 ft^2 to 800 ft^2 size. In many instances, the tethering rigs are used to connect the payload instruments with ground control (Brown, 1989; Matos et al., 2000; Coleman et al., 2016).

Tethered parafoils have been utilized by meteorologists to study climatic parameters. They help to record changes in the gaseous composition of atmosphere. No doubt, agencies dealing with greenhouse gas emissions and atmospheric quality over farming zones could utilize, a series a tethered parafoils. They can collect valuable data about the weather that prevails over crop fields (Jensen, 2000).

Parafoils tethered to trucks or pickup vans could be moved across crop fields. They could be stationed for a while to collect spectral data of crops and the atmospheric data, at each point. Tethered parafoils are portable if attached to an automotive vehicle on the ground. Tethered parafoils can be made to hover around a point in the atmosphere. They could be used as a watch tower over ground installations. Tethered parafoils connected to ground control station computers can monitor terrain, military camps, agricultural farms, crops, irrigation channels, etc. (Langers, 2016). A tethered parafoil fixed to a central location in a farm could provide excellent surveillance, all through the crop season. In addition, it could relay useful data about crop growth, its nutrient status, water stress index, need for irrigation, etc. A tethered system permanently flying over a location within a field can help in detecting disease/pest attacks to crops.

2.3.4   POWERED PARAFOILS (MANNED)

Powered parafoils (PPC) are among the most preferred aerial vehicles. They are relatively easy to assemble, swift in the air, light in weight and safe to use in different regions. Powered parachutes/parafoils meant for use over agricultural and urban locations possess a three wheeled cart (or payload area), 1 or 2 seats for pilots, an engine (petrol or electric) and a propeller placed below the canopy (airfoil). In a powered ram-air parafoil, air rams into front openings (cells). It keeps the canopy stiff and in correct shape. Powered parachutes are relatively sedate and easy to fly. They fly at a roughly constant speed of about 30 mph (27 knots). There are only two controls. The engine throttle, if pressed, makes parafoil go up faster. If we retract from throttle, it makes parafoil slower and helps to descend. The foot-pedals which tug on some of the support lines help in swerving/turning left or right. At the time of landing, we press on both pedals at once. This is to slow the parafoil down just before you touch down. Aerial photography could be done manually, if piloted. If autonomous, we may have to place sensors (visual, multispectral and IR) and control the shutters, to get good spectral images of ground features, topography, crops, irrigation lines, etc. These powered parafoils are typically flown at low altitudes like 1000 ft (300 m). Powered parachutes take-off from large, grassy fields. They cost about US$14,000 for a new one-seater and US$20,000 for a two-seater. (2018 price level). They are small and can be transported in a small trailer or pick-up van and stored in a garage. (Buchanan, 2017; Kolf, 2013; Plate 2.3; Table 2.2).

As stated above, a powered parachute has a motor, propeller and wheels. It has para-wings or air-foil parachute to produce aerodynamic lift. The chute is positioned above the machine and it collects air, as the pilot starts for take-off. Once the parachute is fully inflated, the take-off is achieved by pressing full throttle. Full throttle helps to achieve greater speed on the runway. Maximum speed of a powered parachute in the air is about 25 to 35 km. It can fly low skimming just over treetops and buildings. Otherwise, if needed, it can reach a ceiling height of 10,000 ft above ground level. The powered parachute's endurance depends on the motor and gasoline that it holds in the tank. However, it can drift and stay afloat for a while before descending slowly, in case of exhaustion of fuel or engine brake-down. Powered Parachutes are easy to set-up, take-off or land as many times, quickly and in succession on a runway of 100 ft length. Powered parachutes are safer than fixed-winged drone because of greater stability in the air. Therefore, they are used for sky sports. Usually, it takes about 10–15 hr of training to use powered parachutes. In farming zones, it is said a powered parachute is a highly useful aerial gadget. Farmers can quickly take-off and surveillance large fields. Farmers can get a close-up view by flying low. They may also fix sensors (cameras) of visual, multispectral and infra-red bandwidth. This is to obtain sharp images of the crops below. Farmers can obtain information on growth pattern, leaf chlorophyll content, disease/pest attack, if any, drought affected patches and soil erosion, etc. (Blue Sky Gyros, 2016). A very detailed description of powered parachutes, its various parts, minimum flying standards, regular maintenance and servicing of engine, etc. are provided, in a handbook published by the United States Department of Transportation's Federal Aviation Administration (United States Department of Transportation, 2007).

Powered manned parafoils have already found their way into farming belts in Europe and Australia. Reports suggest that, parafoils models such as 'Buckeye' and 'Paraplane' manufactured elsewhere in USA has been adopted in farms. Parafoils conduct aerial survey of farmland, crops and irrigation channels. The imagery and detailed spectral data are analyzed prior to conducting ground operations, in a farm (Farm Show Magazine, 2018a,b). These powered parafoils are unlike the light aircrafts because they do not crash, if malfunctions occur. In case the engine fails the parafoil still floats and descends slowly with payload and pilot intact. The pilot buggy usually allows one or two pilots. They say, it takes just one hour to learn flying a powered parafoil with the help of a good trainer. Therefore, it is easy indeed to adopt parafoils in farms. The parafoil takes off on a runway strip of 25 ft. and travels up to 26 kmph speed over the crops or urban areas at low altitudes. A five-gallon tank with petrol or diesel keeps the parafoils aloft and moving briskly over fields or towns. The parafoil models can be assembled from scratch in about 20 hrs. We have to use manuals supplied by the manufacturer (e.g., Coldfire Inc., California and Paraplane Inc., New Jersey, USA; see Taylor, 2018).

PLATE 2.3   *(Continued)*

PLATE 2.3   (Top): A powered parafoil in flight. (Bottom): A three-wheeler cart with two-seater including pilot and a pay load area.

Note: Powered parachutes can be guided by a pilot or they could be used as aerial robots. The GPS connectivity and pre-programmed flight path are essential, only, if they are entirely robotic. Powered Parafoils are useful in aerial surveillance and imagery of natural resources. They provide excellent aerial images of farmland, topography, soil type, cropping systems, etc. Spectral data that they provide can be used, to assess crop's-N status, water status, disease/pest attack, drought/flood or soil erosion, etc. Also see: http://www.infinitypowerchutes. com/videos/.

Source: Alvie Wall, Founder President, Infinity Power Parachutes Inc., East Lafayette Rd. Sturgis, MI, USA.

The two-seat powered parachutes (PPC) are light sport vehicles. The person flying PPCs must have at least a sport pilot's certificate issued by the Federal Aviation Agency (FAA), to fly them. A minimum of 10 hours of flight instruction, and 2 hours of solo as a student pilot, is required to obtain this certificate. Powered parachuting is not to be confused with powered paragliding.

The PPCs are often considered to be safer than normal fixed-wing aircraft. This is because of their inherent stability, limited response to control inputs, and stall resistance. There are two primary means to control a PPC: increasing or decreasing engine power (which controls vertical rate of climb) and deflecting the right or left trailing edge of the parachute—typically by moving the steering bars with the feet. It turns the parafoil right or left. If the trailing edge of the wing is pulled in, on both sides at the same time, the parafoil "flares," i.e., slows and temporarily gains additional lift.

TABLE 2.2   Powered Manned Parafoils: A Few Examples

Name: Sky Rascal

The powered parachute, 'Sky Rascal' is basically a recreational sport machine. It has a sleek lightweight design, integral frontal fuselage bars, and powerful engine. It allows farmers to have a solo flight for lengthier durations, above cropped fields.

Specifications:

Stall Speed: Virtually stall resistant; *Max Payload:* 447 lbs; *Engine:* 350 lbs; *Take-Off Roll:* 25 ft. to 300 ft.; *Land Roll:* 10 ft., to 100 ft.; *Sink Rate:* 10 – 12 ft. per second; *Glide Ratio:* 3:1 to 4:1 (rectangular) 5:1 to 6:1 (elliptical); *Air Speed:* 26 – 35 mph; Climb Rate: 350- 1000 ft. pm; *Descent Rate:* 600 ft. per min. (engine off).

Name: Infinity Commander Series 65 HP

The Commander Series Parafoil is available as S-LSA or E-LSA.

Specifications:

Max Speed: 28–32 mph; Fuel Consumption: 4 gal./hr; Stall Speed: Virtually Stall Resistant; Climb Rate: 200–900 ft/min; Cruise Speed: 28–32 mph; Take-off Roll: 150–600 ft.; Glide Ratio: 4/1; Landing Roll: 100 ft.; Decent Rate: 10 ft/.sec; Ceiling: 10,500 ft. Empty Weight: 385 lbs.; Engine: 582 ROTAX DC/DI Two Stroke, Oil Injection, "B" Drive, 65HP; Payload: 600 lbs; Wing Area: 500 ft²; Fuel Capacity: 10 gallons.

Name: Infinity Commander Series 100 HP

The Infinity Commander series is also available with a 100 HP engine.

Specifications:

Max Speed: 28–32 mph; Fuel Consumption: 3 gal./hr; Stall Speed: Virtually Stall Resistant; Climb Rate: 900+ ft/min; Cruise Speed: 28–32 mph; Take-off Roll: 150–600 ft.; Glide Ratio: 4/1; Landing Roll: 100 ft.; Decent Rate: 10 ft/sec; Ceiling: 10,500 ft.; Empty Weight: 485 lbs.; Engine: 912S ROTAX Four Stroke, Four Cylinder, Dual Altitude Compensating Carburettors, Dual Ignition, 100HP; Payload: 600 lbs; Wing Area: 500 ft.; Fuel Capacity: 10 gal.

Name: Airwolf 912ULS

Airwolf is an excellently engineered powered (manned) parafoil. It is a tough and reliable flying machine. It has a 4 stroke 100 HP engine. It can fly over agricultural terrain, urban locations, mountains and carry payload. It has longer endurance per flight. The engine is to be serviced only once in 2000 hr.

Specifications:

Rotax 912ULS 100hp engine; Stainless Steel Exhaust; Dual Quad Hydraulic Spring Over; Suspension – SUPER 912 (EIS); Electronic Information System with altimeter; Electric Fuel Pump; a 10 gallon fuel tank; Choice of Color Powder coat; Rigid Chromoly frame extending full length; 4 Point Harness Seatbelts; Deluxe Dual Seats; Ground Steering Tow Handle; Strobe Light; Standard or Extended Footbars; Dry Cell Battery; Gear Reduction; Electric Starter; Spun Aluminum Wheels; Tundra Tires; 912ULS 100 hp Engine; Oil Cooler; Side Bag; Turf Glide Tundra Tires; 3 Blade 68″ Warpdrive Propeller; Performance Designs PD500 wing; Chute Bag and Line socks. There are several other optional attachments.

Source: Infinity Power Parachutes Inc., East Lafayette Road, Sturgis, Michigan, USA; Buckeye Industries Inc., Cassapolis, Michigan, USA.

The power-off glide ratio of a powered parachute ranges from 3:1 to 6:1. Glide ratio varies depending on the chute size and shape, and the weight that the chute is carrying. Engine-off landings are generally safe provided that the parafoil is within glide range of a suitable landing zone. Also, the pilot is properly trained in the use of proper "flaring technique." This is caused by the reshaping of the rear of the wing or parachute, which is pulled downwards. This results in the entire parafoil rocking on the transverse axis. Thus, slowing it momentarily. It allows an engine-out landing to become less rough within the last few feet off the ground.

The flare is generally used to make fine adjustments in altitude when flying close to the ground. In particular, when the parafoil is landing. "Flaring," in terms of powered parachuting, refers to the usage of the steering bars pushed out simultaneously, causing the parafoil to rock forward. Flaring will allow an engine-out landing to be much smoother than simply relying on the lift of the ram-air canopy, to cushion the landing. Chute collapse is considered by many pilots to be virtually impossible with square wings. The wing is more likely to collapse with elliptical winged parafoils.

Wind is a major hazard. Wind direction and speed need due consideration. The main hazards one faces while flying a powered parachute are associated with wind and obstacles. Flight should not be attempted in winds exceeding 10–15 mph or in gusty conditions. Wind hazards include terrain-induced air disturbances. It is advisable to stay above trees, mountain peaks, and other obstacles that disturb the flow of the wind. Special care must be taken to avoid power lines, trees, and other low-level terrain obstacles.

Recreational flights using PPC are common. It is possible to fly low and slow over a terrain, using PPCs. PPC pilots typically enjoy flying low and slow. The PPC is an excellent platform for sightseeing and photography.

The PPCs are also used in agriculture. It is adopted occasionally by law enforcement agencies. The PPCs do not need an airport to take-off and land. Small strips of land are enough to effect a lift-off. Generally, ultra-light PPCs are not allowed to fly at night. However, a properly equipped PPC could be flown at night or over metropolitan areas, by a private pilot with a PPC rating.

2.3.5   TRIKES OR MICROLIGHTS ABOVE CROP FIELDS

Trikes could be useful in aerial survey, imagery and air dropping of pay loads. Grain producing companies and farmers with large land holdings may find it useful, to conduct overall surveillance of terrain, farms and crops, using trikes. Trikes may help in inspecting and detecting variations in topography, land and soil types, crop growth pattern, disease/pest afflictions on crops, irrigation lines and equipment, soil erosion, etc. In Europe, trike is usually called as 'microlight.' It essentially has a three wheeled cart that hangs from a pivot point on the triangular wing. The cart has 2 or 3 seats for pilots. It has an engine (petrol or electric) and a propeller. The engine throttle helps in regulating the speed of the trike. A steering bar helps in altering the direction of flight. The pilot can pull or push to alter the wing position and its tilt. The steering is attached to wings. We can select the type of wings based on purpose. Smaller wings give higher speeds (60 to 70 knots). They also help in negotiating windy conditions better. The small trike may range in cost from around 13,000 US$ to 30,000 US$ (2018 price level). They can take-off or land on a small paved runway of 500 m or on grass surface. Trikes can be dismantled. Its wings could be rolled, folded and towed on a trailer, if required. Regular radio connectivity, GPS and autonomous flight is also possible (Northwing, 2018; Wikipedia, 2018c, d; Buchanan, 2017; Plate 2.4). An example of microlight with specifications is given below:

The advantages provided by flexible wings (easy storage, light weight, low cost) is a significant criterion. It should be considered prior to procurement of a Microlight (Trike). Such advantages overcome their limited aerodynamic performances.

Microlights are popular wherever short aerial flights are required. Particularly, those needing low altitude surveillance. During the late 1970s and early 1980s, the hang-gliding movement induced many people to purchase powered microlights. The lightweight, slow-flying trikes were subject to minimum regulations. The weight and speed limits differ from country to country. In affluent countries, microlights or ultralight aircraft now account for a significant percentage of the global civilian-owned aircraft (Wikipedia 2018 c,d).

PLATE 2.4 A microlight (trike) hovering over undulated terrain.

Note: Farmers with large holdings may use a microlight flight above a crop field, to obtain an overall aerial image, trace variations in crop growth and soil fertility, identify locations with disease/pest attack if any, spots with flooding, stagnation or even erosion, spots with leaks in irrigation pipes, etc. A microlight offers a 'bird's eye view' of the farm. It offers a quick overall status of the crop field. They are semi-autonomous with an engine, propeller and a pilot to glide them at altitudes over farms. Spectral images (visual, IR and multispectral) could also be obtained by programming the sensor's shutters or manually handling them. Also see https://www.youtube.com/watch?v=1RbeNMBvEdg; https://www.youtube.com/watch?v=d6hR-5i675U.

Specifications of a typical Microlight (Trike) of 80 to 125 lbs. include:

Sail area: 181 ft²; Sail Span:34.3 ft.; Nose angle: 126°; Aspect ratio: 6.3; Stall at 490; Gross speed: 22 mph; Trim Speed: 28–34 mph; Top speed: 45 kmph; VNE: 53 mph; Sink rate: +/–235 ft. per minute; Wing weight: 84 lbs; Wing length: 18.3 ft; Glide ratio (L/D): 10.5:1 +/–; Minimum Gross: 330 lbs. to 480 lbs. gross.

Source: Dr. Kamron Blevins, President, Northwing Inc., Gala Avenue Chelan, Washington State, USA.

2.3.6   *AUTONOMOUS POWERED PARAFOILS (UNMANNED AERIAL VEHICLE SYSTEMS)*

No doubt, there are several types of unmanned aerial vehicle systems (see Krishna, 2019). Drone aircrafts are getting popular due to their applications in the civilian regions. They are also being evaluated extensively in agriculture zones, recently. Yet another set of UAVs are called 'Unmanned Guided Parafoils.' They are often called 'Autonomous Parafoils.' They are autonomous. Yet, it is possible to control their flight path by altering the position of the "aileron" found at the rear side of the parafoil. Their flight path and waypoints are usually pre-programmed, using appropriate computer software (e.g., Pixhawk, e-Motion, Goose Pilot).

These guided parafoils possess a wide range of applications in military, civilian and agricultural realms. They are used in supply of ammunition, fuel, tools and food packets to different military camps. Those fitted with visual, multi-spectral and infra-red sensors are utilized to conduct aerial survey and reconnaissance of city traffic, public places, etc. There are a number of applications possible adopting autonomous parafoils. Autonomous parafoils could be used for aerial photography of farms and cropped fields. Assessment of crop's growth rate, its nutritional status and water requirements is a clear possibility. Air dropping of supplies in disaster affected areas such as floods, drought, quakes, etc. are other possibilities (Veronte Autopilots LLC., 2015).

A very soft and safe landing of autonomous parafoil is a necessity, particularly, if it is carrying delicate and fragile pay load. In general, safe landing is a requirement even in farmer's fields, if the parafoil is carrying a payload of delicate sensors or spray equipment. They say, usually, lack of knowledge about surface wind pattern causes rough landing. Also, it becomes difficult to maneuver the autonomous parafoil, using remote controller or through automatic GPS-guided pre-programmed commands. Often, simulations on a computer screen may be essential. This step will help in regulating the speed and descent rate of the parafoil, at landing. Overall, we may have to understand that, if wind speed and pattern is unknown, then, soft landing is difficult. Therefore, contingency measures are required (Lou et al., 2018).

Autonomous airdrop systems lacking propulsion may be adversely affected by high winds. Strong winds encountered during Draper Laboratory flight testing prevented lightweight parafoil systems from landing accurately (Cheil, 2013). Therefore, multiple guidance strategies are designed to overcome high wind scenarios. A powered guided parafoil may overcome such problems only to a certain extent.

No doubt, autonomous parafoils are useful in relatively accurate air drop systems. They are utilized to pick cargo and deliver it at a distance with greater accuracy. However, autonomous powered parafoils do experience faults. By definition 'faults' are unpermitted deviations from normal functioning of parafoils. Such faults could occur right at the time of flight during transit or air drop. They may firstly affect the landing accuracy, payload safety and parafoil condition. Faults can affect parafoil's performance (Stoeckle et al., 2014). In flight fault corrections are possible. In flight faults may affect behavior of the parafoil. Accurate delivery in farms or urban areas may difficult, if faults occur. Faults generally increase landing errors. Faults may affect long distance movement of cargo via parafoils. The performance of parafoils could be simulated to detect before a payload is carried and air dropped. Generally, early detection of faults either prior to launch or in-flight helps in restoring, accuracy of flight path and landing location. Flight testing has shown that parafoils may suffer from a few common sets of faults. We can group them into faults caused due to motor non-function, defects in parafoil canopy, defects in strings, computer programming and guidance related faults, etc.

During recent years, autonomous parafoils have found utility in space science. Autonomous parafoils have been tested for use during recovery of astronauts and space modules that return to earth. The autonomous parafoils holding astronauts are amenable for careful guidance from control center. They could even be pre-programmed. This way, we can avoid landing with a splash. The descent rate and speed could be regulated using remote control computers (Ward et al., 2001).

Let us consider the prime context of this chapter, i.e., parafoils in agriculture. During recent years, autonomous parafoils have been adopted during precision farming. Also, during general crop husbandry related aerial survey activities. For example, in the European plains, researchers have successfully monitored their farms and crops, using autonomous parafoils. They have obtained consistent data about topography, land and soil characteristics, and water resources. They have used such data to form 'management blocks' during precision farming (Thamm et al., 2013; see Krishna, 2016, 2018). A few other examples suggest that, autonomous parafoils are efficient in collecting data about crops grown in experimental stations. They are used to evaluate and rank the performance of thousands of germplasm lines and cultivars. The crop's characteristics such as growth rate, leaf chlorophyll content, biomass accumulation rate, plant-N status and water stress index could be recorded. They collect data utilizing a set of multispectral and infrared sensors on autonomous parafoils (see Krishna, 2016, 2018).

The following are few examples of autonomous powered parafoils which are used in general aerial survey of terrain and in agriculture:

2.3.6.1   TETRACAM'S HAWKEYE

Hawkeye is an autonomous powered parafoil. It has wide open parafoil towering over its rugged carbon composite frame. The 'Hawkeye' is becoming common over farms and forests. They say, the system's high visibility is an asset as the craft's fully autonomous autopilot can carry the Hawkeye more than 1.5 miles away from its ground control station. Hawkeye is produced by an USA based company known as Tetracam LLC. It is located in Chatsworth, California, USA. The Hawkeye is unique among unmanned aerial vehicles due to its comparative low-cost, ease of operation and its inherent safety. With its chute continually deployed, if a malfunction occurs while the craft is airborne, the Hawkeye is designed to simply float to the ground. Therefore, it protects the craft and its camera cargo from damage (John, 2011; Plate 2.5).

PLATE 2.5   *(Continued)*

PLATE 2.5 Hawkeye autonomous parafoil. Top: Canopy (air foil) and engine attached to propeller; Bottom: Engine fitted to a propeller (no canopy).

Note: The Hawkeye autonomous parafoil is energized by a petrol/diesel engine. It is connected to a propeller. The parafoil has a central processing unit. It receives signals from ground control center. The parafoil's flight path can also be pre-programmed, using appropriate computer programming (e.g., Pix4D Mapper).

Source: John Edling, Tetracam LLC., Chatsworth, California, USA.

Reports by Moscow Aviation Institute, Russia suggest that paragliders that are unmanned, autonomous and take to flight path that has been pre-programmed is being sought. The autonomous powered parafoils are being tested for usefulness and efficiency. The 'unmanned paraglider' is said to fly for 6 hr at a stretch. It takes a payload of 15–150 kg (Akhramovich, 2017). The UAV paraglider has been tested under different atmospheric conditions. It is powered by an electric motor. Steering chords control the paraglider's flight path. Major advantages stated regarding the autonomous paraglider are long flight time, heavy payload weight, low-cost of production, mobility and simplicity of operation, full self-sufficiency and night-flight capability. In case of emergency onboard, the paraglider can perform soft landing and prevent cargo crashing. This is in contrast to many other UAVs. The paraglider requires only 20–30 meters runway. It needs three men, two of them to unfold the wing and the third one to control engine thrust, at the take-off stage. The paraglider is fitted with on-board micro-computer, GPS and GLONASS receivers. Usually, flight path and way-points information are loaded to computers available in the payload area of the parafoil. Parafoil's wing size is an important character. With the wing area of 3 sq. meters, payload weight is 15 kg. A parafoil with 40 m² wing size carries 150 kg payload. The stand-alone flight lasts from 3 to 6 hours, but it depends on wind conditions. The flight program and path are uploaded on the onboard computer. We may note that, similar autonomous paragliders are being tested and adopted by other nations. For example, in Israel, the paraglider has been evaluated for use during spraying plant protection chemicals on crops. This is in addition to usual aerial survey of farm installations, crops and estimation of crop growth, leaf chlorophyll and water stress index.

Swarming of autonomous parafoils is a clear possibility. It hastens the task of aerial survey, acquisition of spectral data and images. It is apt when land surface to be covered in an agrarian region is large. In March 2005, Brooklyn-based defense contractor Atair Aerospace Industries Inc., announced the first successful demonstration of flocking and swarming techniques in a parafoil UAV. They adopted five computer-guided Onyx parafoil gliders in an experiment meant for the U.S. Army. The company says the parafoils can be released from 35,000 ft above ground. They glide autonomously as a flock for 30 miles, then, land together within 150 feet of a pre-programmed target. The system is designed to air-drop troop supplies with previously unachievable precision. Such methods may also find utility in agrarian regions, for example, to drop large packages of fertilizers at different fields.

2.3.7   NANO PARACHUTES AND PARAFOILS

High altitude guided small parafoils are sought by Space Science and Military establishments. Such remotely guided or autonomous parafoils are of great utility in air dropping small pay loads. Researchers at the University of Idaho, in USA are evaluating very small guided parafoils also called 'Nano-Parafoils.' They are capable of air dropping small loads from very high altitudes or space. Such parafoils, in future, may routinely transport small payloads from space crafts and space stations. In fact, National Aeronautics and Space Agency of USA intends to use such small parafoils to regularly drop small payloads from International Space Station (Kologi, 2015). Such high altitude small parafoils are remotely controlled, using wireless connectivity and global positioning system. Yet, there are no known applications listed for such small autonomous parafoils in agricultural settings. However, we may explore several of them, wherever there is need for a small handy parafoil. We may require parafoils that transships and airdrops small payloads, as per computer instructions or wireless guidance. A very small parafoil may have some useful applications in agrarian regions.

2.3.8 GIGA PARAFOILS

Here, we discuss about largest Autonomous Ram-Air Parachute: 'GigaFly Airborne Systems.' Recent reports suggest that, the largest autonomous parafoil capable of carrying a payload of 42,000 lbs (19,022 kg) has been tested. The large autonomous parafoil is named as 'Gigafly.' This ram-air parachute is 10,400 ft^2 in size, with a wingspan of 195 ft. Its size is said to be nearly same as a Being 747 jumbo jet. This new parafoil system has carried 33,000lb (15,000 kg) payload several times. Also, air dropped it from a C-130 cargo plane cruising at 15,000 ft., above ground. It is interesting to note that such large parafoil descends extremely slowly and accurately as per wireless and GPS-based controls. It descends at a gentle 14 ft s^{-1} rate. The accuracy of landing location is <275 m (Airborne Systems, 2018). The Gigafly is designed to air drop large payloads from above 25,000 ft. Such large parafoils are being sought by military establishments. They are classified as 'LVADS (Low Velocity Aerial Delivery Systems).' They are supposedly useful in air dropping large amounts of ammunition and food supplies in combat zones. Of course, we can envisage excellent use for large parafoils in disaster prone areas, emergency food supplies and large equipment supply into areas not reachable easily by road. In agriculture, we can think of innumerable uses for such large autonomous parafoils. It starts with supply of large amounts of seeds, fertilizers, plant protection chemicals and even equipment into far flung agrarian zones. The supplies could be easily carried inside a C-130 and air dropped using GigaFly parafoil. There are a few other models/brands of large autonomous parafoils such as 'MegaFly' (Carter et al., 2005; Airborne Systems, 2018). A few other large parafoils, for example, one known as 'DragonFly' carries 10,000 lb. (4500 kg) payload and 'Firefly' carries 2000 lb (980 kg) payload. The design, development and preliminary adoption of such large autonomous parafoils is said to be a remarkable achievement in aviation history.

2.3.9 'DRONE PARACHUTES' FOR RECOVERY, RESCUE AND SAFETY

Drones referred here are the small autonomous aircrafts. They are used frequently in military for reconnaissance, in civilian settings for aerial surveillance and most importantly in agriculture, to conduct aerial moni-toring of crops. Also, to get spectral data about terrain, soils and crops. Such drones could be recovered using parachutes on a routine basis or when they encounter emergency. Such parachutes are often termed as 'Drone Parachutes.' Parachutes of different specifications have been utilized during recovery of an UAV aircraft. A few specially designed parachutes and para-foils are also in vogue, during drone recovery and rescue. Drone aircrafts may have to be fitted with suitable parachute/parafoil. They should be provided with specific computer-based instructions (pre-programmed) or controlled remotely. The parachutes need to be opened at the correct instance. Usually, parachutes/parafoils are opened at a specific threshold height above ground. Such a threshold depends on the drone model, its flight pattern, parachute, etc. The parachute loaded into drones for rescue are usually selected based on specifications suggested by the drone manufacturer. The parachutes/ parafoils for rescue and recovery are specially designed. They are usually small, compact to fit into cockpit or fuselage of the small fixed-winged or copter drone. Such parachutes are light in weight. They are made of light foil or cloth and strings. Drone parachutes are light in weight and weigh 30–45 g only. Drone recovery parachutes could be circular, ellipsoidal or straight (foil). Drone parachutes are usually well integrated into the drone's computer program (European Aviation Safety Agency, 2017; Schroth, 2017). They open (pop-up) automatically. Let us consider an example of drone parachute called 'Iris Ultra Parachute' (Fruity Chutes Inc., 2018c,d). The Iris Ultra Parachutes are small, compact, and easily packed into drone aircraft. The ultra-compact versions need 35% less space inside the drone's body. The parachutes are compatible with wide range of drone models, both fixed-winged or copters. They are available in different shapes and sizes. A drone parachute, particularly, its size is usually selected by first ascertaining the size and weight of the drone to be recovered. Flight pattern of the drone should also be considered. The height at which the parachute should open (pop-up) and descent speed are other important considerations. Parachute parts like tethers to attach drone, deployment bags, recovery harness, nylon webbing are carefully selected. It is based on the drone model that must be recovered or rescued. Let us consider a few more examples. Aeromao is a flat fixed-winged drone manufactured by a company known as 3D

Robotics, California, USA; Delta-M fixed-winged drone by AVASYS Inc, Krasnoyarsk, Russia; AII Textron and Quest UAV are the good examples. They adopt parachute-aided landing of small drones (See Krishna, 2019; AVASYS LLC. 2016; Plates 2.6 and 2.7).

PLATE 2.6   Drone recovery and rescue parachutes/parafoils.

Note: Drone parachutes are utilized to recover several different models of drones. Agricultural drones that are small are also fitted with small, light-weight compact parachute/parafoil. It could be usual round shaped or ellipsoidal parachute.

Source: Gene Engalgou, Fruity Chutes Inc., Monte Serena, California, USA; AeroMao Inc., California, USA.

Delta-M UAV system is another good example for parachute-aided recovery and/or rescue of small fixed-winged drone. This drone is attached with a compact 'parachute launcher' assembly of very small weight of 30–35 g. The parachute pops out when the drone is going awry out of path or based on pre-programmed instructions. The parachute opens (pops) when it is intended to land the drone after completing the tasks or to re-charge the batteries. It is a product designed by Siberian Federal University and produced by AVASYS Inc, Krasnoyarsk, Siberia, Russia (AVASYS LLC., 2016; see Krishna, 2019).

EOS-mini UAS is yet another good example for parachute-aided recovery of agricultural drone. EOS-mini UAS is a small UAV. It is launched either using catapult or by hand. It delivers clear images of the surface features including a range of agricultural aspects. Regarding military, it is used to monitor platoon movement, installations in a camp, and international borders. Major advantages attributed to EOS-mini UAS are over 2 hr of endurance, low noise and acoustic signatures, excellent aerodynamics, fully stabilized gimbal and photographic equipment. It offers excellent video depiction of surface features and sharp aerial maps of crops (Threod Systems, 2017). *EOS-mini is often recovered using a small parachute*. Hence, it does not crash land on belly or on wheels. Instead, it floats slowly and descends to earth.

PLATE 2.7   AeroMapper EV2 UAV being guided to land, using a popped parachute.

Note: AeroMapper is an example of parachute-aided recovery of drone: https://www.youtube.com/watch?v=GFSyG9kdy5g.

Source: Mauricio Ortiz, Aeromao Inc., Mississauga, Canada.

As stated earlier, 'Parachute Launcher' on a drone is an important component during regular landing of the drone. Also, in times of emergency. Let us consider a typical parachute launcher assembly that is fitted into a drone aircraft (fixed-winged or copter). The function of 'parachute launcher' is to open in response to computer-based instructions or as pre-programmed, at the time of landing of drone aircraft. There are several designs of parachute launchers. They are all usually very light. The canopy of parachute is made of light fabric. The drone ejects the parachute far from the airframe, like a bullet. This makes parachutes effective even at very low altitudes or high-speed spin situations. For example, a California-based company Fruity Chutes Inc., has introduced the 'Harrier Drone Parachute Launcher.' It is a spring-loaded parachute launcher meant for fixed-winged drones and multi-copters. The Harrier parachute launcher features a thin wall carbon fiber canister, lightweight nylon components, and a high energy compression spring. The canister has a new line of parachutes called the "Iris Ultra Zero." The parachute (IFC-30-SUZ) weighs just 27.7 grams and is 30 inches width. The complete 40 mm Harrier Parachute launcher assembly with the IFC-30-SUZ parachute weighs just 81 grams and is only 103 mm long (Fruitychutes, 2018d; Aero News Network, 2018).

The safety of autonomous aerial robots such as drones is important. This is irrespective of cruising altitude, size of the drone, or area above which it is flown and the purpose it serves the farmers and other types of customers. At the same time, it is not uncommon for drones to break down due to various reasons related to model, engine, batteries, computer programming, flight path, and weather conditions. Drone safety and careful retrieval is essential in farmland. Autonomous drones should be fitted with parachutes/parafoils that get activated in few seconds, immediately after a defect or a critical failure is detected. Computer programs should detect defects quickly. There are parachute and compatible computer software that can be integrated into several different models of agricultural drones (ParaZero Safe Air Inc., 2018). We should note that, if drone is a low flying copter, then, parachute must open very quickly before drone loses height. If the drone is at a high altitude, then, parachute should perform safely until the drone touches the earth. It must withstand wind currents and drift accurately. There are several reports of disasters pertaining to small drones used in the farming regions of USA and Canada. The reasons quoted most commonly related to malfunction of electronic circuitry, wrong pre-programming, incorrect guidance from ground control computers and sometimes erroneous functioning of basic parts of aircraft body. The crash landing could be deleterious to drone body. Hence, parachutes are fitted to drone's fuselage that pop-up in times of emergency. Many of the farmers are taking precautions by fitting drones with parachutes that open-up automatically, if electronics go awry. For example, 'Nexus Parachute Launcher' helps in releasing a 'drone parachute' in times of distress automatically (Moore, 2018). There are indeed several companies that make drone rescue parachutes. They rescue the drone when they stray outside beyond the visual light of sight (BLVOS). For example, a recent report suggests that an Austrian company has offered two new models of drone rescue parachutes known as DR-5 and Dr-10. They can help farmers in rescuing their drones, if and when they go awry from pre-determined flight path.

2.3.10   AGRICULTURAL PARAFOILS

At present, both powered parafoils guided by a pilot and entirely autonomous ones meant for general aerial survey purpose are being utilized in agricultural regions. There are no models of powered parachutes/parafoils or autonomous parafoils meant exclusively for use in aerial survey of crops or even to spray plant protection chemicals. Specialized agricultural parafoils are not yet manufactured. However, a few parafoil models may eventually evolve into being called 'agricultural parafoils.' The parachute models listed and discussed below such as Buckeye, SUSI 62, Hawkeye and Pixy are meant for general aerial survey. They have been already adopted to conduct aerial photography of agricultural experimental stations, its topography, soil and water resources. They also collect spectral data about crops. One of the brands of autonomous parafoils has been utilized to spray plant protection chemicals on crops (e.g., Buckeye) (Taylor, 2018). Perhaps, such parafoil models may eventually evolve into totally an 'agricultural parafoil.' Sooner, we may develop larger number of specialized powered manned as well as autonomous parafoils, exclusive to farming. They would be able to conduct several specialized operations above farmland. Krishna (2019) has discussed a couple of parafoil models that could be utilized in aerial survey of experimental crop fields, detection of water and nutrient (plant-N) status of crops, NDVI etc. Now, let us consider a powered manned parafoil about which there are reports that, it could be utilized to spray crops with plant protection chemicals or liquid fertilizers. The Buckeye parafoil is powered and piloted. It has facility for either one or two pilots to survey crops. The parafoil has a pay load area for pilots plus to store a large quantity of fluid or granular formulation. The formulations are eventually to be sprayed on to crops. This parafoil can be piloted at low altitude over the crop's canopy, may be 3–5 m above crop. So, it allows pilots to distribute liquid or granular formulations accurately over the crops. It avoids wind caused drifts. The buckeye powered manned parachutes are also provided with visual, multi-spectral and infrared cameras. Hence, they could be adopted, to obtain data about crops. Buckeye powered parafoil is among the earliest models to be tested and adopted to spray crops with plant protection chemicals.

2.3.10.1 BUCKEYE POWERED PARAFOILS (MANNED)

The Buckeye Powered Parachutes are capable of low altitude flights over crop's canopy. Hence, they are of great utility during aerial photography of crop's growth status. Buckeye powered parafoil has also been used to spray fungicides and foliar liquid fertilizers (Taylor, 2018; Buckeye Industries, 1997). There are two versions of Buckeye powered parafoils that are used in agricultural settings- a two-seater and single seater with a payload area for cameras. There is also facility to attach spray nozzles and storage tanks of 20–25 lt capacity. This is to hold plant protection chemicals. The specifications as quoted by the company for a single seater Buckeye parafoils (e.g., 'Eagle') usable in farming sector are as follows; Its empty weight (i.e., without payload) is 220–230 lb (100–105 kg). It can carry a payload of 285 to 310 lb (130–145 kg). wing is 450 sq.: ft. Fabric used for parafoil is 1.1 oz Soar coat. Its fuel storage capacity is 5 gallons. So, the fuel load decides its endurance in the air as powered parafoil. Although, it can float, if fuel is exhausted. It is fitted with a Rotax engine of 45 hp.

Regarding performance characteristics; the maximum attainable speed while in flight is 30 kmph. Its glide ratio is 3.8/1. and descent rate is 10 ft s^{-1}. Fuel consumption is 2 gallons h^{-1}. Its rate of climb is 650–700 ft min^{-1}. Its ceiling height is 10,500 ft above ground level. This Buckeye powered parafoil could be adopted to carry several of the tasks above crop fields. Almost, on a routine basis all through during the crop season. It is quick to assemble. We can gather useful data about farms/crops. A few different models of Buckeye parafoils are available in the market. They differ in their specifications and purpose. A few examples are 'Buckeye-Eagle,' 'Buckeye-Falcon,' 'Buckeye-Eclipse,' and 'Buckeye-Endeavor' (Buckeye Industries, 2018; Taylor, 2018).

2.3.10.2 HAWKEYE PARAFOIL: AN UNMANNED AERIAL VEHICLE

The 'Hawkeye Autonomous Parafoil' is among the earliest of parachute models utilized effectively in the realm of agriculture. This is an autonomous parafoil fitted with petrol engine. It is guided using pre-programmed launch and flight path. The Hawkeye autonomous parafoil is versatile. It flies at low altitude above crop's canopy and picks sharp imagery of land, its topographic details, soil and crop's growth status and health. Specific computer programs such as 'Goose autopilot' or 'Pixhawk' could be used, to set the flight path and way points. Hawkeye agricultural parafoil is usually fitted with a set of sensors that operate at visual, multispectral, infrared and red edge wavelength bands. The sensors provide farmers or farming companies with the necessary spectral data and aerial imagery. Most importantly, they provide data about crop's nutrient and water status. We can also calculate crop's growth rate. Aerial surveys of large farms allow farmers to judge the extent of soil damage due to erosion. Hawkeye is supposedly a low cost autonomous aerial vehicle when compared with UAV aircrafts. It is portable. Hawkeye could be operated by farmers with little training in controlling UAVs. In case of emergency or engine troubles, the parafoil floats until it is set right. A few specifications of this autonomous parafoil has been discussed earlier under Section 2.3.6 that deals with autonomous unmanned parafoils (John, 2011; see Krishna, 2019). Regarding major agriculture related uses, Krishna (2019) lists the following: Hawkeye is mostly utilized to conduct aerial survey of large crop production zones. It is utilized to monitor farm activities and send instantaneous reports to GCS. The sensors offer detailed images of crops. The spectral data is highly useful to farmers (see Plate 2.8). The data collected includes NDVI that depicts crop growth pattern and canopy growth rate. The sensors offer leaf chlorophyll (crop's-N) and CWSI (crop water stress index) (Hunt et al., 2014, 2018). Non-agricultural Uses: These pertain to recording public events, surveillance of mines, transport vehicles, buildings, dams, lakes, riverine regions, etc.

Specifications of a Hawkeye Autonomous Parafoil (Tetracam Inc., 2018):

- Endurance – 10 – 30 min depending on batteries and payload.
- Speed – 10 knots per hour nominal.
- Parachute size – 120 ft. x 32 ft.x 26.6 ft.
- Vehicle length – 3ft.; Vehicle Width – 16 ft.; Vehicle Height – 20 ft.; Vehicle Weight (RTF/no payload) – 8lbs.
- Payload Capability – 3 lbs.
- Autopilot to GCS Range – 1.5 miles (extensible to > 4 miles).
- Manual RC Spektrum Control Range – 800 yards (line-of-sight).

Hawkeye Airframe Includes: Payload control box; Goose autopilot and GPS; Wiring harness; Adjustable steering system with Titanium gear high torque servo; Brushless motor with propeller adapter; Motor vibration isola-tion mount; Brushless motor speed controller; APC 15x6 propeller (Qty 2); 2.4Ghz 7CH RC receiver – (Spektrum AR7000); 5S1P 4400 mAh Lithium Polymer flight batteries (Qty 2) 3S1P 2200 mAh Lithium Polymer payload battery (Qty 1); 2S1P 2200 mAh Lithium Polymer RC system battery (Qty 1); Parachute (120 ft. x 30 ft. – Bright Green, Bright Orange).

Ground Control Station Includes: Goose Ground Station Software (Windows O/S); RC Transmitter (Spektrum DX7 with Charger); Lithium Polymer Battery Charger. All electronics including Goose Autopilot, RC transmitters and speed controllers are pre-programmed. They are flight tested prior to shipping. Users need to provide their own Windows-based laptop for ground control software.

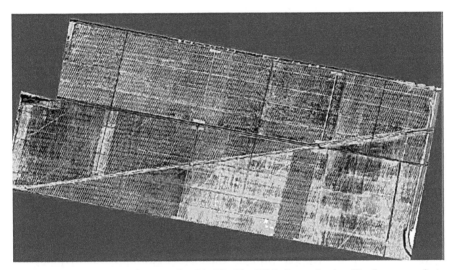

PLATE 2.8 Citrus trees in an orchard in Florida, USA, imaged using Hawkeye parachute fitted with visual and multi-spectral sensors, in Florida, USA.

Note: The image depicts variations in NDVI across the orchard. Hawkeye autonomous parafoil has also been utilized to assess crop's nitrogen status (leaf chlorophyll) and water stress index (infrared imagery).

Source: John Edling, Tetracam Inc., Chatsworth, California, USA.

2.3.10.3 PIXY ABS AUTONOMOUS MOTO-GLIDER

Pixy ABS is an unmanned aerial vehicle (UAV). It is frequently called as 'Moto-glider.' It is manufactured at Montpellier, in France, by French Institute for Development. It is a compactly foldable and portable autonomous parafoil model. Pixy ABS is made of thin plastic/aluminum foil. The payload area has an aluminum basket. The basket accommodates a motor and sensors such as visual, multispectral and infrared cameras. The petrol tank holds 3–5 lt of petrol. Hence, the endurance is short. Just a period of 30–45 minutes. It gains height using only a short runway. Pixy ABS has been already tested and routinely utilized in Eastern European farms. For example, in Poland, it has been adopted to assess the performance of wheat genotypes grown in experimental farms (Lelong et al., 2008). InCzech Republic, it has been adopted for regular surveillance of crops such as wheat, barley and lentil.

Pixy ABS parachute is a lightweight UAV. It has ability to gain height quickly. It only needs a short runway. It has a propeller attached to brushless motor. The motor is powered by Lithium Polymer batteries. Pixy ABS is easily foldable. Therefore, it is easily portable to any place in the farm or an urban location. Pixy has a usual complement of sensors placed in the payload area. It offers a range of spectral data very useful to farmers. Such data helps while deciding inputs and agronomic procedures. 'Pixy' has been already tested in farms across many locations in the European plains (Lelong et al., 2008).

Regarding its specifications and performance characteristics: Pixy ABS is made of toughened plastic foil and aluminum. Its size is 35 ft. length, 120 ft. width, 32 ft. height. It weighs 6–8 kg including a payload of 2–3 kg. The engine is connected to a propeller that has 3 plastic blades. It needs a runway of 10–15 m length to gain height. It reaches a speed of 20 km, but it can float at much low speed of 3–5 kmph during aerial photography of terrain and crops. the Pixy ABS autonomous parafoil is guided, using remote control computers. Otherwise, the entire flight path and way points could be pre-programmed, using a software such as 'Goose Pilot' or 'Pixhawk' or 'eMotion.' The autonomous parafoil has 'auto return home' facility. It floats freely if engines stop/fail. The photographic accessories include Sony DSCS F828 or Canon EOS visual and multispectral cameras. The imagery is done from an altitude of 20 to 700 m above crop's canopy.

Now, let us consider uses of Pixy ABS in agricultural zones. It is useful in aerial scouting of terrain, crop fields and water resources. These are most common functions in the agrarian regions. Pixy is used to draw digital data about crops' growth (NDVI) and leaf chlorophyll (crop's-N) status. The thermal imagery provides data on crop's water stress index. It can also be used to collect air samples, in the atmosphere above the crops. In other words, it can be used to assess gaseous quality of atmosphere above crop's canopy. Pixy ABS can be used to monitor general farm activity. Pixy ABS has been utilized in agricultural experimental stations. It can be used to collect periodic data during experimental evaluation of crop species and genotypes (Pudelko et al., 2012; Lelong et al., 2008; see Krishna, 2019). Regarding non-agricultural uses; Pixy ABS parachute could be flown above installations, roads and public places to monitor human/vehicle activity. It is used to surveillance mines and ore transporting vehicles. It helps in providing aerial view of disaster prone or affected areas. Pixy ABS can also be used to surveillance natural resources such as water bodies, dams, rivers, etc.

2.3.10.4   SUSI 62: AN AUTONOMOUS PARAFOIL

SUSI 62 is an UAV parafoil. It is one of the series of SUSI parafoils manufactured by a German company. SUSI 62 is a multipurpose parafoil. It is used in urban locations to monitor traffic. Several models of SUSI are utilized during recreational gliding. SUSI 62 has been utilized in farming sector too. The aerial imagery obtained using this autonomous parafoil has been utilized effectively, in precision farming. The images of terrain, water resources and soils have been used to demarcate the 'management blocks' (Thamm, 2011; Thamm et al.,2013; Krishna, 2019). SUSI 62, with its full complement of sensors such as R, G, B, IR, NIR and LIDAR has been utilized, to survey agricultural land, crops and water resource in the German Plains and other agrarian locations within Europe (Thamm, 2011; Thamm et al., 2013). SUSI 62 has provided excellent data about soil fertility variations, biomass accumulation patterns, vegetation and its species diversity. SUSI 62 is a parachute developed to withstand certain levels of vagaries in weather parameters. It has been designed to carry a payload. It can accommodate different kinds of objects, parcels or photographic equipment. It is a portable parafoil UAV. SUSI 62 has been used to conduct variety of tasks. In Germany, it has been used for recreation purposes, in the city of Cologne. This is primarily, to carry visitors above the city and other locations (Thamm, 2011). In Africa, it has been utilized to photograph migrants and immigration routes of the tribes. In agrarian regions, SUSI 62 has provided excellent data about soil fertility variations, biomass accumulation patterns, vegetation and its species diversity. Hence, it could be adopted to survey land resources and mark 'management blocks' during routine crop management (Thamm, 2011; Thamm and Judex, 2006; Krishna, 2019).

Regarding specifications; SUSI 62 has a canopy made of plastic fiber foil and an aluminum frame. It has four light carbon fiber wheels. They are wheels that help on runway to launch and land. They withstand rough landings. It is easy to assemble this autonomous parafoil for use anywhere in a farm location. All major parts of the SUSI 62 are easy to dismantle. It can be re-assembled in minutes. SUSI 62 allows a payload of 8 kg. The payload area accommodates a set of sensors such as visual, multispectral, infrared and red-edge cameras. the parafoil could be flown up to a height of 3500 ft above crop's canopy. During aerial survey and sharp close-up photography, it is flown at low altitudes of 5 to 10 m above crop's canopy (Krishna, 2018).

SUSI 62 has a 5.5 hp two stroke engine as source of power. The engine is attached to propeller with three plastic blades. It also has rechargeable Lithium polymer batteries. The LiPo batteries support its electronic circuitry and computers. It is basically a free-floating parafoil. It has long endurance in flight. The remote-control computers can program the flight path, way points, start and off the engine, alter direction and speed of the parachute, also frequency of photographs and shutter activity of the sensors. The autopilot mode allows the parachute to trace the path and reach starting point, in case of emergency or upon completion of aerial survey. SUSI 62 is a stable parachute. The parachute can also be managed in the manual mode. It just requires two days of training for the farmer, to learn how to control the flight path and sensors, in order to obtain aerial photography. Optical sensors include mid-range SLR (e.g., Nikon 300 D or a sophisticated Canon D5 Mark 11). It also carries a thermal range sensor (Hasselblad 60 megapixel). (Krishna, 2018, 2019).

Regarding agricultural uses, the parachute UAV (SUSI 62) is used to conduct aerial survey of land, decipher land use pattern, detect soil type variation, cropping systems, water resources, etc. The sensors offer data about crop growth, canopy size, its reflectance and temperature. They offer data about NDVI, leaf chlorophyll (crop's-N status), crop's water stress index, disease/pest attack, disasters due to flooding, droughts and excessive soil erosion, etc. SUSI 62 is also used to monitor farm activity and animals in the pastureland. Aerial images from SUSI 62 has been effectively used, to mark 'management blocks,' during precision farming (Thamm, 2011).

Regarding non-agricultural uses, SUSI 62 parachute can be left to float above buildings and other installations. This is to surveillance and monitor activities on ground. In game sanctuaries, it has been used to monitor animal activity. In agrarian regions, it has been used to visualize and get pictures of drought or disaster affected areas. It has been used to monitor wood logging and movement of vehicles out of forest plantations.

2.3.11   CARGO PARACHUTES AND PARAFOILS

Parachutes have been used extensively for airdrop of personnel and cargo, by the military establishments of different nations and humanitarian agencies.

As stated earlier, the two most common types of parachutes used are round parachutes and rectangular shaped parafoils. The circular parachutes capture air and that provides a drag force to decelerate a payload and achieve steady descent velocity. The latter, i.e., parafoil, by virtue of its design with air inflated air-foil shaped cells, provides a lift force as well as a drag force. Therefore, parafoils can be steered to deliver payloads accurately, on the ground. Based on this capability, parafoils are currently actively pursued by the military for precision aerial delivery of cargos. The development and introduction of ram-air parachute (parafoil) with facility to steer it and control its rate of descent is an important event. It induced great interest in cargo transport and airdrop. Further, during recent years, GPS-guided transport using sophisticated electronic and computer-based commands has allowed for greater accuracy, Particularly, during airdrop and landing of parafoils (Yakimenko, 2016). It has been forecasted, that, in due course, precision air drop of large cargo using autonomous parafoils may become common.

Parachutes (circular types) meant for cargo transport need certain special modifications. They need to be sturdy, easily navigable, and possess larger pay load capability and area. The parachute cloth, its strings too may have to be specifically selected and evaluated. A study by Niccum et al. (1965) has shown that cross type parachutes are well suited for both drag and cargo transport purposes. Let us consider characteristics of a circular flat parachute. For example, the G12 cargo parachute was developed, to use it, during deceleration and air drop of bulk cargo.

Some of the characteristics and specifications of the G 12 cargo parachute are as follows: Shape: Flat Circular; Diameter (nominal): 64 ft. (19.5m); Gore material: 2.25 oz, type I, nylon parachute cloth; Number of suspension lines: 64; Suspension Line Material: Type IV, PIA-C-7515,1000 lbs. (450 kg) tensile strength; Length of Suspension Line: 51 ft. (15.5m); Suspension Riser Assemblies: 2; Suspension Riser Length: 60 ft. (18.28m); Complete Assembly Weight: 125 lbs. (57 kg); Maximum Payload Capacity: 2,200 lbs. (997.9 kg) Pilot Chute Shape: Flat Octagonal; Diameter (nominal): 68 inches (1727 mm); Canopy Material: 1.1 oz PIA-C-7020, type I ripstop nylon parachute cloth (Mills Manufacturing Inc., 2018a,b). There are indeed numerous companies that produce several types of circular parachutes. They could differ in size, weight, payload area, deceleration rate and accuracy of landing, etc. Of course, their choice depends on purpose.

Luders et al. (2015) have pointed out that ability to accurately and consistently deliver supplies into difficult, complex terrain is important, during airborne delivery. Particularly, when systems such as parafoils are utilized. Since environmental disturbances occur constantly while parafoils are in flight and during landing, robustness is essential. Similarly, accuracy during landing is important. Often, accurate and soft landing is a necessity in military and civilian cargo transport. Therefore, we should plan trajectory and landing site of an autonomous parafoil as accurately as possible. Landing should be smooth. Again, robustness is a primary concern, given that environmental wind disturbances are often highly uncertain. As stated above, parafoils are subject to uncertain and variable wind environments. They should be compensated to arrive at correct location on ground surface. If such deviations are not compensated, it results in large errors in landing. We should also note that, many applications often have tight landing restrictions. Missing the target location even by a small distance, can lead to unintended collisions with natural or man-made hazards. Precise delivery is essential to avoid loss of supplies or unacceptably dangerous recovery efforts (Luders et al., 2015; Yakimenko, 2005; Hur, 2005). During past few years, improvement in air drop accuracy and delivery of cargo has been rapid. It has been attributed to GPS and electronic control of the parafoils.

'Affordable Guided Airdrop System – AGAS 2000' is one such air drop system. It is unique among precision airdrop parachute systems. AGAS 2000 integrates a relatively inexpensive autopilot controller with a round cargo parachute. AGAS 2000 can deliver a one-ton payload to within 211 m circular error probable (CEP) of its target. Its programmed mission profile should be based on forecasted winds. It lands within 38 m CEP when based on a current wind profile provided by a system such as the Air Force's Precision Airdrop System (PADS) (Jorgensen and Hickey, 2005). In such airdrop systems, parafoils or guided round parachutes with GPS are used, to control the falling direction. It helps in precise landing within a few meters of the designated target (Jorgensen and Hickey, 2005). Jian et al. (2011) have reported the development of a parafoil based cargo delivery. The parachute uses the background data and active disturbance rejection system. This helps in smooth and accurate landing of payload.

At present, several variations of cargo parafoils are available in the market. Some of them are tailor-made for specific tasks. For example, there are 'Precision -guided Parafoils to drop air cargo. These aerial vehicles pick, transport and drop air cargo with great accuracy. They are relatively more accurate and safer compared to simple parachutes. Let us consider an example. The parafoil plus a parachute combined unmanned aerial system called ONYX is produced by Atair Aerospace Inc., of New York State, USA (Atair AerospaceInc., 2018). Such autonomous powered precision guided 'parachutes + parafoils' systems are used, to drop air cargo. They accomplish it with unprecedented accuracies, both, in flight and landing. Some parafoil enthusiasts compare them to 'smart bombs' used by military establishments. This is because, each UAV system fulfills the precision airdrop requirements stipulated under the U.S. Army's Joint Precision Airdrop System (JPADS) program. Each system is deployed from the air using fixed-wing and/or rotary aircraft, such as a C-130 or C-17. The parafoil can be deployed at altitudes up to 35,000 ft. and at speeds up to 150 KIAS. The system autonomously glides over 44 km and lands cargo within 100 m of its selected target. Onyx systems offers capability of strategically, positioning equipment and supplies for special operations forces. The capabilities of delivering payloads range from 0 lbs. to 2200 lbs. (1000 kg). Of course, we must adopt different sized UAVs. Using real-time collision avoidance and swarming/flocking flight algorithms. Therefore, Onyx can operate where multiple units are deployed in the same air space (Atair Aerospace Inc.2003).

As stated above, Onyx is a two-parachute hybrid. It has a high-efficiency, ram-air elliptical parafoil for autonomous guidance. The system uses a round recovery parachute to provide a soft landing for sensitive cargo. The on-board guidance, navigation and control system incorporates an integrated GPS and navigation system. It helps in continually adjusting to the flight characteristics of the system, as well as the weather conditions, throughout the flight. It helps to control the system to a pre-programmed altitude and position. The second, non-guided round 'Recovery Parachute' deploys just prior to landing, for a soft touchdown at its programmed point of impact (Atair Aerospace Inc., 2003, 2018).

Some of the key features for this hybrid parafoil/parachute system are as follows:

a) It is available in three payload configurations: Micro-Onyx (0 to 20 lbs.), Onyx-500 (0 to 500 lbs.), and Onyx-2200 (500 to 2200 lbs).

b) Its accuracy of landing with payload is 100 m.

c) This patented two-parachute ("hybrid") system uses a high-speed elliptical parafoil for autonomous guidance, and a round recovery parachute for a reliably soft landing.

d) It has flocking/swarming (formation flying) and active in-air 'Collision Avoidance' for simultaneous deployment of up to 50 Onyx systems.

e) It has 'Adaptive Control,' an advanced self-learning method for flight control. It enables gross variances in cargo weights to be airdropped.

f)   High glide ratio over 4.5:1 provides a horizontal standoff of 44 km
     from an altitude of 35,000 ft.
g)   It is deployable from military fixed-wing and rotary aircraft moving
     at speeds up to 150 KIAS. Ultra-fast flight speed (80 kts) increases
     accuracy and reduces vulnerability, to wind-induced errors.
h)   Continuously dis-reefed guidance parafoil provides for lowest
     opening shock and high-speed deployment capabilities. It is a rigger-
     friendly, recoverable, modular, and reusable system.
i)   Onyx 2200 is fully compatible with A-22 CDS bundles.

2.3.12   PARAFOILS WITH SENSORS TO STUDY WEATHER PARAMETERS AND ATMOSPHERIC POLLUTION

Previous reports indicate that parafoils have been utilized in various tasks
related to military and civilian intelligence, surveillance, telecommunication net-working and air drop (Pellone et al., 2017). We should note that
parafoils have also been utilized to study weather parameters immediately
above the farm or urban areas. They can also be utilized to collect data
useful in general weather forecast. Autonomous parafoils have been
utilized in pollution assessment, fire detection, and radiation monitoring.
Normally, such parafoils are fitted with sensors that detect atmospheric
conditions, gases, chemical pollutants and dust particles. A parafoil is also
used to assess visibility, etc. Let us consider an example. The parafoil used
had 1.6 m sail span, 2.4 m^2 sail area and weighed 5 g. The weather sensors
could be lowered and exposed to collect atmospheric sample, once the
parafoil has reached the desired altitude above ground. The sensors are
usually relatively very small. Their size ranges from 3–4 mm to 19 mm. A
recent report from Italian Aerospace Research Organization suggests that,
parafoils could be effectively utilized to monitor environmental parameters, record and store them in the central processing unit (CPU) on the
parafoil itself. If not, parafoils may relay the data to ground stations. Such
parafoils have been used to obtain information on accumulation of non-
methane volatile gases, methane, CO_2, NO_2 emissions, toxic gases (SO_2),
smoke and carbonaceous particles, etc. (Pellone et al., 2017; Jensen, 2000).
Huang et al. (2013) have stated that, small parafoils with a series of sensors
are apt to detect atmospheric pollution. Atmospheric measurements could
be focused on getting values for particulate content. Also, to get data

on gaseous composition such as CO_2, NO, N_2O, methane, etc. Specific chemical detectors (probes) could be attached to parafoils.

It is interesting to note that autonomous parafoils with fixtures to spray pollution remedying chemicals are being sought in China. The intention is to reduce atmospheric smog over their major cities and other locations affected by smog. The China Meteorological Administration has evaluated the large payload carrying guided autonomous parafoils, to solve weather related aspects. They have used parafoils to spray and distribute chemicals that supposedly reduce smog and fog in the atmosphere. The parafoil with spray nozzles and a large payload tank of chemicals (700 kg) is able to spray, in one stretch, up to 5 km area. The parafoil was built based on specifications offered by the meteorological agency. They tested the efficacy of applying smog reducing chemicals into atmosphere just above the busy Beijing Airport. The intention was to clear atmosphere and improve visibility above the airport. A few city municipalities, it seems, have decided to use parafoils above their cities in order to clear the atmosphere (Woolsten, 2014). Now, extrapolate this situation to agricultural settings, it is clear that we can use parafoils with spray nozzles and payload tank of plant protection chemicals, to spray the crop. It supposedly takes less time and effort compared to human skilled workers. Parafoil sprayer keeps farmers away from contact with harmful chemicals.

We may note that UAVs have been used to house sensors in the payload area. Several different types of sensors have been fitted and utilized to detect spectral characters of crops and chemical variations in the atmosphere, above the crop. Pobkrut et al. (2016) report that, irrespective of platforms, chemical sensors attached on UAVs could also be utilized to detect gases such as ammonia (NH_3) (Lorwangtragool et al., 2011), volatile compounds such as toluene, ethylene (Siyang et al., 2013), etc. Special type of sensors known as composite gas sensors that possess electronic nose for gases are installed on the platforms. Such a platform could be a parafoil flying just above the crop. In agriculture, for example, a parafoil with these gas sensors could detect NH_3 emissions from rice fields, cattle shed, and other cropped fields. Farmers can obtain a map of greenhouse gas emissions above their fields. Above a plantation, for example, a parafoil flying low over the fruit trees can detect concentration of ripening gas like ethylene. It helps in judging the ripeness of fruits on trees (see Pobkrut et al., 2016). Of course, Pobkrut et al. (2016) have utilized a low flying copter drone, but it could always be a different platform such as a parafoil (e.g., SUSI 62 or Pixy ABS). Most important component is the gas sensors.

They are made of polyvinyl chloride or polyvinyl pyrrolidone, polyvinyl alcohol, polyvinyl isobutyl/methyl mixture, etc. In addition to agricultural uses, such sensors can detect volatile compounds emanating from stored explosive dumps/stores. They could also detect volatile gases emanating into atmosphere immediately after an explosion. Pesticide residue in the atmosphere too could be detected.

2.3.13. RECREATIONAL PARACHUTES/PARAFOILS AND PARAGLIDERS

Parafoils form part of aerial sport and related recreational activities. There are also parafoils that land and take-off from water bodies such as lakes, canals, rivers and dam sites. Let us consider an example. 'Explorer' is a parafoil which is powered by a fuel engine and guided by a pilot. It allows the pilot to take-off or land on water. Initial reports on its performance suggests that water surface provides extra lift to the parachute when it takes-off. Such parafoils need no new or extra skills to fly out or land on water surface. The float system has a fixed main gear and a retractable nose gear system. It allows pilots to make water options anytime during flight or water taxi operations. The Buckeye Parafoil, for example, has amphibious capability if we opt for landing on ground surface. It is produced by Buckeye Industries located in Cassopolis, Michigan, USA. Incidentally, the same parafoil with certain modifications is also used in agricultural settings. The parafoil is utilized to spray plant protection chemicals on crops (Buckeye Industries, 1997).

There are several other models of powered recreational parachutes (Plate 2.9). Again, a few with suitable modifications allow the parafoil to take-off from water bodies like lakes, large waterfront, etc. They serve recreational purposes. For example, Balena (2005) reports about a parachute that takes-off from water surface, using powered propellers and canopy. Usually, such a parachute can be 14–15 ft long and 6.5 ft wide and may weigh up to 265 kg. It can also accommodate a pilot, a passenger and a few cameras for aerial photography. Such powered parachutes are fitted with a 65 hp motor. This motor is attached to a propeller. Generally, it requires about 25 hr flight training to master take-off from water and glide into air for a few hrs, say 9–10 hr. The parafoil can reach a height of 10,000 ft. above ground surface. It can be maneuvered over difficult terrain or large agrarian farms. During recreational flights the parafoil is usually flown at 200 to 1000 ft. altitude. Obviously, it could be used over a farm to monitor crop fields and detect maladies, if any. We could detect maladies such as flood or drought affected spots, disease/pest attacked zone, nutrient dearth in soil, etc. The best time to fly this parafoil is early mornings when wind velocity and distraction is low. The parachute canopy and tethers are easily packed into a pouch after use (Balena, 2005).

PLATE 2.9   Recreational parafoils.

Source: Till Gottbrath, Nova Performance Glides, Terfens, Austria; https://www.nova.eu/en/company/; https://www.nova.eu/en/my-nova/.

Recreational parachuting began sometime early 1970s. At present, there are indeed several parachuting clubs in each and every nation. Commensurately, there are several companies that produce a wide range of parachute and parafoil models. SkyRunner is yet another example of recreational parafoil. At the same time, it could find greater applications in agricultural surveillance and scouting of crops. It may also find easy acceptance during transit of people and cargo. Particularly, if we encounter rough, difficult to navigate terrain by road, e.g., hill country villages, crop fields in hills or delta areas, etc. Regarding recreation, it is supposedly a 'state of the art' adventure parafoil. It allows the rider and passenger excellent maneuverability. It is known as a good light 'Sport Aircraft.' It seems 'SkyRunner' fuses off-road engineering with proven aviation technology to create a new generation of sports recreation vehicle. It is designed for both work and play (SkyRunner, 2018). The Sky Runner powered guided parafoil costs less than 50 US$ per hr in flight. Hence, it is efficient compared to helicopters (SkyRunner, 2018; Plate 2.10). We need only 8 days of training, to obtain a pilot's license. Hence, it could become a sought after aerial agricultural vehicle. Farmers may adopt it to conduct crop scouting. However, at present, this aerial guided parafoil costs about 140,000 US$ per unit.

2.4   USES OF PARACHUTES AND PARAFOILS

The usefulness of parachutes/parafoils has increased enormously since their first demonstrations to public. Yet, there are many more ways by which

PLATE 2.10   *(Continued)*

PLATE 2.10   A recreational and transport parafoil.

Note: Such parafoils have immense relevance to agricultural setting. They are useful in trans-porting farm inputs, farmers and farm managers from one location to another. Also, during regular surveillance of crop fields and terrain.

Source: Till Gottbrath, Nova Performance Glides, Terfens, Austria; https://www.nova.eu/en/company/; https://www.nova.eu/en/my-nova/.

parafoils can be useful to farming and urban communities. The earliest, and still most common usage of parachutes/parafoils is in air drop of personnel and cargo. Like, many other inventions, gadgets and contraptions, parachutes/para-foils too were largely confined to military establishments. Therefore, much of the improvements in the technique and sophistication in material were aimed at aiding the military purposes. Parachutes drew greater attention from space scientists. At present, we know that parachutes and guided or autonomous parafoils have many uses, during spacecraft recovery and landing. Parachutes/parafoils have been tested and utilized in a series of situations related to civilian conditions such as urban surveillance, monitoring public activities, traffic on highways, etc. Parafoils, in particular, are being tested and new models are being generated. So that, they could be effectively used in agrarian regions. In the agrarian regions, parafoils could find regular use in drawing aerial images, thermal maps of crops, monitoring irrigation channels, monitoring disease/pest attack, evaluating crop genotypes in farm, etc.

2.4.1   MILITARY AND CIVILIAN USES OF PARACHUTES AND PARAFOILS

Records about drone (Unmanned Aerial Vehicles) usage, in general, suggests that during recent years, they are being frequently utilized by military and civilian agencies. It includes different types of drones such as small flat-winged aircrafts, copters and parafoils. At present, guided powered parafoils are utilized in good number in the surveillance of military camps, convoys and ammunition dumps. The logbooks of powered parafoils show frequent flights and landings by the military personnel (Lynch, 2012). So, it is clear that, powered and autonomous parafoils/paragliders could gain greater acceptance as a military surveillance drone.

Parachutes are utilized in military mainly to conduct air drop of personnel and cargo. Mostly, round or circular parachutes find utility during cargo airdrop. A military cargo parachute lowers supplies safely from the aircraft onto the ground. Parachutes are also effective when road traffic gets hindered or when the terrain is difficult for surface transport. Parachutes were first utilized in a significant measure during 1930s by Russian (USSR) and German military. Around 1950s, improvement in transport of military personnel and cargo was perceptible. It was possible due to larger parachutes. They could drop 1 to 2-ton cargo from an aircraft. They could drop the cargo at pin-point locations, from aircraft (Kovacevic et al., 1998). By 1970s, military cargo transport aircrafts regularly air-dropped cargos sized 16 to 20 t, using very large parachutes.

Parafoils and other air drop mechanisms are said to be of great utility, when military personnel and cargo have to be shifted, from one ship to another, in high seas. Armament airdrop into ships are possible, using guided parafoils (Committee on Naval Expeditionary Logistics, 1999; Puranik et al., 2006).

Military establishments of different nations have been constantly striving to improve the performance of parachutes/parafoils. Basically, they aim at swift and accurate transport and air drop of different types and sizes of cargo. For example, parafoils of 4200 ft^2 have been tested and deployed by the United States Army. The aim is to air-drop cargo from large aircrafts. The cargo is lowered slowly. Parafoils land accurately at pre-determined spot (Bennet and Roy, 2005). Flight tests and routine use has shown that parafoils can safely air-drop large sized cargo, up to a size of 15,000 lbs (6500 kg) of military supplies. The air drop could be performed right at the military camps or at pinpoint spots near the combat zone.

No doubt, payload delivery is an important aspect of parafoil technology. Military establishments, now a days are opting for autonomous parafoils, to achieve accurate air drops. The autonomous parafoils are guided using GPS connectivity to improve accuracy. An autonomous parafoil is a highly preferred air-drop vehicle, because, it avoids involvement of army personnel. Further, Klerman (2011) states that, helicopter and trucks are often the major methods of food and ammunition supply. However, these could be vulnerable. Under such conditions autonomous parafoils are useful, so that we do not risk human life in the combat zone.

In some nations or regions, military establishments have adopted autonomous parafoils to get aerial imagery of military convoys. They have also been used to detect areas prone to drug trafficking. The parafoil either transmits or stores imagery of terrain and details. The video imagery of movement of military convoys have been highly useful in planning (Kogler et al., 1995).

Parafoils are used to air drop cargo at pre-determined locations as accurately as possible. They are of utility when cargo has to be slowly lowered at a point that is otherwise very difficult to transit on surface/roads. A different kind of problem is to be negotiated when parafoils are to be lowered and cargo transported, on to the deck of a moving ship, on high seas. They say improvements in identification of target location and its velocity could help use of parafoils. It is pertinent when cargo transport has to be handled using parafoils, say, either at harbor or on high sea (Hewgley et al., 2009, 2011, 2014a 2014b). In future, parafoils could be used at harbor and Naval Dockyards, to shift heavy load from one place to another or even to a moving ship.

Now, let us consider a few civilian uses of parachutes/parafoils. Wind power generation by a parafoil is also a possibility. It has been duly explored (Kim and Park, 2009). The para-wing on ships pulls it using wind energy. The wind velocity increases with altitude. Hence, para-wings placed higher can offer more energy for ships to move on high seas.

Powered Parachutes and Parafoils could form an important aspect of emergency supplies, relief and rescue acts in case of a disaster. Let us discuss a recent example, where in, parachutes have been used shrewdly, to supply food relief and rescue material, to the affected people. It relates to a disaster in the Caribbean region. This region was hit by a typhoon. Here, aeroplanes carrying food supplies were also loaded with parachutes. The food packages were attached to parachutes that alighted slowly in the devastated regions. So, air drop of food and other supplies was efficient and safe (Price, 2018). Aid packages delivered by parachute have saved lives in the storm-hit Caribbean. Let us consider yet another example. 'Hurricane Irma' caused devastation to the people of Puerto Rico. They found themselves recovering from a disaster again when 'Hurricane Maria' struck the island in September 2015. Yet, many families were able to begin piecing their lives back together with aid delivered. They used a revolutionary device developed in a farm shed in Kent, United Kingdom (Price, 2018). The Air Drop Box uses a parachute to gently send supplies thrown out of light aircraft. It provided humanitarian organizations a cheap and safe method of getting food and clothing to those in need, quickly.

Forecasts and preliminary evaluations suggest that parachutes/parafoils that pop-up from a drone carrying a medical supply could land at a pre-determined location. Such locations are usually, say, near a hospital or a temporary medical center in difficult terrain. The airdrop is done swiftly and accurately. Fixed-wing drone aircrafts are typically used to cover longer distances and carry a payload of 1.5 to 4 kg medicines payload. They can fly at speeds around 100 km/hour and cover a round trip of 150 km. These drones operate in conditions with up to 50 km/hour winds. A major drawback of fixed-wing drones is that they are not designed for vertical take-off and landing (VTOL). Fixed-winged drone aircrafts are less suitable to pick up samples for diagnostic services. Instead, parachute delivery of emergency medical supplies to health centers in hard-to-reach areas seems apt.

The Zipline cargo drones (ZIPs) are operated out of medical operation centers. Medical centers are of the size of a shipping container. Such containers are placed close to the public medical warehouses. The drone aircrafts lift about 1.5 kg of cargo. The UAV drops into a marked "mailbox" area, using a parachute before it returns without landing, to the ground control station. Such drones may conduct about 15 deliveries per day. Use of parafoils for air drop of medicine from drone aircrafts will depend on efficiency and accuracy. Also, on success and economics of fixed/copter drones that carry the medicine and parachutes used to air drop. Evaluation is needed.

Powered autonomous parafoils also those piloted could find greater use with surveillance and law enforcement agencies, in urban areas. Let us consider an example. A powered parachute is being utilized by police department of one of the small towns called Rippon, in Canada. Parafoils have been adopted during suspect captures, river rescues, critical infrastructure over-flights, crime scene photos, narcotics enforcement and crime suppression. They say, parafoil costs small fraction of the cost of helicopter. It seems parafoils have also been utilized by police and law enforcement departments in USA.

2.4.2   AGRICULTURAL USES OF PARACHUTES AND PARAFOILS

So far, fixed-wing airplanes, helicopters, multi-rotor copters, remote controlled kites, balloons, gliders, and motorized parafoils all have been used for low altitude remote sensing (LARS) and imagery (Pudelko et al., 2012; Krishna, 2018, 2019). Here we are concerned with parafoils. A parafoil must be equipped with imaging system, software for autonomous flight, stabilized sensors that include GPS (Global Positioning System), etc. The desirable resolution is about 1–2 centimeters. A parafoil offers relatively better image resolution than that of any satellite-based images.

The parafoils that could be either guided (piloted) or those autonomous in flight have several uses in a farm setting. They offer certain clear advantages to farmers over other modes of crop surveillance and data collection. Also, in general farm maintenance. The usage of a parafoil can begin even before land development, right at the stage of site selection. Let us list a few uses of parafoils in agrarian regions:

Parafoils are utilized to study and map topography of an agrarian location. This is done prior to initiating land development activity. Using parafoils, we can obtain aerial photography of topography. Such images could be consulted in demarcating farm blocks, land leveling, planning of planting schedules, laying irrigation pipelines, etc. (Thamm et al., 2013; Pudelko et al., 2012). The parafoils can be used to conduct aerial scouting of crops during mid-season.

Aerial imagery of soils using parafoils helps in the formation of 'management blocks' based on soil types. Along with ground data from previous seasons, aerial maps can be used to map soil fertility variations. We can overlay maps to obtain borders of 'management blocks' that are based on soil fertility variation (see Krishna, 2016, 2018; Thamm, 2011; Thamm et al., 2013; Pudelko et al., 2012).

Parafoils are used to judge fields right from seeding. Aerial images of seed germination trend, monitoring crop stand establishment, identification of seedling gaps and plant stand evaluation are possible, using parafoils. Farmers can accomplish field scouting to fill seedling gaps. This procedure could be done relatively rapidly compared to field workers walking all through the large fields. Parafoils could be used to obtain aerial imagery (visual band width) to assess water resources, presence of irrigation channels, ponds, dams, etc. in a given location. Such data helps in formation of management blocks and organizing water resources efficiently.

Parafoils are used to surveillance crops and collect NDVI data, using multispectral sensors. Such data helps in judging the growth pattern, biomass accumulation and grain yield. Autonomous parafoils can also be flown over crop fields periodically at different stages of crop growth. This is to obtain an estimate of NDVI, leaf chlorophyll (crop's-N status), panicle/grain formation trends. Consequently, we can compute and forecast grain yield, using data collected from parafoils.

Parafoils fitted with visual, NIR and IR sensors are used to collect data about crop's water stress. The infrared imagery helps farmers to judge variations in soil moisture and map the crop's water stress index. Eventually, IR imagery got using parafoils helps in deciding quantum and frequency of irrigation (Pudelko et al., 2012).

Autonomous parafoils could be programmed to fly above large areas of crop fields. This is to obtain aerial imagery about crop's health. The visual and multi-spectral imagery indicates spots affected by crop diseases. Of course, it depends on the type of symptoms generated by the pathogen or insect pest. Piloted parafoils with facility for multispectral imagery could also be used, to detect several types of crop maladies and major diseases. Maps showing spots with disease/pest attack can be treated, selectively. So, it helps in reducing the need for farm labor and plant protection chemicals. Also, it helps in thwarting the crop loss due to disease/pest right, at an early stage. For example, in Russia, autonomous powered parafoil was utilized to detect disease affected zones on crop. Also, to spray with plant protection chemicals (Akhramovich, 2017).

Tracking spots affected by natural disaster such as floods, soil erosion, rills and gully formation, loss of seedlings, crop lodging, drought is possible using aerial imagery. Autonomous parafoils with multi-spectral sensors could be programmed to fly past disaster zones. The ground control stations may be relayed with accurate data about the intensity of natural disaster and crop loss. Once, the data is available, remedial measures are usually channeled swiftly.

Recently, parafoils that fly at low altitudes above agricultural experimental stations have been sought. Such parafoils help in keeping a general vigil and in monitoring farm activity. Such parafoils with sensors helps us in gathering crop physiological data useful for plant breeders. Agronomists evaluating large number of crop varieties and genotypes can rely on parafoil collected data. Parafoils help in assessing agronomic performance of crops and their genotypes in experimental stations (Pudelko et al., 2012; Akhromovich, 2017). Parafoils connected with ground control station via wireless and possessing GPS connectivity are used to monitor farm activity, farm vehicle movement, and farm borders/fence.

2.4.3 USES OF PARACHUTES AND PARAFOILS IN SPACE SCIENCE

Parachutes are most commonly used as aerodynamic decelerator systems. They are utilized in slowing down aircraft during landing. Also, while recovering rocket pay loads. Such non-guided parachutes differ in size, shape and efficiency. Selection of a non-guided parachute depends on the purpose of decelerator system. Generally, parachute-based decelerators are easy to fit. They are low cost systems. There are a few design features specific to high performance non-guided parachutes. They are usually evaluated for a series of features. Some of the features evaluated are: determination of parachute area and mass; drag stability and stress of the parachute; filling time and flight dynamics calculation; wind tunnel testing; flight and landing (impact) testing; material and manufacturing process involved. Essentially, tedious selection of materials, manufacturing process and testing is necessary, before deciding on a non-guided parachute model/brand, for deceleration (Koldaev and Moraes, 2005; Knacke, 1991; Peterson, 1990; Deweese et al., 1978; Machin et al., 2002).

As early as in 1960s, a guided parafoil system was intensively tested. It had successfully demonstrated the feasibility to land a large vehicle. The parafoil drop test program used subscale and full-scale parafoils. Over 300 drops were conducted using subscale parafoils having a wing area of 750 ft^2. Using full-scale parafoils, the program conducted twenty drops adopting 5,500 ft^2 parafoils and thirteen drops adopting 7,500 ft^2 parafoils (Stein and Strahans, 2018). A series of flight tests were conducted, using the Buckeye powered parachute test vehicle. The Buckeye was modified by Southwest Research Institute to enable the vehicle, to fly autonomously or via remote commands from the ground. As a result of the successful demonstration of NASA's autonomous, guided parafoil system, Natick Soldier's Center was interested in forming a cooperative effort with NASA, to build and test a guided parafoil system for the U.S. Army. The aim was to deliver 10,000 lbs (4500 kg) of useable payload (Stein and Strahans, 2018; NASA, 2014; Corelfield, 2017). Overall, we may note that space agency tested parafoils of different sizes.

Further, we may note that miniature, lightweight, autonomous parafoil systems are used in high-altitude weather experiments. Balloon payloads routinely carry parachutes for safe recovery after the bursting of the balloon. Light weight miniaturized parafoils/parachutes are commonly placed with high altitude balloons. They are also used in the recovery phase of spaceships, modules and personnel, if it is a manned flight. Parachutes are often the best bets for recovery of space vehicles and personnel returning to earth's gravitation (Benton and Yakimenko, 2013). They retard and slow down the descent to ground surface. In some cases, such as space shuttles, they are used to retard the shuttle during landing and bringing it to a halt.

During the past 4–5 decades, National Aeronautics and Space Agency has tested several models of parafoils, to recover space craft and astronauts involved in space journey. They have used parafoils at the final stages, to steer the payload slowly and to exact location. Such ram-air parachutes (parafoils) usually possess GPS connectivity and facility to steer them. A few of the models are autonomous parafoils that could be pre-programmed regarding air-drop rate and location. Parafoils with ability to airdrop payload from 10,000 ft. above ground were tested way back in 1993 (Sim et al., 1993; Murray et al., 1994). Currently, parafoils are being evaluated for safe recovery of space craft payloads right from 69,000 ft. above ground.

Parachutes have been utilized to slow down space vehicles. Often, such parachutes undergo severe testing before being adopted on actual space missions. For example, a prototype of large parachute, a composite of three parachutes were tested for their reliability. These parachutes were evaluated for their ability to decelerate, when the aircraft was travelling at 250 mph. The mock test was conducted at 25,000 ft. above ground. Such parachutes are of great utility in decelerating and halting space crafts, upon landing (Chow, 2013). Parachutes that decelerate the space shuttle are among the best examples of parachutes used in space science.

Adopting autonomous parafoils to drop a payload from 60,000ft. on to a defined point is said to be a great leap forward. Further, the use of a steerable aerofoil parachute to bring the payloads back to earth is also a neat and logical extension of technology. Reports from NASA states that, cellular aerofoils will be utilized, to land payload from high altitude of over 60,000 ft. They say, such parafoils will be able to recover payloads without having to chase balloons. Further, they say, parafoils will be able to steer the payload to desired spots on land or sea.

Reports suggest that NASA's X-38 has tested the largest parachute known till date. This large parachute is part of recovery system of X-38 space craft.

The intention is to develop a few International Space Stations' crew return vehicles, for much less cost than the space shuttle orbiter. In future, such huge parafoils may find routine use. The parachute tested in Arizona is almost as large as a Boeing 747 jumbo passenger plane. It has a span of 143 ft. and total area is 7500 ft². In its initial trials it has dropped an 18,000 lb cargo from 21,500 ft above the desert surface. The parachute, despite its large size opened swiftly in 30 sec. This parachute is called lifeboat of International Space Station.

Parafoils could be an appropriate technology when International Space Station is visited more often by the space crafts from earth. Autonomous parafoils could be apt when use of personnel is to be avoided. Reports from NASA suggest that, parachutes seem to have found a highly important use during landing and exploration of robots on Mars. They say, like on earth, slowing down the descent and careful landing of craft on mars surface is important. Therefore, space scientists are now trying out new very large supersonic parachutes, to land space crafts on Mars. At present, they are testing the low-density supersonic decelerator, at 34 km away from earth. Such parachutes may find more common use in landing cargo on other heavenly bodies too (NASA, 2014).

In the Mars's canyon, while surveying the shallow sloped hills, it seems, astronauts could use controlled parachutes/parafoils. Autonomous parafoils could be used to speed down-hill, just 2–3 feet above the Martian surface (Bhosri et al., 1999).

2.4.4 USES OF PARACHUTES AND PARAFOILS IN ARCHAEOLOGICAL STUDIES

Aerial imagery of an archaeological site is not a new proposition. It has been conducted using several different types of platforms. A few of the platforms utilized are static and located at a vantage point high above the ground features. A few others could be flown over the sites, swiftly (e.g., fixed-winged or copter drones). The flight path and imaging points could be pre-programmed. There are many studies that demonstrate the potential of aerial visual and thermography, to reveal surface and subsurface cultural features. Yet, technological and cost barriers have prevented the widespread application of thermal imaging in archaeology (Cesana et al., 2014). However, during recent past, aerial vehicles such as small drone aircrafts, parafoils and microlights have been utilized, to make high resolution aerial imagery of archaeological sites. For example, Cesana et al. (2014) report that, UAV could be utilized to make detailed aerial imagery of historic sites in the 'Chaco area' of New Mexico. They could easily obtain well processed photos from ortho-images. Aerial imagery using thermal sensors have been improvised. At present, they offer increased spatial resolution and thermal sensitivity. At the same time, advances in photogrammetric image processing software packages, such as Pix4D Mapper, AgieSoft's PhotoScan and Microsoft's Image Composite Editor (ICE), now make it easier. They provide high resolution processed photos of archaeological sites. Parafoils that fly slowly and at low altitudes over the historic remains offer clear images from vantage points. Powered piloted parafoils and microlights (trikes)too are utilized to conduct aerial photography. No doubt powered (motorized) autonomous parafoils have also been effectively utilized, to surveillance and study archaeological sites (Hailey, 2005).

Hailey (2005, 2008) utilized a Destiny 2000 powered microlight to conduct low altitude aerial imagery. It is a two-seater piloted microlight aircraft. A broad band thermal sensor was used to obtain infrared imagery of archaeological sites. The thermal camera was able to detect temperature difference of 0.2 degree centigrade. The positions were marked using TrimbleProXRS GPS Positioning system. Both, vertical and oblique shots were obtained about the archaeological sites. Tethered parafoils and aerostats too have been utilized, to study archaeological sites. For example, in Greece and Turkey, they used tethered parafoils/aerostats fitted with cameras to get photogrammetric details of several archaeological sites (Whittelesey, 1970).

2.5 PARAFOILS AND PARACHUTES IN AGRICULTURAL CROP PRODUCTION

2.5.1 AERIAL SURVEY OF NATURAL RESOURCES RELEVANT TO AGRICULTURE

Thamm (2011) and Thamm et al. (2013) believe that in due course autonomous parafoil (e.g., SUSI 62) could help in obtaining accurate topographical maps of agricultural terrain, land holdings, soil types and cropping systems in vogue in a given area. Farmers may choose appropriate land and soil management practices, contouring, laying irrigation pipelines, etc., based on aerial photography got using parafoils. Planning land conservation measures and managing natural vegetation could be accomplished using aerial imagery derived using autonomous parafoils with sensors. One of the reports clearly shows how management blocks could be marked, using parafoil-derived aerial images of a sloppy and undulating land. They consulted aerial imagery derived using SUSI 62.

Reports from Great Britain and Australia suggest that powered parafoils are being increasingly sought by farmers. Powered manned parafoils are being utilized to conduct series of aerial surveillance and imagery tasks. They are usually done by parafoil while flying at low altitudes above crop fields. In addition, these parafoils are also used to watch functioning of irrigation lines and movement of field vehicles. Grazing pattern of cattle is also monitored using powered parafoils. There are a few models of powered parachutes such as the Buckeye's Power Paraglider which is produced by the Buckeye's 'Coldfire Systems Inc., situated in California, USA. Yet another parafoil model sought is named 'Paraplane' produced by ParaPlane Corporation. Both of them are utilized by farmers for aerial survey of terrain, open fields, soil conditions and crops. They say, what started as a hobby flying of paragliders has been adopted in farming zones. These parafoils are used mainly to survey crops, conduct spectral analysis and decide agronomic inputs (Farm Show Magazine, 2018b).

Agricultural drones have been adopted to conduct aerial survey of water resources in farmland. Aerial images got using visual band sensors offer accurate data about water flow in channels and rivers. Seasonal changes in water flow and storage can be monitored accurately, using aerial images. Recent reports from Utah State University (AggieAir) suggests that, we can assess water resources in a farm using aerial imagery obtained, by autonomous small aircrafts. It has to be followed by improved image processing. Further, we have to adopt appropriate computer programs and water flow calculating methods (King and Nelson, 2018). No doubt, autonomous or powered guided parafoils too could be flown at low altitude over crops. Then, aerial imagery-based methods could be adopted to assess water resources. In fact, studies that compare the usefulness, cost effectiveness and rapidity of data procurement about water resources, using drone aircrafts, parafoils and blimps may be necessary. Incidentally, autonomous parafoils have been examined in experimental farms to conduct aerial survey of soil and water resources (Thamm, 2011, Pudelko et al., 2012; Plate 2.11).

PLATE 2.11   A Powered Manned Parafoil above agricultural fields.

Source: Alvie Wall, Founder President, Infinity Power Parachutes Inc., East Lafayette Rd. Sturgis, MI, USA.

2.5.2   *PARAFOILS TO STUDY GROWTH AND MAP NUTRITIONAL STATUS OF CROPS*

There are several reports about small light aircrafts being utilized to conduct aerial survey of crops, to ascertain nutritional and water status. They have also tried small aircrafts to spray fertilizer-based nutrients. Foliar fertilizer application may not be accurate and uniform, if we utilized aircrafts. Plus, during precision farming we require good control over the release of nutrients from the nozzles. Accurate regulation of foliar nutrient release may not be possible at the speed with which the small aircraft flies over the crop. At best, only blanket prescriptions of foliar nutrient application could be achieved. Precision farming requires application of foliar nutrients at different rates based on variation in soil fertility and crop's nutrient status. The aircraft-aided aerial photography and foliar application needs low flying platforms. Also, platforms such as copters or parafoils are needed for accurate photography and spectral analysis. Fertilizer trials have been monitored and assessed, using small aircraft-aided multi-spectral photography (Lebourgeois et al., 2012). However, it could be worthwhile to try and standardize methods involving parafoils.

Several models of UAVs have been adopted, to obtain useful data about crop growth and productivity. Unmanned aerial vehicles such as fixed-winged (Sensefly Inc., 2018), helicopters (e.g., RMAX. 2015), multi-copters (e.g., AGRAS MJ-1; DJI Inc., 2016; e.g., MD 3000; Microdrones, 2018), parafoils (SUSI 62; see Thamm, 2011; Thamm et al., 2013; e.g., Hawkeye; Tetracam Inc., 2018) fitted with visual and multi-spectral sensors have offered data about NDVI, chlorophyll content (i.e., nitrogen), biomass, crop height, water stress index, etc. (see Krishna, 2016, 2018, 2019). Recently, Olga et al. (2018) have reported use of copter drones to obtain critical data about spring wheat crop's nitrogen status. There are other reports that deal with estimation of nitrogen (i.e., chlorophyll content) of a crop (for example, potato (*Solanum tuberosum*) using parafoils (Para wings) (Hunt et al., 2018). They have estimated leaf chlorophyll which is directly and proportionately related to crop's N status.

Let us consider the details about an autonomous moto-glider (e.g., Pixy ABS) that was utilized at the agricultural experimental station, Montpellier, France. They have adopted Pixy ABS to assess the performance of crops, such as wheat grown in experimental plots. They conducted an experiment, wherein fertilizer-N was applied to wheat plots, at five different levels. The idea was to assess the performance, i.e., response of wheat to fertilizer-N, using spectral analysis. They used a parafoil (Pixy ABS) fitted with visual (Canon 8 gigapixel red, green, blue and cyan band width), multi-spectral (SONY DSC multispectral camera), and IR (Canon Infra-red camera) sensors. They aimed at recording NDVI, leaf chlorophyll index (i.e., crop's-N status)and biomass accumulation trends. About 17 genotypes of wheat were evaluated for their response to fertilizer-N. They compared the aerial photography, spectral data and ground truth data for accuracy. They stated that spectral data correlated well with those of ground truth, particularly, for traits such as leaf chlorophyll, leaf area index, nitrogen uptake and GNDVI. Such trials were conducted for different seasons to confirm the procedures that utilized parafoil, sensors and spectral data. Aerial photography of growth and nitrogen uptake trends of different genotypes of wheat were obtained. Later, the trends could be mapped and analyzed. We may note here that, regular monitoring of experimental stations, the installations, farm vehicles, crop's performance in the outfield, evaluation of crops found in experimental plots, all could be done at low-cost and efficiently, by adopting parafoils. Further, it is said parafoils with sensors could be utilized, to conduct aerial survey to form 'management blocks,' in case, precision farming is practiced. Rapid accrual of maps depicting variations in growth, nutrient uptake, disease/ pest attack and water status is almost essential, if precision farming is to be adopted. Parafoils flying at low altitudes over crops can be immensely useful in such situations.

Autonomous parafoils such as SUSI 62 have been utilized to obtain detailed aerial images of crops, at various stages. Then, a digital surface model (DSM) is prepared using appropriate computer programs. These DSMs are highly informative regarding variations in crop growth and nutritional status. Such DSM could be consulted quickly by farmers. Then, appropriate inputs and remedies could be decided. Let us consider an example from tropical Africa (Tanzania). The parafoil (SUSI 62) has been effectively utilized to prepare management blocks. The 'management blocks' are based on topography and soil fertility of fields. During the crop season, aerial photography has been utilized to get DSMs. They say, ortho-images and DSMs of crops are useful, while deciding fertilizer inputs and irrigation scheduling (Thamm et al., 2013; see Krishna, 2018).

2.5.3   AERIAL SURVEY TO MEASURE AND MAP WATER STATUS INDEX (CWSI) OF CROPS IN FIELDS

There are reports that parafoils such as SUSI 62 or Tetracam's Hawkeye are highly useful autonomous UAVS, in assessing crops. In particular, they are good aids in detecting crop's water status (Thamm, 2011; Thamm et al., 2013). Such parafoils are usually fitted with infra-red and red-edge cameras. In addition, visual and multi-spectral sensors are also fitted. The thermal imagery (maps) show details about variations in crop's water stress index. Such maps can be over-layed with those obtained using visual bandwidth sensors. Using appropriate computer programs, we can apply irrigation when required and in apt quantities. Such CWSI maps could also be consulted to apply irrigation, using center-pivot sprinklers. Basically, digital data obtained using sensors on parafoils has to be relayed to variable-rate sprinklers.

Parafoils such as SUSI 62 fitted with a full complement of sensors (visual, RGB, IR and LIDAR) have been effectively utilized, to monitor irrigation in large farms and experimental station. This parafoil has also been used to obtain accurate thermal (IR) maps of experimental crops. Such maps help in knowing the variations in the water stress index of a genotype. So, it helps in deciding quantity and the timing of irrigation events. Water stressed genotypes show higher temperature (Thamm, 2011). Overall, monitoring experimental plots and irrigation lines could be accomplished, using parafoils. It is said a large farm of few thousand hectares could be surveyed quickly using such low flying autonomous parafoils.

2.5.4   AERIAL IMAGERY AND SPECTRAL ANALYSIS TO DETECT AND MAP DISEASE AND INSECT PEST ATTACK ON CROPS

Aerial survey using low flying autonomous drones such as parafoils, small copters or fixed-winged aircrafts has several advantages. The multi-spectral imagery derived from vantage heights above crop's canopy could be utilized in automated detection of crop diseases/pests. The data bank with spectral signatures of crops that are healthy, and those patches affected by microbial pathogens or insects are basic requirements. This is because the computers have to refer and compare the spectral data (from data bank) to decide the disease affliction, its cause and intensity. Computer software programs to calculate the amount of foliar spray required at each spot are required. In near future, parafoils with appropriate sensors and specific computer software should be available to farmers.

The low flying powered parafoils move slowly at an altitude of just 10–15 ft. above the crop's canopy. Such parafoils offer excellent advantages to farmers. They allow farmers to survey the fields in detail for pests/diseases and drought affected patches. At that low altitude, it gives a clearer image of soil erosion or damage to irrigation pipelines, if any. For more general view of the crop's canopy, the pilot can maneuver the parafoil to 1000 ft. above ground level and obtain multispectral images. Then, identify the spots that need remediation. Such parachutes have also been used to keep a vigil on borders of a farm and identify cattle damage, if any. (Buckeye Industries Inc., 2018)

Screening crops for disease tolerance should be possible, using parafoils with multispectral sensors. Adoption of parafoils helps in reducing need for human scouts, to assess crop damage due to disease/pest. Human fatigue and error in judgment can add to inaccuracies. Such inaccuracies are removed, if automated spectral analysis steps are followed. Further, parafoils could be used to manage experimental plant breeding farms. Mainly, parafoils could be used, to monitor disease situation and identify genotypes/lines tolerant to disease. Lipson (2015) states that, spectral methods has been adopted effectively, to survey and identify maize crop affected by northern leaf blight (Lipson, 2015; Dechant et al., 2017).

Parafoils have been utilized to fly above large farms in Pennsylvania, USA. The aim was to surveillance crops and collect data about them, at different stages during production of grains. Such parafoils have a payload that includes sensors such as visual, multispectral and NIR and IR cameras. The aerial videography is later used to identify locations in crop field that show up nutrient deficiencies, insect attack and water stress. This technology helps farmers, to apply remedial measures swiftly and accurately. Remedial measures could be confined only at the spots that are affected. At present, private companies and farm extension agencies in USA are demonstrating the use of parafoils, to obtain aerial photography of cereal crop fields (Cedar Meadow Farm, 2010).

Let us consider a recent example. Aerial survey of pine trees aimed mainly to detect pine beetle attack is possible, using different platforms. For example, we may use a small UAV (fixed-winged or copter) fitted with sensors. Then, trace the attacked trees on maps. We can also use piloted aircrafts to detect beetle attacked trees. Pine trees attacked by beetle usually show loss of crown foliage. The color of foliage changes to yellow, orange and eventually red. In the present context, we may note that a parafoil either autonomous or guided or a piloted microlight could also be used, to survey and detect pine beetle attacks (Billings and Ward, 1984). Periodic survey of forest stands for pest attack using parafoils is a good idea. A parafoil fitted with visual, multi-spectral sensors could be useful. However, detailed studies directed towards assessing parafoils, for weed detection and control are needed. The advantages and disadvantages of parafoils need to be ascertained, in experimental fields and farmers' field.

2.5.5   *SPRAYER PARACHUTES/PARAFOILS FOR PLANT PROTECTION*

Firstly, we have to identify the fields with crops that have been under stress due to disease or pest or weed infestation. Also, those fields that are nutrient stressed and need foliar fertilizer application. Next, in case of disease/pest, we have to identify the causal agent and extent of damage. The remedial measures have to follow up in equal measure and as accurately as possible. Until now we know that aerial robots such as autonomous fixed-winged small aircrafts, copters and helicopters have been successfully utilized, to first scout for disease/pest or weeds (see Krishna, 2016, 2018, 2019; Buckeye Industries, 1997, 2018). Then, to apply plant protection chemicals as a blanket prescription. In case, the farmers are adopting precision farming methods, then, formulations are applied, using drone sprayers fitted with variable-rate applicators (nozzles). Variable-rate nozzles are electronically controlled. The nozzles and flow are dictated by computer commands. The computer decisions are based on aerial images and spectral data got using various sensors. We ought to realize that all of these steps in crop husbandry could also be accomplished, using parafoils fitted with sensors and sprayers. However, we need apt modifications on parafoils, so that, they can be operated as specialized 'farm parafoils' (see Akhromov, 2017; Buckeye Industries, 2018; Taylor, 2018).

Reports suggest that, powered parafoils could be utilized efficiently, to spray plant protection chemicals. They could be manned using a one-seater or two-seater payload area made of light aluminum. Of course, the parafoil could be made entirely autonomous, using appropriate computer and program ground control computers. Parafoils could be pre-programmed regarding flight path and way points. Let us now consider use of a powered piloted parafoil above crops and plantations, in the Kissimmee county of Florida, USA. A powered parascender produced by Parascender Inc., was utilized by the farmer (Farm Show Magazine, 2018a). The parafoil was powered using a 52 hp twin engine (Rotax engine). It was fueled by petrol. The engine was connected to a propeller with plastic blades. There were a few modifications done on the pilot area of powered parafoil. One of the seats was removed to accommodate a plastic tank that holds 30 gallons of pesticide/fungicide. The powered piloted parafoil was also fitted with 5.2 ft. spray bar with 4 nozzles. The nozzles could be regulated electronically. This way, most appropriate flowrate could be decided. In all, this parafoil was able to carry a payload of agricultural chemicals, sensors and personnel totaling 520 lb (230 kg). Answers from farmers suggest that, such a parafoil sprayer was utilized mostly, to spray insecticide, to control root worms that affect corn. Also, to control weevils that affect wheat, hay and soybeans. They have opined that, these powered parafoils are safer, more accurate and economically efficient, if compared with conventional rapid transit small aircraft. They say, such parafoil sprayers with 30-gallon pesticide tank were typically selected, if the area to be sprayed with insecticide is about 300 acres. Interestingly, reports state that powered parafoil sprayers were well suited to conduct aerial spray of post-emergent 2.4-D herbicide, on a standing crop of maize or wheat (Farm Show Magazine, 2018a). Field evaluations suggest that, compared to piloted aircraft, a powered parafoil is smaller at 5–7 ft size. It is easy to handle the parafoil while landing. It can be packed and transported swiftly. Regarding economics, a powered parafoil, like the one used above the crops in Kissimmee costs 12,000 US$ (September 2018 price level). The parafoil is quick to assemble. It requires only few hours of pilot training.

Framers in Oklahoma have stated that a sprayer parafoil covers large areas of cropped land per flight. Forecasts in late 1990s, suggested that low flying, slow parafoils that are guided can be of immense help to farmers. Farmers with large land holdings and widely scattered fields may require aerial transport and spraying ability, using parafoils or any other means. Aerial spraying is most efficient method for applying plant protection chemicals. They say, powered guided parafoils are excellent in locations where drone air crafts pose problems, in terms of endurance and pay load. Let us consider an example of parafoil which is an agricultural sprayer. Parafoil with an ability to carry a payload of 30 gallons plant protection chemical and fly at altitudes, say 10 m above the crop canopy is available (Buckeye Industries, 2018). The sprayer parafoil has an aluminum structure. The payload area has a 2- seater facility along with 2 tanks, 15 gallons each, to store plant protection chemical. The sprayer system has jet nozzles and atomizers.

Huang et al. (2013) point out that light autonomous parafoils such as Tetracam's Hawkeye, could be definitely utilized, to conduct aerial photography of crops found in experimental stations. To do this, parafoils are attached with visual, multispectral, hyperspectral and IR sensors. They are provided with laser pods to assess growth of crops accurately. In addition, they could be fitted with sprayers, to conduct aerial sprays on crops. This possibility is being explored with different models of parafoils. Clearly, parafoils with ability for both spectral analysis and spraying of foliar fertilizers or plant protection chemicals are evolving. They may become popular and routine, in due course.

2.5.6   PARAFOILS IN PRECISION FARMING

Parafoils could be touted as good aerial vehicles to conduct spectral analysis of crops and to spray inputs, during precision farming. Let us consider an example. The 'Hawkeye' is an autonomous parafoil with facility for spectral analysis of crops. It can also be fitted with spray bar and variable-rate nozzles to apply inputs at variable rates. The resulting NIR/Visible light images can be analyzed to determine the relative health of plants, in the area surveyed. Hawkeye system can also be used to identify unique spectral signatures indicative of specific plant species and diseases. We can distinguish weeds from crop plants and spray herbicides accordingly. Each mission may be pre-programmed via it's 'Goose Autopilot' and Ground Control Software. The Hawkeye's GCS software enables users to pre-plan the mission's flight path, identify camera trigger points and monitor the crop.

As the Hawkeye arrives at the designated waypoint, the 'Goose autopilot' triggers the integrated Tetracam multispectral imaging system, to capture an image. At the same instant, the GPS coordinates are recorded by the camera. It correlates with the captured image with its geographical location. The ground control software works on any computer with Windows program. The Hawkeye may also be flown manually via its Spektrum RC controls.

The craft has been designed to be a docile platform. So that, it is easy to operate, for someone who has only little experience with flying radio-controlled aircraft. The Hawkeye has only two controls: Power which controls speed and altitude, and the roll servo which turns the parafoil right or left. The parafoil flies at approximately 10 knots per hour. The forward speed of the Hawkeye is determined by the parafoil wing and wind conditions only. In general, a Hawkeye cannot be stalled like a conventional winged aircraft.

'Precision Agriculture' necessitates use of aerial imagery of terrain and its topography. Special features on land has to be known. These features could be utilized to prepare and mark a 'management block.' Soil fertility variations should also be known (mapped). Such management blocks are required to conduct different sets of agronomic procedures and apply inputs at variable- rates (Krishna, 2012). For example, a management block with soils that are hard and not ploughed for few seasons, need thorough deep ploughing and turning of soil. While a field under no-tillage may be directly sown with seeds. Fields differing in soil characteristics such as texture and mineral/organic fraction may have to be treated accurately, using precision farming principles. Agronomic procedures, fertilizer and irrigation inputs may have to vary with each management block. Knowledge about soil fertility, water status of crops, NDVI and leaf chlorophyll content (crop's-N status) is necessary. Small drone aircrafts with a full complement of 4 or 6 sensors are well suited, to conduct such aerial survey, rapidly. We can also obtain similar details of fields/crops, using a parafoil (e.g., Hawkeye).

Tetracam's Hawkeye parafoils equipped with all sensors (visual, multi-spectral, hyperspectral, Infrared and red edge) are available. They could be of utility during precision farming. The aerial imagery depicts a posse of characteristics useful to farmers, while deciding inputs and procedures. If farmers are able to reach data banks with spectral features of disease/pest attacked and healthy crops, then the same imagery can help in detecting and mapping the damage to crops. Such maps could be effectively used during application of plant protection chemicals at variable rates(see Krishna, 2018; Tetracam 2018, John, 2011; Krishna, 2019).

Thamm (2011), Thamm et al. (2013) and Thamm and Judex (2006) have clearly shown that autonomous parafoils (e.g., SUSI 62) fitted with a full set of sensors, which are useful in precision farming. The parafoils carry visual (R, G, and B), multi-spectral, infrared and red edge sensors, and also a lidar pod. Such sensors could offer excellent imagery and spectral data, that can be useful during precision farming. They have utilized spectral data depicting topography, water resources and soil types to demarcate appropriate 'management blocks.' The agronomic procedures and crops planted were dependent on aerial images. A parafoil with tank to hold plant protection chemicals and a nozzle bar could be utilized, to spray the crops. Variable-rate nozzles have to be fitted, if precision spraying is to be achieved. Precision spraying is based on maps depicting variations in disease/pest attack intensity (see Buckeye Industries, 2018; Krishna, 2018, 2019).

There is an underlying trend among farmers and farming companies in different agrarian belts. The aim is to adopt precision farming techniques. In other words, they intend to first detect variations in the crop growth, its water and nutritional status. This involves tedious efforts by farm scouts. They have to map the variations accurately using visual scores or leaf meter readings. Usually, human scouts do not have the opportunity for an overview of large farms. Therefore, during recent years, farmers are being advised to obtain aerial maps, using drones and optical sensors. Parafoils could also be used to get aerial images. The digital data depicting variations is then utilized in variable-rate applicators (nozzles) or planting vehicles (devises). A few different types of unmanned aerial vehicles have been used, to detect such variations in crop growth, soil nutrient and moisture distribution. No doubt, low altitude slow moving autonomous parafoil with requisite sensors could be put to use, during precision farming (Unal and Mehmat, 2016; Krishna, 2019; Thamm, 2011; Pudelko et al., 2012). Clearly, there is a need to integrate parafoils into farming systems activities.

2.5.7 PARAFOILS ARE USEFUL IN MONITORING AGRICULTURAL EXPERIMENTAL STATIONS

As stated earlier, agricultural drones include different types of small autonomous aircrafts, parafoils, blimps, and aerostats. Among them, drone aircrafts have already gained acceptance among farmers and farming companies. Farmers have used them to obtain aerial photography. The spectral images are then analyzed (processed) using appropriate computer programs. This helps to decipher the crop growth trends. Also, to judge crop's water stress or to trace any disease/pest attack. Further, drone aircrafts have also been adopted during evaluation of crop genotypes, in large number, and grown in experimental plots (see Krishna, 2016, 2018). However, within the context of this chapter on Parafoils, we are more concerned with Powered Parafoils and Autonomous Powered Parafoils that could be utilized, to monitor agricultural experimental stations. We are concerned with parafoils that are fitted with a set of sensors such as visual, multispectral, IR and Red edge cameras. Indeed, there are very few studies, if any, that deal with parafoils and their utility in monitoring an entire agricultural experimental station. Reports about use of parafoils to collect useful data about growth of large number of genotypes are sparse. The genotypes of a crop species could be evaluated for parameters such as growth, NDVI, leaf chlorophyll content, drought tolerance, disease /pest attacks and resistance, etc. Indeed, use of parafoils to collect data from large number of germplasm and elite lines needs greater impetus.

Rydberg et al. (2007) have clearly stated that, interest in using UAVs (e.g., parafoils) in farms is increasing. Particularly, to overview crops in experimental stations. They further state that powered parafoils that are guided by pilots to fly at low altitude over the crops are comparatively better. This is because, they offer accurate and clear images. They state that, moto-gliders offered high quality images of fields. The sensors used should be of high resolution. We have only to fly the parafoil low above the crops.

Pudelko et al. (2012) state that autonomous parafoils such as 'Pixy ABS' is useful in monitoring and obtaining aerial images of terrain, soils, general vegetation, crops, weeds, pests and mechanical damage if any to crops. The high-resolution aerial photography obtained using sensors on 'Pixy ABS' parafoil helps in judging crop's response to different fertility factors. It provides spectral data useful in detecting spatial variations in crop growth.

Thamm (2011) states that, SUSI 62- a parafoil with its' set of visual and multispectral cameras helps farmers, in detecting damage caused to plantations. Such damage could be due to floods, drought and even stray animals that may trample standing crops. This aspect is of utility in an agricultural experimental station, where in, round the clock aerial surveillance is essential.

Now, let us consider an example from Oregon State Agricultural Experimental Station. (Hermiston Agricultural Research Station) (Hunt and Horneck, 2013; Hunt et al., 2005, 2018). They have evaluated potato (Cultivar: Ranger Russet) crop's response to 4 different levels of fertilizer-N. The crop was actually being tested for its response to sulfur coated urea. The data was collected extensively and repeatedly in small intervals, using a parafoil fitted with sensors. A Tetracam's Hawkeye parafoil was used. It was fitted with a full complement of sensors (visual, multi-spectral, IR, NIR and LIDAR. The sensors used to collect data was known as 'Agricultural Digital Camera Light version.' The flight path, way points and shutters of camera were pre-programmed, using appropriate computer programs. We may note that, parafoil's flight path is not very easy to control, particularly, if it is windy. The ortho-mosaics were stitched using GPS data and a computer program such as PIX 4D Mapper. Crops' nitrogen status was measured using aerial pictures. The chlorophyll distribution pictures were obtained, using Konica Minolta SPAD 502. Further, leaf chlorophyll was also measured, using leaf disc method and chemical estimation. Interestingly, leaf chlorophyll content measured using aerial spectral data corroborated with other methods, namely, SPAD readings, petiole nitrogen, and leaf -N (1 cm leaf disks) adopting chemical estimation (Hunt et al., 2014). The spectral data of potato crop was measured and calculated using procedures described by Hunt et al. (2005, 2014). Incidentally, McCollum et al. (2018) have utilized NDVI measurements using UAVs to judge N status of sugarcane.

Daughtry et al. (2000) say that it is easy to adopt parafoils. Parafoils may become more common above crop fields, in future. Apart from the above studies and a few other reports dealing not with parafoils but with drone aircrafts, our usage of autonomous aerial robots is still feeble. We have to adopt parafoils in a much wider sense to monitor the activities of farm personnel, agricultural vehicles and progress of agronomic procedures. More importantly, they need to be adopted while assessing performance of crops. We could adopt them regularly to assess cultivars (genotypes) in open fields and experimental plots. We ought to note that, parafoils are low cost, slow flying autonomous vehicles. They are portable and deployable quickly. Parafoils have the potential of offering high resolution imagery, since, they fly at very low altitude over the crops.

Production of UAVs, in general, including parafoils has increased in recent times. Concurrently, production of different types of sensors too has increased. Sensor models and brands are flooding the market. After all, they say, more than platforms or their models it is the sensors that are key to obtaining high resolution images of crops. A few other uses that we can imagine for parafoils flying above the agricultural experimental stations are in evaluating greenhouse gas emissions (e.g., NH_3, SO_2, methane, etc.). We can use chemical sensors placed in the payload area (Pobkrut et al., 2016; Phillips, 2017; Scenteroid Inc., 2017; Hugenholtz and Barchyn, 2016; Krishna, 2018, 2019). Overall, a parafoil that is either tethered or powered and piloted or an autonomous one is an excellent aerial contraption. It helps to watch the agricultural experimental station.

Let us consider an example related to use of autonomous powered parafoil (an UAV), to conduct aerial survey of crops grown in experimental station. The autonomous parafoil (Pixy ABS) is a light UAV at 6 kg weight. It picks a payload of cameras and CPU of 2–3 kg. The cruising speed of this parafoil is just 15 to 20 kmph. So, it allows for detailed imagery. The endurance in flight is 40 minutes per flight. The autonomous parafoil (Pixy ABS) is attached with visual, multi-spectral and infrared sensors. This parafoil is utilized to conduct aerial photography of crops, at various stages of evaluation. We may note that, such trials with autonomous parafoils were actually conducted for 3 consecutive years between 2007 to 2010 (see Pudelko et al., 2012). The aerial images were obtained from 20–700 m above the crop canopies. Crops such as wheat, spring barley and maize grown in plots were assessed, using parafoils. They have concluded that autonomous parafoil such as Pixy ABS with full set of sensors is good enough, to obtain high quality sharp imagery of experimental station, its installation and crops. The low altitude imagery of crops has clear advantages over that obtained using aircrafts or satellites. The digital single reflex sensors provide excellent images. The aerial images derived using autonomous parafoil was amenable for ortho- rectification. Appropriate computer-based image processing programs had to be used. For example, Pix4D Mapper. Pudelko et al. (2012) further state that, aerial imagery provided insights on spatial variations of land, its topography, surface details, crops, their growth pattern, chlorophyll content and water status. So, a parafoil such as Pixy ABS could be very useful, in managing and obtaining detailed data about crop genotypes. A lacuna related to use of parafoil to conduct aerial surveillance and spectral analysis is that, parafoil could be affected by wind blowing beyond 20 kmph. Thermal currents may affect parafoils' movement (see Rydberg et al., 2007; Pudelko et al., 2012; Krishna,

2018, 2019). We can fly the parafoil over plantations, field crops and forest plantations too, to measure growth rates, logging trends, disease/pest attack, if any, and water status. Lelong et al. (2008) have shown us how autonomous parafoils could be deployed over an agricultural experimental station. Also, how crops could be monitored and evaluated for their performance. They say, a parafoil such as Pixy ABS is good to conduct regular surveillance of crops for growth traits, nutrients status and pest/disease attacks, if any.

2.5.8   PARACHUTES AND PARAFOILS IN AGRICULTURAL CARGO TRANSPORT

Agricultural cargo transport adopting autonomous and powered guided parafoils is a concept worth exploring. We should note that, sky above cropping expanses has least obstacles and disturbances, if any, to the movement of parafoils. Autonomous Parafoils could be immensely cheaper compared to surface transport of large goods carriages. Autonomous parafoils with cargo could be pre-programmed, to fly the safest routes above crop fields. Otherwise, they could be carefully controlled, using ground control station (iPad). Parafoils may reach locations hitherto impossible, difficult or costly to do so, if we adopted conventional trucks. Small parafoils could transport useful inputs from one farm to another. So far, parafoil aided farm cargo transport has not been explored. Parafoils could be of immense use when the terrain is difficult (e.g., in terrace farming conditions or in mountainous conditions). However, reports, if any, about small parafoils and cargo transport in agrarian regions are even difficult to trace.

Agricultural cargo transport using drones and parafoils is an important aspect that needs to be improved. The cargo could be either inputs or harvested grains/fruits that have to be briskly transported to different locations. Agricultural cargo transport needs special attention if the farming terrain is difficult to reach, such as mountainous locations, semi-arid belts without proper surface transport facilities or into areas affected by disasters such as floods, earthquake, wetlands, etc. Transport of products such as grains/fruits or other perishable vegetables too needs greater attention. Parafoils that are autonomous may carry only a certain small quantity of payload. Yet, they could be of advantage in supplying farm inputs, to locations within a farm. Large parachutes/parafoils, of course can air drop large quantities of agricultural inputs. Particularly, goods such as fertilizers, seeds, pesticides and herbicides needed. Precision guided parafoils with ability to carry payloads for farms could be of great utility (see Cacan et al., 2017; Thamm, 2011). Such airdrop systems could utilize GPS for navigation. There are several computer programs that allow us to decide flight path, way points and landing location (e.g., Pixhawk, *eMotion*, Goose Pilot, etc.). We may also use radio frequency-based control from a ground control station (Cacan et al., 2017). Large parachutes could also be of utility. Large machinery such as tractors, planters, weeders, fertilizer applicators may have to be air-dropped, into farming zones that are isolated or far-off from highways. A C-130 aircraft with loads of agricultural cargo can deliver them safely, by using a parafoil. It could be either machinery, bulk farm inputs (fertilizers, pesticides) or large quantities of grain harvests. Large quantities of food grains are often trans-shipped via large cargo airplanes such as Antonov or C-130. In that case, they could after all use parachutes to air drop cargo.

2.6 ECONOMIC AND REGULATORY ASPECTS OF PARAFOILS AND PARACHUTES UTILIZED IN AGRICULTURAL CROP PRODUCTION

At present, this is a topic that needs greater attention. Rules for usage of powered parafoils, both piloted and autonomous ones need to be highlighted. Several aspects of safety for use of autonomous parafoils may need thorough review and discussion by concerned committees in different nations. Overall many rules that apply to autonomous small aircrafts obviously apply to autonomous parafoils too.

There are many rules and regulations to follow while managing and controlling a parachute/parafoil. A regular training for few hours to days are required before one qualifies as a pilot of a powered parafoil. Equally so, a good knowledge and training are essential to control and program autono-mous parafoils. Many of the rules and regulations for use of parafoil over rural, agricultural and urban areas are listed in a handbook. Also, usually, the parachute or powered parafoil comes with handbooks and manuals to handle them safely. In many countries, fines are stipulated to those overlooking rules for use of parafoils. Also, we should note that autonomous parafoils could be treated on par with small aircraft drones such as fixed-winged or helicopters or multi-copters regrading rules for their use. Permissions from appropriate agencies are needed (see Krishna, 2018).

In most countries, parafoil pilots operate under simple rules that spare them certification requirements. Those laws, however, limit where they can fly- specifying that pilots avoid congested areas and larger airports to minimize risk to other people or aircraft. U.S. pilots operate under Federal Aviation Administration regulation, Part 103 (Wikipedia, 2018c).

2.7 FUTURE OF PARACHUTES/PARAFOILS IN AGRARIAN REGIONS

Parachutes (not parafoils) of wide range of sizes and ability for payload are available. Their role in agricultural settings may get confined to cargo drop and aerial photography. This is because they are not easily navigable in the sky. Parafoils, with their versatility for transit and pay load carrying ability may rule the agricultural skies to a certain extent. Of course, parafoils will have to share the agricultural sky with several other aerial robots, such as fixed-winged drone aircrafts, helicopters, multi-copters, blimps, aerostats and kites. Yet, the prediction is that parafoils and microlights (manned) may offer extra advantages to farmers. Particularly, when they prefer to assess their cropped fields, using aerial photography. Also, when adopted to apply inputs, i.e., using sprayers. Parafoils are not excessively costly. They are not difficult to maintain and service. They can be easily stored and initiated quickly into flying in the air. Parafoils are versatile and are known to have innumerable uses, in addition to agricultural operations. Parafoils are relatively slow, particularly, the autonomous ones. Parafoils offer excellent high-resolution imagery. The digital data that parafoils offer are of use to various professionals. Regarding, economics, parafoils/parachutes are useful to farmers. However, detailed studies that stringently compare the utility/economics of parafoils/parachutes is a necessity. Since the competition among various aerial robots is on, there is a need for urgent and correct assessment of advantages/ exchequer accrued to farmers, due to adoption of parafoils/parachutes. Comparative studies examining the preferences of farmers and economic advantages are highly relevant, at present. Regarding production of various types of parafoils/parachutes, there are specialized agencies that relay information about the market for para-foils and aerial photography and need for these aerial vehicles. Industries that produce parafoils may consult the agencies periodically. Regarding training personnel to use parafoils in agriculture, it is said that a farmer can learn to fly a parafoil in 8–10 hr. and learn to ride a microlight in about 3–7 days of 5 hr. per day training. Therefore, forecast is that, to a certain extent, parafoils and micro-lights may garner the space in the agricultural sky.

Parachutes/parafoils are utilized regularly in military and general cargo transport. The demand for sophisticated models and those capable of cargo transport and drop may increase. There is no reason to believe that demand for parachutes/parafoils would reduce, in the near future. Almost all nations seem to be using parachutes/parafoils in one way or other, in their military establishments and in civilian organizations. In fact, predictions on global market for parachutes/parafoils indicate a significant increase during the period from 2019 to 2025 (Sargar, 2018). The above discussions make it clear that parachutes/parafoils have innumerable applications through their ability for lifting cargo, their transport and drop. They are sought during aerial surveillance and sentry services.

Market reports for previous years and forecasts about parachutes and parafoils are available (Rohan, 2018; Morder Intelligence Ltd. 2018). Major users of parachutes/parafoils are military establishments, cargo companies, sports and rescue mission organizers, in that order. Right now, in 2019, the military establishments of different nations are dealing with a market worth 1.21 billion US$ per year, for parafoils/parachutes. The military segment is supposed to experience increased demand for parafoils. Regarding type of parachute, the reports indicate that highest demand is for ram-air parachutes (i.e., parafoils), followed by round and ribbon types. Agricultural sector too may opt for more of parafoils in future. At present, use of parafoils in farming is too feeble. Region-wise, North America is the most prolific user of parachutes/parafoils. This region is followed by Europe, Asia Pacific and Latin America in that order. Globally, 47% of parachutes/parafoils are utilized in North America.

However, in future, perhaps most prolific use of parafoils could be in agricultural crop production zones. The need for incessant aerial images, crop surveillance for nutrient dearth, water shortages and mapping variation in growth pattern may make farmers to adopt them, in greater number. So, we believe that, in due course we may come across innumerable, highly advanced and sophisticated parafoils with ability for autonomous flight patterns, cargo lift, aerial imagery and digital data collection of crop production zones.

Manned Parafoils and Microlights that are semi-autonomous need detailed evaluation for their ability, to offer useful spectral data about crops. They are yet to be examined in different agrarian regions. Economic feasibility studies are needed, if we have to impress the farmers and companies to adopt parafoils, during crop production. If we could make purchase and regular use of parafoils above crop lands less-costlier, then, these aerial vehicles have potential to initiate a revolution, in agrarian regions. The sprayer parafoils may attract greater attention from farmers. Aerial spraying could become easier to accomplish. The way we assess crops and mange agronomic procedures in farms may altogether change and get easier, if we adopt parafoils/parachutes.

KEYWORDS

- central processing unit
- circular error probable
- crop water stress index
- digital surface model
- high altitude high opening
- high altitude low opening
- image composite editor
- infra-red
- joint precision airdrop system
- low altitude low opening
- low altitude remote sensing
- semi-rigid deployable wing
- vertical take-off and landing
- zipline cargo drones

REFERENCES

Aero News Network, (2018). *New Harrier Parachute Launcher Designed for Drones* (p. 1). http://www.aero-news.net/index.cfm?do=main.textpost&id=3758535f-ed27-4490-a24a-8da9fc160439 (accessed on 30 July 2020).

AeroVironement Inc., (2018). *Tethered Eye* (pp. 1–6). https://www.avinc.com/resources/press_release/u.s.-combattingterrorism-technical-support-office-evaluating-new-aeroviron. (accessed on 30 July 2020).

Airborne Systems, (2018). *World's Largest Autonomously Guided ram-Air Parachute: GigaFly* (pp. 1, 2). https://www.copybook.com/companies/airborne-systems/articles/worlds-largest-autonomously-guided-ramair-parachute (accessed on 30 July 2020).

Akhramovich, S., (2017). *Russia to Create UAV Paraglider* (p. 103). MIL. Press. http://mil.today/2017/Science4/ (accessed on 30 July 2020).

Atair Aerospace Inc., (2003). *Atair Aerospace Transforms Precision Airdrop Delivery with Hi-Tech Solutions* (pp. 1–7) https://www.businesswire.com/news/home/20030616005636/en/Atair-Aerospace-Transforms-Precision-Airdrop-Delivery-Hi-Tech (accessed on 30 July 2020).

Atair Aerospace Inc., (2018). *Atair Aerospace Awarded 3.2 million US $ Contract by US Army to Supply Onyx Precision-guided Parafoil System* (pp. 1–18). https://www.prweb.com/releases/2006/09/prweb433958.htm (accessed on 30 July 2020).

Australian Parachute Federation Ltd, (2017). *The Early History of Parachuting* (pp. 1–7). Australian Parachuting Federation ltd, Sydney, Australia, https://www.apf.com.au/

APF-Zone/APF-Information/History-of-the-APF/Early-History-of-Parachuting/default. aspx/ (accessed on 30 July 2020).

AVASYS LLC, (2016). *Delta-M UAV Systems* (pp. 1–8). Autonomous aerospace systems-geoservice. Krasnoyarsk, Siberia, Russia http://uav-siberia.com/en/content/delta-m-uav-system?gclid=CNKdmZ3–0NACFdi (accessed on 30 July 2020).

Balena, G., (2005). Powered parachute takes off from water. *Farm Show Magazine, 29*, 4.

Bellis, M., (2017). *History of Parachute* (pp. 1–4). Thought Company Inc., https://www.thoughtco.com/history-of-the-parachute-1992334 (accessed on 30 July 2020).

Bennet, J. P., Rhodes, D. J., & Lewis, L. W., (2015). *Remembering Historic Camp Claiborne* (p. 120). Louisiana, USA. US Department of Agriculture, Baltimore, USA.

Bennet, T. W., & Roy, F., (2005). *Design Development and Flight Testing of the US Army 4200 sq. ft Para Foil Recovery System* (pp. 1–7). NASA Technical Reports Server (NTRS). https://ntrs.nasa.gov/search.jsp?R=20070022338 (accessed on 30 July 2020).

Benolol, S., Zaparin, S., & Ramassar, P. I., (2018). *The Fast Wing Project: Parafoil Development and Manufacturing* (pp. 1–6). http://www.cimsa.com/pdf/aiaa_paper.pdf (accessed on 30 July 2020).

Benton, J. E., & Yakimenko, O. A., (2013). On Development of Autonomous HAHO Parafoil System for Targeted Payload Return. *Proceedings of the 22nd AIAA Aerodynamic Decelerator Systems Technology Conference*. Daytona Beach, FL. http://hdl.handle.net/10945/35310 (accessed on 30 July 2020).

Bhosri, W., Cojanis, P., Gupta, M., Khopkar, M., Kiely, A., Myers, M., Oxnevad, K., et al. (1999). *The Exploration of Mars: Crew Surface Activities* (pp. 1–15). University of Southern California Aerospace Engineering. https://www.lpi.usra.edu/lpi/HEDS-UP/usc.pdf (accessed on 30 July 2020).

Billings, R. F., & Ward, J. D., (1984). How to conduct a southern pine beetle aerial detection survey. In: *Southern Pine Beetle Hand Book* (p. 22). Texas Forest services, Texas A and M University, College station, USA Extension Service Circular 267.

Blue Sky Gyros, (2016). *What is a Powered Parachute* (p. 17). https://www.blueskiesppc.com/powered-parachute-overview (accessed on 30 July 2020).

Borneman, A. M., (2006). *Proud to Send Those Parachutes Off: Central Utah's Rosies During World War II* (p. 133). Brigham Young University BYU Scholars Archive, Brigham Young University-Provo, Utah. https://scholarsarchive.byu.edu/etd (accessed on 30 July 2020).

Bosman, R., Reid, V., Vlasblom, M., & Smeets, P., (2013). Airborne wind energy tethers with high-modulus polyethylene fibers. In: Ahrens, U., Diehl, M., & R. Schmehl, R., (eds.), *Airborne Wind Energy* (pp. 563–585). Berlin: Springer. Berlin, Germany.

Braunberg, A. C., (1996). Parachute guidance empowers programmed payload placement. *Signal, 50*, 83–85.

Brown, G., (1989). Tethered parafoil test technique. *10th Aerodynamic Decelerator Conference, Aerodynamic Decelerator Systems Technology Conferences* (pp. 1–7) https://doi.org/10.2514/6.1989–903 (accessed on 30 July 2020).

Buchanan, S., (2017). *Fun Things to Fly: Powered Parachutes, Trikes, and Gyroplanes* (p. 121) https://letterstocreationists.wordpress.com/2017/02/10/fun-things-to-fly-powered-parachutes-trikes-and-gyroplanes/ (accessed on 30 July 2020).

Buckeye Industries, (1997). *Introducing the Aquatically Enhanced Buckeye Explorer* (p. 14). http://buckeyedragonfly.com/contact.htm (accessed on 30 July 2020).

Buckeye Industries, (2018). *Summary of Aircrafts and Powered Parachutes* (pp. 1–3). https://en.wikipedia.org/wiki/Buckeye_Industries (accessed on 30 July 2020).

Buhler, W. C., & Wailes, W. K., (1989). *Development of a High Performance Ringsail Parachute Cluster*. AIAA Paper No. 73, https://vdocuments.site/american-institute-of-aeronautics-and-astronautics-10th-aerodynamic-decelerator-58532edf2dbe9.html (accessed on 30 July 2020).

Cacan, M. R., Castello, M., & Scheureman, E., (2017). Global positioning system denied navigation of autonomous parafoil systems using beacon measurements from a single location. *Journal of Dynamic Systems and Measurements.* doi: 10.1115/1.403765Histry (accessed on 30 July 2020).

Carter, D., George, S., Hattis, P., & Singh, L., (2005). Autonomous guidance, navigation, and control of large parafoils. *Proceedings of the 18th AIAA Aerodynamic Decelerator Systems Technology Conference and Seminar* (pp. 1–16). AIAA 2005–1643.

Cedar Meadow Farm, (2010). The 8th Annual Farm field Day. The Lancaster Chamber's Agriculture Services Coordinator, Haltwood, Pennsylvania, USA, web address.

Cesana, J., Kantner, J., Wiewel, A., & Cothren, J., (2014). Archaeological aerial thermography: A case study at the Chaco-era blue j community, New Mexico. *Journal of Archaeological Science, 45,* 207–219.

Cheil, B., (2013). *Autonomous Para Foil Guidance in High Winds* (p. 163). Boston University, Massachusetts, USA https://hdl.handle.net/2144/21117 (accessed on 30 July 2020).

Chow, D., (2013). NASA Tests Orion Space's Parachute with Mock Glitch (p. 15). National Aeronautics and Space Agency, USA, https://www.space.com/21155-orion-parachute-test.html (accessed on 30 July 2020).

Coleman, J., Ahmad, H., & Toal, D., (2016). Development and testing of a control system for the automatic flight of tethered parafoils. *Journal of Robotics, 34,* 519–538.

Committee on Naval Expeditionary Logistics, (1999). *Naval Expeditionary Logistics: Enabling Operational Maneuvers from the Sea* (p. 104). National Academy Press, Washington D.C.

Corelfield, G., (2017). *NASA Extends Trial of Steerable Robo-Stunt Kite Parachute: Why Chase Balloons for Hundreds of Miles When You Can Drop the Payload Outside?* https://www.theregister.co.uk/2017/02/21/nasa_stunt_kite_aerofoil_tech_trial/ (accessed on 30 July 2020).

Daughtry, C. S. T., Walthall, C. L., Kim, M. S., Brown, D. C. E., & McMurtrey, III. J. E., (2000). Estimating corn leaf chlorophyll concentration from leaf and canopy reflectance. *Remote Sensing and Environment, 74,* 229–239.

DeChant, C., Wiesner-Hanks, T., Chen, S., Stewart, E. L., Yosinski, J., Gore, M. A., Nelson, R. J., & Lipson, H., (2017). Automated identification of northern leaf blight-infected maize plants from field imagery using deep learning. *Phytopathology, 11,* 1426–1432.

Delurgio, P. R., (1999). *Evolution of the Ring Sail Parachute* (pp. 1–12). Irvin Aerospace Inc., Santa Ana, California, American Institute of Astronautics and Aeronautics. 99–1700, https://airborne-sys.com/wp-content/uploads/2016/10/aiaa-1999–1700_evolution_of_the_ringsail.pdf (accessed on 30 July 2020).

Deweese, J. H., Shultz, E. R., & Nutt, A. B., (1978). *Recovery System Design Guide*. Technical Report AFFDL-TR-78–151, California. USA, https://docplayer.com.br/43391013-Analise-por-demodulacao-aplicada-ao-monitoramento-de-falhas-em-engrenagens.html (accessed on 30 July 2020).

Dhobrokhodov, V. N., Yakimenko, A., & Junge, C., (2003). Six degree of freedom model of controlled circular parachute. *Journal of Aircraft, 40,* 233–245.

DJI Inc., (2016). *Above the World* (p. 238). UAS Magazine. http://www.uasmagazine.com/articles/1591/dji-explains-new-book-of-UAV-captured-images (accessed on 30 July 2020).

Donaldson, P., (2009). *Unmanned Vehicles Handbook 2008: The Concise Industry Guide* (p. 258). Shepard. www.uvonline.com (accessed on 30 July 2020).

Dunker, S., Huisken, J., Montague, D., & Barber, J., (2015). Guided parafoil high altitude research (GPHAR) Flight at 57,122 ft. The American institute of aeronautics and astronautics, *Proceedings of Aerodynamic decelerator Systems technology Conferences* (pp. 1–12).. Daytona Beach, Florida, USA.

Eric, H., & Chun-Chi, H., (2016). *Parachutes Designed and Invented* (pp. 1–3). Historylines http://www.historylines.net/history/15th_cent/parachutes.html (accessed on 30 July 2020).

European Aviation Safety Agency, (2017). *Concept of Operations for Drones: A Risk-Based Approach of Regulation of the Unmanned Aerial Vehicles* (p. 11). European Aeronautic Space Agency, Cologne, Germany.

Ewing, E. W., (1972). *Ring Sail Parachute Design* (p. 380). AFFDLTR-72–3, Air Force Flight Dynamics Laboratory, Air Force Systems Command, United States Air Force, Washington D.C. USA.

Farm Show Magazine, (2018a). *Parascender Powered Parachute* (pp. 1, 2). https://www.farmshow.com/a_article.php?aid=12632 (accessed on 30 July 2020).

Farm Show Magazine, (2018b). *Powered Parachutes Catching on Fast* (Vol. 15, p. 36). Farm Show magazine. https://www.farmshow.com/a_article.php?aid=8140 (accessed on 30 July 2020).

Fields, T., Yakimenko, O., LaCombe, J. C., & Wang, E. L., (2014). Lower stratospheric deployment testing of a ram-air parafoil system. In: *2014 AIAA Atmospheric Flight Mechanics Conference* (pp. 1–7). American Institute of Aeronautics and Astronautics. 0193. https://my.nps.edu/documents/106608270/107784480/Fields+-+Lower+Stratospheric+Deployment+Testing+of+a+Ram-Air+Parafoil+System.pdf/be8ab409–185b-495b-8a18-5546be31b288. Report number 0193 (accessed on 30 July 2020).

Fruity Chutes Inc., (2018c). *Professional Aerospace Recovery Systems*. Location and web address, Fruitychutes LLC., Monte Seen, California, USA. (pp 1–7), https://fruitychutes.com/parachute_recovery_systems.htm' (August 7th, 2020)

Fruitychutes LLC, (2018a). *Online Descent Rate Calculator* (pp. 1–5). Fruity chutes, Monte Seen, California, USA https://fruitychutes.com/genes_blog/online-descent-rate-calculator/ (accessed on 30 July 2020).

Fruitychutes LLC, (2018b). *World Class Parachute Recovery for Rockets Drones, UAV, Research* (p. 10) https://fruitychutes.com/parachute_recovery_systems.htm (accessed on 30 July 2020).

Fruitychutes LLC, (2018d). Ultimate Parachute System for all Drones, Multi-copters and UAS (p. 1–7). https://fruitychutes.com/uav_rpv_drone_recovery_parachutes.htm p (accessed on 30 July 2020).

Global Security. Org., (2018). *Airborne Operations in the Cold War* (p. 16). https://www.globalsecurity.org/military/ops/airborne2.htm (accessed on 30 July 2020).

Gromov, M. M., (1939). *Across the North Pole to Americas (in Russian)* (p. 38). Foreign Language Publishing House, Moscow, Russia.

Hailey, B. S., (2008). *An Investigation of New Philadelphia Using Thermal Infrared Remote Sensing Center for Archaeological Research University of Mississippi* (pp. 1–8). USA http://faculty.las.illinois.edu/cfennell/NP/2008aerial.pdf (accessed on 30 July 2020).

Hailey, T. I., (2005). *The Powered Parachutes AS AN Archaeological Aerial Reconnaissance Vehicle* (pp. 1–17). Archaeological prospection. Northwestern State University, Natchitoches, Louisiana, USA. https://doi.org/10.1002/arp.247 (accessed on 30 July 2020).

Hartsfield, J., (2018). *NASA'S World's Biggest Parafoil* (pp. 1–14). Johnson Space Center, Houston, Texas., USA National Aeronautics and Space Agency, Washington D.C. USA. https://members.kite.org/resources/Documents/kiting_2000_v22_i2.pdf (accessed on 30 July 2020).

Hewgley, C. W., (2014a). *Pose and Wind Estimation for Autonomous Parafoils* (p. 133). Naval Postgraduate School, California, USA. Dissertation http://hdl.handle.net/10945/43926 (accessed on 30 July 2020).

Hewgley, C. W., Cristi, R., & Yakimenko, A. O., (2009). Precision Guided Airdrop for Vertical Replenishment of Naval Vessels. *20th AIAA Aerodynamic Decelerator Systems Technology Conference* (pp. 1–13). American Institute of Aeronautics and Astronautics Inc, 2009 AIAA 2009–2995. https://pdfs.semanticscholar.org/61cb/31fc3c8474797c794ccee6ff01197419 d28e.pdf (accessed on 30 July 2020).

Hewgley, C. W., Cristi, R., & Yakimenko, O. A., (2014b). Visual pose estimation for shipboard landing of autonomous parafoils. *Proceedings of the 2014 IEEE/ION Position, Location and Navigation Symposium, PLANS 2014* (pp. 1301–1308.). Monterey, CA, United States of America.

Hewgley, C. W., Yakimenko, A., & Slegers, N. J., (2011). Shipboard landing challenges for autonomous parafoils. *21st AIAA Aerodynamic Decelerator Systems Technology Conference and Seminar 2011* (pp. 1–9). AIAA 2011–2573 https://pdfs.semanticscholar.org/2384/ e429383adf48d0d9a0476dfd800bd3da6fbb.pdf (accessed on 30 July 2020).

Historylines, (2016). *The Parachute Was Invented by Ancient Chinese-Not Leonardo Da Vinci* (pp. 1–4). Ancient History facts. http://www.messagetoeagle.com/the-parachute-was-invented-by-ancient-chinese-not-leonardo-da-vinci/ (accessed on 30 July 2020).

https://pdfs.semanticscholar.org/475d/8ab1efaa43142ac4beb2b919ea923583172e.pdf (accessed on 30 July 2020).

Huang, Y. B., Thomson, S. J., Hoffmann, W. C., Lan, Y. B., & Fritz, B. K., (2013). Development and prospect of unmanned aerial vehicle Technologies for agricultural production management. *International Journal of Agriculture and Biological Engineering, 6*(3), 1–10.

Hugenholtz, C., & Barchyn, S. C., (2016). *A Drone in Search of Methane* (pp. 1–6). http:// https://ventusgeo.com/wp-content/uploads/2018/02/UAS-Methane-Gas-Detection-White-Paper.pdf (accessed on 30 July 2020).

Hunt, E. R. Jr., Donald, A. H., Charles, B. S., Robert, W. T., Alan, E. B., Daniel, J. G., Joshua, J. B., & Philip, B. H., (2018). Monitoring nitrogen status of potatoes using small unmanned aerial vehicles. *Precision Agriculture, 19,*314–343. doi: 10.3965/j.ijabe.20130603.001.

Hunt, R. E. Jr., Cavigelli, M., Daughtry, C. S. T., McMurtrey, III. J. E., & Walthall, C. L., (2005). Evaluation of digital photography from model aircraft for remote sensing of crop biomass and nitrogen status. *Precision Agriculture, 6,*359–378.

Hunt, R. E., & Horneck, D., (2013). *UAS/Precision Agriculture Experiment* (pp. 1–24). At Hermiston, Oregon, USA https://www.ars.usda.gov/research/publications/ publication/?seqNo115=301005 (accessed on 30 July 2020).

Hunt, R. E., Daughtry, C. S. T., Mirsky, S. B., & Hively, W. D., (2014). Remote sensing with simulated unmanned aircraft imagery for precision agriculture applications. *IEEE Journal of Selected Topics in Applied Earth Observations and Remote Sensing, 7,* 1–12 doi: 10.1109/JSTRS.2014.2317876.

Hur, G. B., (2005). *Identification of Powered Parafoil-Vehicle Dynamics from Modeling and Lift Test Data* (p. 225). Texas A&M University College Station, Texas, USA, Dissertation.

Jensen, M. L., (2000). *Tethered Lifting System for Measurements in the Lower Atmosphere* (p. 130). PhD dissertation, University of Colorado PhD dissertation, https://cires1.colorado.edu/science/groups/balsley/people/jensen.html (accessed on 30 July 2020).

Jian, L., Sun, Q., Kang, X., & Liu, Z., (2011). Autonomous homing of parafoil and payload system based on ADRC. *Control Engineering and Applied Informatics, 13*, 1–27.

John, P., (2011). *Tetracam's Hawkeye UAV: Flying Multispectral Camera Platform* (p. 1) https://www.youtube.com/watch?v=ERSAm7k6clI (accessed on 30 July 2020).

Jorgensen, D. S., & Hickey, M. P., (2005). *The AGAS 2000 Precision Airdrop System* (pp. 1–3). doi: 10.2514/6.2005-7072.

Kerman, B., (2012). *A Brief History of the Parachute* (pp. 1–17). http://www.popularmechanics.com/flight/g815/a-brief-history-of-the-parachute/ (accessed on 30 July 2020).

Kim, J., & Park, C., (2009). Wind power generation with a parawing on ships, a proposal. American Institute for Aeronautics and Astronautics AIAA paper 2009. *Energy, 35,* 1425–1432.

Kim, T. W., & Song, Y., (2017). *A New Guidance Algorithm for a Powered Ram Air Parafoil System Under Wind, International Journal of Applied Engineering 12*(18), 7558–7565 https://www.ripublication.com/ijaer17/ijaerv12n18_61.pdf (August 8th, 2020)

King, S. B., & Nelson, B., (2018). *Aerial Imagery Gives Insight into Water Trends* (pp. 1–5). Utah State University, Water Science Center. https://engineering.usu.edu/news/main-feed/2018/aerial-view-of-water-trends (accessed on 30 July 2020).

Klerman, S. B., (2011). Path planning for autonomous parafoils using particle chance constrained rapidly-exploring random trees in a computationally constrained environment Massachusetts institute of technology. *Master's Thesis*, 1–58, https://dspace.mit.edu/bitstream/handle/1721.1/85396/870304756-MIT.pdf; sequence=2 (accessed on 30 July 2020).

Knacke, T. W., & Madelung, G., (1992) *Parachute Recovery Systems Design Manual* (p. 188). 1sted., Para Publishing, Santa Barbara, California. USA.

Knacke, T. W., (1991). *Parachute Recovery Systems Design Manual* (pp. 1–511). Recovery Systems Division. Naval weapons Center, China Lake, California, USA paper No NWC TP 6575.

Kogler, K. J., Sutkus, L., Troast, D., Kisatsky, P., & Charles, A. M., (1995). *Simulation of Parafoil Reconnaissance Imagery* (pp. 1–18). NASA Astrophysics Data System (ADS). https://ui.adsabs.harvard.edu/abs/1995SPIE.2622..580K/abstract (accessed on 30 July 2020).

Koldaev, V., & Moraes, P. Jr., (2005). *Non-guided Parachute System for Recovery of Small Orbital Payloads*. Instituto de Aeronáutica e Espaço/Centro Técnico Aeroespacial 12228–904 São José dos Campos, SP, Brasil, https://www.ipen.br/biblioteca/cd/conem/2000/LC8830.pdf (accessed on 30 July 2020).

Kolf, G., (2013). *Flight Control System for an Autonomous Parafoil* (p. 117). Faculty of Engineering at Stellenbosch University, South Africa, thesis, http://scholar.sun.ac.za/handle/10019.1/85757 (accessed on 30 July 2020).

Kologi, M., (2015). *Guided Parafoil System* (pp. 1–7). High Altitude Balloon Program, University of Idaho Space Engineering Department, San Jose State University and University of California, Internal Report.

Kovacevic, D., Bazijanic, E., & Jurum-Kipke, J., (1998). Airdrop of armament and military equipment from aircraft. Transport for military purposes: A Review. *Traffic. 10*, 93–99.

Krishna, K. R. R., (2013). *Precision Farming: Soil Fertility and Productivity Aspects* (p. 179). Apple Academic Press Inc., Waretown, New Jersey, USA.

Krishna, K. R., (2016). *Push Button Agriculture: Robotics, Drones and Satellite Guided Soil and Crop Management* (p. 451). Apple Academic Press Inc., Waretown, New Jersey, USA.

Krishna, K. R., (2018). *Agricultural Drones: A Peaceful Pursuit* (p. 394). Apple Academic Press Inc., Waretown, New Jersey, USA.

Krishna, K. R., (2019). *Unmanned Aerial Vehicle Systems in Crop Production: A Compendium* (p. 675). Apple Academic Press Inc.

Kurt, C., (2018). *Flight Junkies: Parts* (pp. 1–22). http://poweredparaglidingfreetraining.com/powered-paragliding-equipment/parts/ (accessed on 30 July 2020).

Langers, N., (2016). *Tethered Drone Provides a Hovering Watchtower to Troops on the Ground* (p. 18). https://newatlas.com/tether-eye-drone/43520/ (accessed on 30 July 2020).

Lebourgeois, A., Begue, S., Labbe, S., Houles, M., & Martine, J. F., (2012). A light weight multispectral aerial imaging system for nitrogen crop monitoring. *Precision Farming, 13,*525–541.

Lee, A., (2015). *The Amazing Red Devils: A History of the Elite Group* (pp. 1–3). https://www.express.co.uk/life-style/life/585992/Amazing-red-devils-history-elite-group (accessed on 30 July 2020).

Lee, C. K., (2005). Geometric properties of parachutes using 3-D laser scanning. *Journal of Aircraft 73*, 1–7 https://www.researchgate.net/publication/239414574_Geometric_Properties_of_Parachutes_Using_3-D_Laser_Scanning doi: 10.2514/1.18387Q1 (accessed on 30 July 2020).

Lelong, C. C. D., Burger, P., Jubelin, G., Roux, B., & Baret, F., (2008). Assessment of unmanned aerial vehicles imagery for quantitative monitoring of wheat crop in small plots. *Sensors (Basel) 8,*3557–3585.

Lingard, S. J., (1995). Ram-air parachute design. *13th AIAA Aerodynamic Decelerator Systems Technology Conference* (p. 163). Clearwater Beach, FL. USA, http://citeseerx.ist.psu.edu/showciting?cid=286902 (accessed on 30 July 2020).

Lipson, H., (2015). *Automated Disease Detection in Crops Using Drones and Deep Learning* (pp. 1–7). Columbia Tech Ventures. http://techventures.columbia.edu (accessed on 30 July 2020).

Lorwangtragool, P., Wisitsoraat, A., & Kerdcharoen, T., (2011). An electronic nose of Amine detection based on Polymer/SWNT-COOH Nano composite. *Nanoscience and Nanotechnology, 11,*10454–10459.

Lou, S., Tan, P., Sun, Q., Wu, W., Luo, H., & Chen, S., (2018). In-flight wind identification and soft-landing control for autonomous unmanned powered parafoils. *International Journal of Systems Science, 49*, 23–29 https://www.tandfonline.com/doi/abs/10.1080/00207721.2018.1433245/ (accessed on August 8th, 2020).

Luders, B., Sugel, I., & How, J. P., (2015). *Robust Trajectory Planning for Autonomous Parafoils under Wind Uncertainty*. Aerospace Controls Laboratory, Massachusetts Institute of Technology MIT Laboratory. American Institute for Aeronautics and Astronautics. USA. doi: 10.2514/6.2013–4584.

Lynch, J., (2012). *Newly Released Drone Records Reveal Extensive Military Flights in United States of America* (pp. 1–3). https://www.eff.org/deeplinks/2012/12/newly-released-drone-records-reveal-extensive-military-flights-us (accessed on 30 July 2020).

Machin, R. A., Iacomini, C. S., Cerimele, C. J., & Stein, J. M., (2002). Flight testing the parachute system for the space station crew return vehicle. *Journal of Aircraft, 38*(5), 786–799.

Matheson, M., (2017). The rebirth of aerial delivery. *Canadian Military Journal, 2001*, 1–13, 44–46 http://www.journal.forces.gc.ca/vo2/no1/doc/43–46-eng.pdf (accessed on 30 July 2020).

Matos, C., Mahalingam, R., Ottinger, G., Klapper, J., Funk, R., & Komerath, N., (2001). Wind tunnel measurements of parafoil geometry and aerodynamics. American Institute for Aeronautics and Astronautics. *36th AIAA Aerospace Sciences Meeting* (p. 15). AIAA Paper 98–0606, https://arc.aiaa.org/doi/abs/10.2514/6.1998–606/ (accessed on August 8th, 2020).

McCollum, G., Johnson, R., & Fastie, C., (2018). *Correlating Nitrogen Application Rates in Sugarcane with Low Cost Normalized Difference Vegetation Index* (p. 45). Sustainable Agriculture Research Education/ USDA, University of Maryland Final Report No. FS 14–282.

McKinney, J., & Lowry, C., (2009). Mars precision landing using guided parachutes. *20th AIAA Aerodynamic Decelerator Systems Technology Conference* (p. 2983) American Institute of Aeronautics and Astronautics.

McWilliams, P., (2013). *A summary of Different Parachute Types* (p. 17). AWE 365.com https://awe365.com/a-summary-of-different-parachute-types/ (accessed on 30 July 2020).

Meyer, J., (1985). *An Introduction to Deployable Recovery Systems* (pp. 1–8). History Reviews: Sandia Reports SAND85–1180. https://www.scribd.com/document/45703473/An-Introduction-to-Deploy-Able-Recovery-Systems/ (accessed on August 8th, 2020).

Microdrones, (2018). The Heavy Lifting Drone md4–3000 https://www.microdrones.com/en/mdaircraft/md4–3000/ (accessed on 30 July 2020).

Mills Manufacturing Inc., (2018a). *Cargo Parachutes* (p. 19) MillsManufacturing.com (accessed on 30 July 2020).

Mills Manufacturing Inc., (2018b). *G12 Cargo Parachute Assembly* (pp. 1–3). http://www.millsmanufacturing.com/products/g-12-parachute/ (accessed on 30 July 2020).

Mines, F., (2016). *Draft History of Parachuting in Australia: A Draft History of Parachuting in Australia up to the Foundation of Sport Parachuting in 1958* (pp. 1–32). https://www.apf.com.au/APF-Zone/APF-Information/History-of-the-APF/Draft-History-of-Parachuting-in-Australia/default.aspx (accessed on 30 July 2020).

Moore, J., (2018). *Unmanned Aircrafts* (pp. 1–3). AOPA Institute for Air Safety. Aircraft Owners and Pilots Association. https://www.aopa.org/news-and-media/articles-by-author/jim-moore (accessed on 30 July 2020).

Morder Intelligence Ltd., (2018). *Global Parachute Market: Segmented by Type (Round Parachute, Cruciform Parachute, Ram Air Parachute, others), Application (Military, Cargo, Sports, others) and Geography-growth, Trends Progress and Challenges (2019–2024)* (pp. 1–17). Mordor Intelligence Ltd. Hyderabad, India. https://www.mordorintelligence.com/industry-reports/parachute-market (accessed on 30 July 2020).

Murray, J. E., Sim, A. G., Neufeld, D. C., Rennich, P. K., Norris, S. R., & Hughes, W. S., (1994). Further Development And Flight Test Of An Autonomous Precision Landing *System* Using A *Parafoil*. NASA Technical Reports Server (NTRS). (accessed on 30 July 2020).

NASA STAFF, (2017). *A Canopy of Confidence: Orion's Parachutes* (pp. 1–8). http://www.nasa.gov/exploration/systems/mpcv/canopy_of_confidence.html (accessed on 30 July 2020).

NASA, (1967). Rogallo Wing-the story Told by NASA (pp. 1–7). NASA History Office. History.nasa.gov. (accessed on 30 July 2020).

NASA, (2014). *NASA is About to Test a Mega-Parachute at the Edge of Space* (pp. 1–3). The Atlantic Daily. https:/www.THEATLANTIC. COM/TECHNOLOGY/ARCHIVE/2014/06/NASA-ISABOUT-TO-TEST-A-MEGA-PARACHUTE-AT-THE-EDGE-OF-SPACE/372030 (accessed on 30 July 2020).

NASA, (2018). *Deployable Aerodynamic Decelerator Systems* (p. 90). Report No. NASA 8066 https://ntrs.nasa.gov/archive/nasa/casi.ntrs.nasa.gov/19710021827.pdf (accessed on 30 July 2020).

National Aeronautics and Space Agency, (2017). *NASA Space Shuttle Crew Escape Systems Handbook* (p. 168). Periscope Film LLC, USA.

Needham, J., (1965). *Science and Civilization: Physics and Physical Technology* (p. 816). Part 2 Mechanical Engineering. Cambridge University Press, England.

Niccum, R. J., Haak, E. L., & Gutenkauf, S., (1965). *Drag and Stability of Cross Type Parachutes* (p. 111). University of Minnesota, Minneapolis, Technical Report No FD TDR-64–155 http://www.dtic.mil/docs/citations/AD0460890 (accessed on 30 July 2020).

Nicoloides, J. D., & Tragarzz, M. A., (1971). *Parafoil Flight Performance* (pp. 1–14). US Air Force Flight Dynamics. Notre Dame, Indiana, USA.

Northwing, (2018). *Northwing ATF Soaring Trike* (p. 15). http://northwing.com/atf-trike.aspx (accessed on 30 July 2020).

Olga, S. W., Sanaz, S., Juliet, M. M., Chad, J., McClintick-Chess, J. R., Steven, M. B., Kristin, S., et al. (2018). *Assessment of UAV Based Vegetation Indices for Nitrogen Concentration Estimation in Spring Wheat* (pp. 71–90). doi: 10.4236/ars.2018.72006.

Onyx Systems, (2018). *Onyx-Autonomously Guided Parafoil System* (pp. 1–3). https://defense-update.com/20070513_onyx-pads.html (accessed on 30 July 2020).

ParaZeroSafeAir, (2018). *ParaZero Drone Safety Systems* (pp. 1–7). https://parazero.com// (accessed on August 8th, 2020).

Patel, S., Hackett, N. R., & Jorgensen, D. J., (1997). *Qualification of the Guided Parafoil Air Delivery System: Light (GPADS-light)* (pp. 234–243). American Institute of Aeronautics and Astronautics. Report No A97--31298-AIAA-F7–1993.

Patworks, (2014). *Early History of Parachutes*. Skydiving museum and hall of fame. http://works-words.com/NSM-WIKI/WP/wordpress/wiki/skydiving/early-history/history/early-history-of-parachutes/ (accessed on 30 July 2020).

Pellone, L., Salvatore, A., Favaloro, N., & Concilio, A., (2017). SMA-based system for environmental sensors released from an unmanned aerial vehicle. Italian aerospace research center-CIRA. *Aerospace, 4,* 1–22. doi: 10.3390/aerospace4010004.

Peterson, C. W., (1990). High Performance Parachutes. *American Scientific Aircraft Journal,* 108–116.

Phillips, A., (2017). *University of Bristol uses UAV Technology for Atmospheric Research* (pp. 1–4). http://www.bristol.ac.uk (accessed on 30 July 2020).

Pobkrut, T., Kerdcharoen, T., & Emsaard, T., (2016). *Sensor Drone for Aerial Odor Mapping for Agriculture and Security Services* (pp. 1–6). Conference paper https://ieeexplore.ieee.org/document/7561340 doi: 10.1109/ECTICon.2016.7561340 (accessed on 30 July 2020).

Potvin, J., (2018). *Calculating the Descent Rate of a Round Parachute* (pp. 1–8). Parks College Parachute Research Group. https://www.pcprg.com/rounddes.htm (accessed on 30 July 2020).

Powerchute Educational Foundation Inc., (2018). *E.L.L.A.S.S Emergency Low Level Search and Surveillance* (pp. 1–3). www.ellass.org (accessed on 30 July 2020).

Poynter, D., (1991). *The Parachute Manual: A Technical Treatise on Aerodynamic Decelerators* (Vol. 1, pp. 416). Para Publications, California, USA.

Price, C., (2018). *The Company Which Invented Aid Packages Dropped by Parachute Which Saved Lives in Hurricane Hit Caribbean* (pp. 1–5). http://www.kentonline.co.uk/

kent-business/county-news/air-drop-box-invents-aid-packages-dropped-by-parachute-on-farm-wrotham-tested-headcorn-aerodrome-159556/ (accessed on 30 July 2020).

Pudelko, R., Stuczynski, T., & Borzecka-Walker, (2012). The suitability of an unmanned arial vehicle (UAV) for the evaluation of experimental fields and crops. *Zemdirbyste-Agriculture 99*,431–436. UDK 631.5.001.4:629.734.

Puranik, A., Parker, G., Passerele, J., Dexter, B., Yakimenko, O., & Kaminer, A., (2006). *Modeling and Simulation of a Ship Launched Glider Cargo Delivery System* (pp. 5332–5341). AIAA Guidance, Navigation, and Control Conference American Institute of Aeronautics and Astronautics Inc.

Rippon Police Department, (2012). Rippon Police Department celebrates 3 years of powered parachute flight (pp. 1–3). http://www.riponpd.org/?page_id=964 (accessed on August 8th, 2020).

RMAX, (2015). *RMAX Specifications* (pp. 1–4). Yamaha Motor Company, Japan. https://barnardmicrosystems.com/UAV/uav_list/yamaha_rmax.html/ (accessed on August 8th, 2020).

Rohan, M. C., (2019). *Military Parachutes Market Worth 1.21 Billion US$ by 2020* (pp. 1–27). Markets and Markets Research, Pune, India. https://www.marketsandmarkets.com/PressReleases/parachute.asp (accessed on 30 July 2020).

Ross, J. F., (2014). *Why Pilots Didn't Wear Parachutes During World War 1* (pp. 1–4). The History Reader: Dispatches in history from St Martin's Press. http://www.thehistoryreader.com/modern-history/parachutes-world-war-1/ (accessed on 30 July 2020).

Rumermann, J., (2009). *NASA-Historical Data Book V7 NASA History Series* (p. 1071). National Aeronautics and Space Agency, NASA History Division, Washington D.C. USA.

Rydberg, A., Soderstorm, M., Hagner, O., & Borjesson, T., (2007). Field specific overview of crops using UAV (Unmanned Aerial Vehicle). *Precision Agriculture, 7,*357–364.

Sanchez, P. D., (2017). Study of a methodology for the flight simulation of a ram-air parachute using vortex lattice aerodynamic model (pp. 1–27). Escola Superior d' Enginyeries Industrial Aerospacial Audio-visual de terrasa. Catalunya, Spain.

Sargar, S., (2018). *Global Parachute Market Experience Significant Growth During Period 201–2025* (pp. 1–55). https://www.fiormarkets.com/report/global-parachute-market-research-report-2019–346236.html#tableofcontent (accessed on 30 July 2020).

Scentroid Inc., (2017). *Scentroid: The Future of Sensory Technology* (pp. 1–12). http://scentroid.com/scentroid-sampling-drone/ (accessed on 30 July 2020).

Schroth, F., (2017). *Indemnis Safety Parachutes for Drones* (pp. 1–4). Dronelife.com (accessed on 30 July 2020).

Sensefly Inc., (2019). *eBee X-Map Without Limits* (pp. 1–7). https://www.sensefly.com/drone/ebee-x/ (accessed on 30 July 2020).

Sim, A. G., Murray, J. E., Neufeld, D. C., & Dale, R. R., (1993). *The Development and Flight Test of a Deployable Precision Landing System for Spacecraft Recovery.* NASA Technical reports Server (NTRS).

Siyang, S., Seesaard, T., Lorwangtragool, T., & Kerdcharoen, T., (2013). E-nose based on metallo-tetraphenylporphyrin/SWNT-COOH for alcohol detection. In: *Proceedings of IEEE International Conference on Electronic Devices and Solid-State Circuits (EDSSC)* (pp. 1–5).

SkyRunner, (2018). *Introducing SkyRunner: The World's First Flying Off-Road Vehicle GO-ANYWHERE* (pp. 1–3). https://www.prnewswire.com/news-releases/introducing-skyrunner-the-worlds-first-flying-off-road-vehicle-go-anywhere-300407206.html (accessed on 30 July 2020).

Stein, J. M., Madsen, C. M., & Strahan., (2012). *An overview of the guided parafoil system derived from X-38 experience.* AIAA Aerodynamic Decelerator Systems Technology Conference, Johnson Space Centre, Houston, USA (pp 1–12), https://arc.aiaa.org/doi/10.2514/6.2005-1652/ (accessed on August 8th, 2020).

Stoeckle, M. R., Fejzic, A., Breger, L. S., & How, J. P., (2014). *Fault Detection and Isolation for Autonomous Parafoils* (p. 124). Massachusetts Institute of Technology, Cambridge, MA, USA https://dspace.mit.edu/bitstream/handle/1721.1/90612/891583273-MIT.pdf; sequence=2 (accessed on 30 July 2020).

Sugel, I., (2013). Robust planning for autonomous parafoil. *MS Thesis* (p. 120). Massachusetts Institute of Technology, Cambridge, MA, USA.

Taylor, B., (2018). *Buckeye Dragonfly* (pp. 1–8). http://buckeyedragonfly.com/ (accessed on 30 July 2020).

Tetracam Inc., (2018). *Hawkeye: Autonomous Aerial Imaging* (pp. 1–8). http://www.tetracam.com/ProductHawkeyewindow2.htm (accessed on 30 July 2020).

Thamm, H. P., & Judex, M., (2006). The low-cost drone. An interesting tool for process monitoring in a high spatial and temporal resolution. *International Archives of Photogrammetry, Remote sensing, Spatial information Science* (Vol. 36, pp. 140–144). ISPs commission 7th Mid-term symposium. Remote Sensing: From Pixels to Process. Enchede. The Netherlands.

Thamm, H. P., (2011). SUSI 62 a robust and safe parachute UAV with long flight time and good payload. International Archives of the Photogrammetry, Remote Sensing and Spatial Information Sciences, 38, 19–24.

Thamm, H. P., Menz, G., Becker, M., Kuria, D. N., Misana, S., & Kohn, D., (2013). The use of UAS for assessing agricultural systems in AN Wetland in Tanzania, in the wet season, for sustainable agriculture and providing ground truth for Terra Sar X data. ISPRS International Archives of the Photogrammetry, Remote Sensing and Spatial Information Sciences (pp. 401–406). XL-1/w2: doi: 10.5194/isprsarchives-XL-1-W2–401–2013.

Threod Systems, (2017). *EOS MINI-Unmanned Aerial System* (pp. 1–9). http://www.threod.com (accessed on 30 July 2020).

Unal, I., & Mehmet, T., (2016). *A Review of Using Drones for Precision Farming Applications* (pp. 276–283). Department of Agricultural Machinery, Faculty of Agriculture, Akdeniz University, Antalya-Turkey, http://tarmek.org/bildiriler/A_Review_on_Using_Drones_for_Precision_Farming_Applications.pdf (accessed on 30 July 2020).

United States Department of Transport. Federal Aviation Administration (2013). *Parachute Riggers Handbook* (p. 344). Create Space Independent Publishers LLC. USA.

United States Department of Transportation (2007). *Powered Parachutes Flying Handbook* (p. 163). FAA-H-8083–39 United States Department of Transportation Federal Aviation Administration. Flight Standards Service. Washington D.C.

United States Parachute Association (2016). *The History of Parachutes Incorporated at Jump Town* (pp. 1–17). Jump Town, Orange, Massachusetts, USA. https://www.jumptown.com/about/articles/the-history-of-parachutes-incorporated-at-jumptown/ (accessed on 30 July 2020).

USDA (2018). *First Parachute Years* (p. 1–7). United States Department of Agriculture. US Forest Service. https://www.fs.fed.us/science-technology/fire/people-working-fire/smokejumpers/smokejumper-base-contact-information/missoula-smokejumpers/missoula-history/parachute (accessed on 30 July 2020).

Van, R. P. H., Boertuzen, P. G., Geerdes, J. B., Dekker, G. J., & Udo, R., (1997). Parafoil technology demonstration OBC development. *Proceedings of Third International*

Conference on Spacecraft, Navigation and Control Systems ESTEC (Vol. 381, pp. 183–189), Noordwijk, The Netherlands SP.

Veronte AutopilotsLLC, (2015). *Guided Parafoils and Unmanned Parachutes UPP* (pp. 1–7). Embention. https://www.embention.com/en/news/unmanned-guided-parafoils-upp/ (September 12th, 2018)

Ward, D. T., Pollock, T. C., & Lund, D. W., (2001). *Flight tests of an Unmanned Powered Parachute: A Validation Tool for GN and C Algorithms*. Vehicle Systems and Control Laboratory, Johnsons Space Centre, Houston, USA. (pp 1–3), https://vscl.tamu.edu/research/flight-tests-of-an-unmanned-powered-parachute-a-validation-tool-for-gnc-algorithms/ (accessed on August 8th, 2020)

Whittelesey, J. H., (1970). Tethered balloon for archaeological photos. Photogrammetric Engineering (pp. 181–186). https://www.asprs.org/wp-content/uploads/pers/1970journal/feb/1970_feb_181–186.pdf (accessed on 30 July 2020).

Wikimilli (2018). *Powered Parachutes* (p. 12). https://wikimili.com/en/Powered_parachute (accessed on 30 July 2020).

Wikipedia (2018d). *Powered paragliding: Rules* (pp. 1–3). Wikipedia https://en.wikipedia.org/wiki/Powered_paragliding#Regulations (accessed on 30 July 2020).

Wikipedia, (2018a). *Rogallo Wing* (pp. 1–14). https://en.wikipedia.org/wiki/Rogallo_wing (accessed on 30 July 2020).

Wikipedia, (2018b) *Powered Parachutes* (pp. 1–7). https://en.wikipedia.org/wiki/Powered_parachute (accessed on 30 July 2020).

Wikipedia, (2018c). *Two-Person Trike, AirBorne XT912 Tourer* (p. 13). From https://en.wikipedia.org/wiki/Ultralight_trike (accessed on 30 July 2020).

Woolston, V., (2014). *China Successfully Tests Smog-Fighting Drones That Spray Chemicals to Capture Air Pollution* (pp. 1, 3). Mail online. https://www.dailymail.co.uk/sciencetech/article-2577347/China-successfully-tests-smog-fighting-drones-spray-chemicals-capture-air-pollution.html (accessed on 30 July 2020).

Wright, J., (2017). *Soyuz Landing: Soyuz Page* (pp. 1, 2). National Aeronautics and Space agency, NASA, Washington D.C. USA.

Yakimenko, O. A., (2005). *On the Development of a Scalable 8-DoF Model for a Generic Parafoil-Payload Delivery System Oleg Paper Presented at the 18th AIAA Aerodynamic Decelerator Systems Technology Conference and Seminar* (pp. 1–13). AIAA 2005–1665, https://pdfs.semanticscholar.org/df9e/84fe34db1102fb66859697150fe0b1e6102f.pdf (accessed on 30 July 2020).

Yakimenko, O. A., (2016). *Autonomous Parachute-Based Precision Delivery Systems* (p. 22). http://www.millsmanufacturing.com/products/g-12-parachute/ (accessed on 30 July 2020).

CHAPTER 3

Blimps in Agricultural Crop Production: A Recent Initiative

ABSTRACT

This chapter focuses on aerial vehicles generally denoted as blimps or zeppelins. Initially, balloons devised were not easily navigable. The dirigible or directable blimp with different shapes were produced in the 19th century. Blimps are non-rigid airships, whereas zeppelins possess rigid framework, made of light wood or aluminum. In a directable blimp or dirigible the lift into sky is derived from a lighter-than-air gas. It was first demonstrated by a Frenchman, Henri Giffard, in 1852. A detailed list of historically important events related to blimps/zeppelins such as its design, development, demonstration to public and first or regular use in different aspects related to military, civilian and agriculture has been included. Blimps were used for long distance travel, including trans-Atlantic journey, during the 1920–30s. They were used by military establishments of USA and several European nations (e.g., Germany, Britain, Russia) beginning in the early 1900s. Blimps/zeppelins were rejected owing to their vulnerability to disaster related to inflammable hydrogen gas (lighter-than-air gas). Such historical information offers a better perspective as we dwell into greater detail and uses of airships in the present times.

Firstly, this chapter provides a background about the airships. It encompasses definitions, terminology, explanations and various components (parts) of a blimp. Blimps are known as non-rigid airships. They are defined as aerial vehicles that levitate, float, and transit in the air using a lighter-than-air gas such as helium. Blimps such as the one utilized for advertisement or military cargo transport or general aerial surveillance contain that several parts. They are the 'envelope' that is made of special texture that is leak-proof. Fins on the envelope help in stabilizing the blimps. The envelope is filled with a lifting gas, i.e., helium. A gondola provides space for crew and electronic instruments. A gondola is tightly attached to the envelope. Specifically, it carries instrumentation required for aerial imagery (electro-optical sensors), cargo transit and even human travel. The powered blimps derive thrust through an IC engine attached to propeller.

There are several types of airships available for use. One section in this chapter provides salient features of at least five different types of blimps. The tethered blimps (also known as aerostats) are held fastened to ground control station, using strong tethers. Untethered, i.e., free-floating airships are meant for aerial surveillance, travel and cargo transport. There are also remote-controlled blimps/zeppelins that are guided using radio control. Blimps that are totally robotic follow pre-programmed flight plans prepared using appropriate computer programs (e-Motion, Pixhawk, etc.). Airships designed so far differ in sizes. They could be small or medium sized ones utilized frequently during advertisements (e.g., Goodyear blimp). Larger blimps are used in travel and military cargo transport. There are also very large blimps known as 'Giga blimps.' A recent design comprises of a hybrid blimp and copter. Such blimps known as 'Plimps' are stable due to the copter component.

Thus far, blimps have found innumerable uses in variety of aspects of human endeavor. Major uses of blimps are in military. They are used for surveillance of camps, vehicle convoys and cargo transit. They are used as sentinels over military camps and missile sites. Blimps have found use as providers of surveillance data related to border security. Blimps are now being tried for their possible use in space science. Particularly, to achieve safe landing and continued relay of telemetry messages from heavenly bodies. Blimps hovering over important geographical sites have offered excellent aerial spectral data for long durations. Incidentally, blimps offer better flight endurance than other aerial vehicles.

The role of blimps in agricultural crop production forms the major section of this chapter. Topics dealt include aerial photography and survey of natural resources, vegetation, land and soil resources, cropping systems, irrigation lines, progress of agronomic procedures, etc. Blimps are utilized during phenotyping of crops in agricultural experimental stations and in a large farm. Crop's phenological data is useful to agronomist and plant breeders. The utility of blimps in offering spectral data relevant to identification of disease/pest attack, soil erosion, drought or floods has been discussed. Blimps could find excellent use in maintenance of agricultural experimental stations, tracing farm activity and in monitoring crop growth.

3.1   INTRODUCTION

3.1.1   AIRSHIPS (BLIMPS AND ZEPPELINS): A HISTORICAL PERSPECTIVE

This chapter focuses on airships that are colloquially called 'Blimps' and 'Zeppelins.' Blimps are the non-rigid airships inflated with lighter-than-air gas. A blimp collapses if deflated. Zeppelins are lighter-than-air aircrafts provided with a rigid aluminum or wooden framework. Let us begin with history of the words 'Blimp' and 'Zeppelin,' their etymology and how the terms were coined. Origin of the word blimp is related to military usage. Initially, lighter-than-air airships were designated as type A-Limp ('Limp Bags'). A second version of airships were designated type-B-Limp. For short they were called 'B-limp or Blimp.' There are a couple of more suggestions about the derivation of the word 'blimp.' It seems Commander of the Capel-le-Ferne Airship Station, R. N. Cunningham went inspecting airships in 1915. The airship produced a sound 'blimp' whenever he flipped his fingers on the body of the ship. Hence, they gave a nickname to non-rigid airships as 'blimps' in 1915. Next, it seems anti-submarine airships were first called as 'blimps,' during 1915. The Oxford Dictionary lists the word 'Blimp.' It is explained as an airship that is non-rigid and is inflated with lighter-than-air gas (History Forum, 2018; Editor, 2018a; Vaeth, 1992). The rigid airships filled with lighter-than-air gas is known as 'Zeppelin.' Zeppelin is a word coined sometime in 1930s. It is derived from its inventor, Ferdinand Zeppelin.

Balloons produced were originally *not navigable*. Balloon designers and producers were not able to control or manipulate the transit route of balloons. They tried to change the shape of the balloon to see, if they could control balloons in the air, when they float. The Dirigible (dirigeable or directable) was first developed by a Frenchman Henri Giffard, in 1852. The first flight of the dirigible, i.e., a directable blimp is said to have occurred on September 17th, 1852. Henri Giffard's blimp was inflated with a lighter-than-air gas, i.e., hydrogen. It had propulsion and a steering. The large propellers were attached to a steam engine. During its flight, Henri Giffard's blimp covered about 17 miles distance at the rate of 5 mph (Bellis, 2017; Sharp, 2012; History Forum, 2018; Greens, 2013; Table 3.1).

An airship (blimp) with gasoline powered engine was first developed by a Brazilian Alberto Santos-Dumont, in the year 1898. He demonstrated the navigable gasoline powered blimp, in Paris, in 1898. During early 1890s, United States Army tested blimps (dirigibles), to transport cargo. For instance, Thomas Baldwin developed a blimp (53 ft. long). The first powered blimp was developed for US Army, in the year 1904. It had a 20 hp gasoline fuel engine (see Table 3.1).

TABLE 3.1 History of Blimps: Important Dates, Inventors, and Their Contributions to Blimp Technology, During Recent History (1600 AD to 1900 AD)

- 1670 Francesco Lana de Terzi (Father of Aeronautics): He published a description of 'Aerial ship' supported by 4 copper spheres that were evacuated.
- 1783 Jean Baptiste Marie Meusnier: He developed designs and depicted blimps of 260 ft. in paintings. The airship had three propellers to obtain thrust.
- 1785 Jean-Pierre Blanchard: He crossed English Channel using an airship supported by hand-powered propellers.
- 1820s William Bland: He flew the first steam engine powered blimp above Paris.
- 1872 Paul Haenlien: He flew the first airship supported by internal combustion engine. The IC engine was run using coal gas as fuel.
- 1874 Micajah Clark Dyer: He filed the first patent for an 'Apparatus navigating the Air.' It made many flights between 1872 and 1874.
- 1883 Gaston Tissander: He conducted the first flight in a blimp powered by electric engines. The blimps had Siemen's Electric motor.
- 1884 Charles Renard and Arthur Constantin Kerbs: They flew the 'La France' airship for 23 min. In all the blimp made seven flights during 1884 to1886.
- 1888 Peter Campbell: He designed the 'Campbell Airship.' It was built by Novelty Airship Company.
- 1888–97 Ferdinand Wolfert: Built 3 ships powered by Daimler petrol engines.
- 1897 David Schwarz: He developed blimp with an aluminum envelope.
- 1897–99 Konstantin Danilewsky: A Ukrainian, he built hand-powered blimps.

Source: Excerpted from Wikipedia, 2017, https://en.wikipedia.org/wiki/Airship; Colvin, 2017; History Forum, 2018; Valiulis, 2014; Abbot and Walmsley, 1998; Wrags, 2008; Brooks, 1992; Laskas, 2016; Greens, 2013.

Next in the series of airships had a rigid frame. It would not collapse like the blimp, if deflated. Suchan airship was developed in Germany. They were called 'Zeppellins.' Zeppelin was the name given to airship with aluminum framework around it. It was named after its inventor Count Ferdinand von Zeppelin (1838–1917). The rigid-framed airship floated and moved in the atmosphere on November 3rd, 1897, above a field near Berlin, Germany. This zeppelin was designed by David Schwarz and had a 12 hp Daimler gasoline engine. Later, an untethered rigid frame zeppelin was successfully flown above Lake Constance in Germany, in June 1900. A school to teach the designing and production of Zeppelins was initiated in 1908, in Germany. These early zeppelins reached a height of 1300 ft. above ground surface. They could transit untethered for 3–5 miles (Bellis, 2017; Editor 2018a; Sharp, 2012).

Airship 'Enterprise (NC-16A)' manufactured by Goodyear Tire and Rubber Inc., first took to sky, on August 23rd, 1934. It was floated in sky above Washington D.C. and New York until end of 1941. It was later shifted to Wingfoot lake, Akron, Ohio. Here, it served as a blimp that could train pilots. However, during early stages of World War II, it patrolled the sky over Ohio. It was used to surveillance urban areas for total black out. It would identify areas that did not comply with black out signals (Crouch, 2017; Plate 3.1).

Russian military produced several models of blimps during 19[th]and 20th century. In 1892, they purchased German zeppelins to counter the French attack. In early 1910, they purchased zeppelins for use in travel and military. During 1920 to 1947, Russian engineers designed and produced a series of airships, mainly blimps, for civilian and military purposes. In 1944, they produced a blimp known as 'Pobeda.' It was meant for transport of cargo. Like other nations in Europe, Russia too experienced disasters related to zeppelins and blimps that had hydrogen in the envelope. One of the last passenger airships produced in Russia was known as 'Patriot' that was de-commissioned in 1950s. Again, Russia too is experiencing a renewed interest in airships and blimp technology, since past 2–3 decades, i.e., 1990-till date.

Erstwhile Soviet Union's military regularly used several blimps of different sizes. They were meant both for surveillance and transport of military personnel. They utilized both semi-rigid and non-rigid airships or blimps. For example, CCCP-V6 was one of the largest blimps used by Soviet Military. It could travel long distances because of extended endurance of 130 hr. This blimp crashed out in 1938 resulting in slowing and dismantling of the blimp program. Despite it, Soviet blimp program lasted till 1950s. In the later period, airships in general were not used for travel or transport of military personnel or civilian passengers. Blimps were confined to adver-tisement, cargo transport, data collection about ground features and general surveillance (Valiulis, 2014).

Regarding history of blimps in Great Britain, it is said that, most of the airships (blimps) meant for World War I were built at Cardington. However, the program was disbanded in 1930. It was due to a crash in which over 50 military personnel perished (Lawless, 2018). It is said that, this accident along with major disaster related to Hindenburg in New Jersey, in 1937 had strong impact on blimp programs. They literally sealed out use of airships or blimps in civil transport and military cargo transport. British blimps were used for scouting, mine clearance and submarine attack duties. Regarding other European nations, both, France and Italy continued use of airships throughout the war.

The prospect of airships as bombers had been recognized in Europe well before they were up to the task. The Italian forces became the first to use dirigibles for a military purpose, during the Italo-Turkish war. However, that also marked the real debut of the airship as a weapon. Germans, French and Italians all used airships for scouting and tactical bombing roles, early in the war. Later, they all learnt that the airship was too vulnerable for operations over the front.

Blimps or non-rigid airships are indeed large aerial vehicles with lighter-than-air gas filled in, to obtain thrust and float. They were utilized frequently for travel of civilian and military personnel, during early 20th century. Transport of cargo across long distance was possible. For a few decades since 1900s, they ruled the sky in terms of transport of pay load. They were predominantly used to ship military equipment and ammunition. Blimps were found to be most practicable vehicles for large scale transport of personnel. Travel using blimps or zeppelins became more common in 1920s (Wiesenberger, 2017; Botting, 1980; Lord and Kolesnik, 1982; Ric, 1994; Dick and Robinson, 1992).

As stated earlier, the first rigid airship was built by a German named David Schwarz, in 1895. It seems his design served as the precursor and model for the Zeppelin series that followed soon, during early 1900s. The zeppelins used 15 hp Internal Combustion engine, for thrust. This effort enthused British to build similar rigid airships and non-rigid blimps. The military called them class-blimps. These blimps were predominantly meant for use by military.

The passenger blimps were developed during 1920s by engineers in USA, Germany and Britain. However, it seems, the first passenger blimp was demonstrated by a Brazilian named Alberto Santos Dumount. As stated earlier, there was an abrupt end to usage of blimps and zeppelin due to disasters in different nations. The resurgence of blimps seems to have begun, since 1990–2000 AD. There are now innumerable blimp/zeppelin producing companies all over world. They are adopted by both military and civilian agencies.

The 'Golden Age of Airships' began in 1900 with the launch of the Luftschiff Zeppelin Lz1. Zeppelins were huge – the Lz1 was 128 m long, while the last one – Graf Zeppelin was 236 m long (Valiulis, 2014). During 1920s, the US Navy constructed blimps with hydrogen as lighter-than-air gas. The first rigid airship was called 'ZR-1.' There were many others developed similarly with hydrogen as the filling gas. However, the US Navy experienced a series of disasters because hydrogen is an inflammable gas. The blimps caught fire and were destroyed. Notably, in one year, there were at least four disasters related to blimps with hydrogen. Later, it became clear that blimps had to be filled with helium, a non-flammable lighter-than-air gas. A blimp called 'Roma' that caught fire in 1923 killed about 45 men. This prompted rejection of hydrogen totally as a filling gas. Instead, strictly, helium was to be used in blimps and zeppelins (Colvin, 2017; Grossnick 1987). Interest in blimps and zeppelins suffered a sharp decline after the disaster of the large airship Hindenburg in 1937. The German Zeppelin collapsed and burnt at Lakehurst, New Jersey. This disaster killed 35 of 97 passengers and all crew on May 6th, 1937. The Hindenburg disaster reduced production and usage of blimps to its lowest levels. The interest in blimps for aerial photography and travel almost died off. Perhaps only a few companies producing dirigibles were still enthusiastic, but only feebly.

The US Navy had recognized the usefulness of blimps early during the World War I era. During World War II, Navy ships were mostly supported by blimps to gather information. At the beginning of World War II, the US Navy had 10 blimps to help the frontline navy personnel and their ships. By the end of the war, the number of blimps had increased to 167. An exclusive blimp base was inaugurated by the US Navy in New Jersey, in 1942. It suggests the importance bestowed on blimps by the US military (Historylines, 2012). This situation was similar with many European military establishments.

During World War II, blimps were used for surveillance of the both west and east coasts of USA. It seems that the use of blimps almost subsided by 1962. Many of the blimp-hangers constructed during war went empty. Particularly, hangers situated at Lakehurst in New Jersey; Tillmook, Oregon; Orange county California; and Moffet Airfield in California were abandoned.

Colvin (2017) points out that during the past few decades, despite revived interest in blimps, many of the projects in USA and Europe never made it beyond the drawing board. A few remained at prototype stage. These were not commercialized. A few examples of Blimp projects that remained as mere prototypes that were developed between 1970 to 2010 are: Helistat prototype built in New Jersey in, 1970; Aeron – a hybrid blimp developed in, 1970; Cyclocrane – a hybrid blimp (blimp plus rotocraft) built in, 1988; CL-160 – a large semi-rigid blimp built in Germany in, 2000; CargoLifter, built in 2002, in Berlin, Germany; and Walrus, a project that became defunct very early in 2005. There may be many more blimp projects that never got reported but were shelved. However, it is possible that some of these prototypes or designs may get selected for large scale use. It depends on the requirements of the clientele.

Historically, there were a few noteworthy expeditions conducted adopting blimps. Travel and expedition using airships attracted many during 1900s. Zeppelins, i.e., a rigid blimp was used to make several trips to Arctic region. The aim was to explore the arctic region for geological features and other useful natural resources. 'Graf Zeppelin' a German airship made expedition to Arctic circle in 1931 (Thiesen, 2010; Prentice et al., 2005). About 40 crewmen conducted an 8000 km journey into arctic. It took them 136 h of nonstop transit.

There is also a new dimension to the use of blimps. During past decade, there seems to be a resurgence of interest in blimps. Blimps may find popularity with farming enterprises who like to conduct round-the-clock aerial survey of crops, i.e., during the crop season. Blimp usage got revived only recently (1990s – 2005) through the demand for aerial survey, imagery of natural resources, observing agricultural zones and monitoring of public events. The need for policing of urban areas has also induced demand for blimps. Blimps are also preferred because of their low carbon footprint (Wiesenberger, 2017).

Colvin (2017) opines that hot-air balloons were fore-runners of the present-day airships. It was Germany that started with passenger zeppelins in 1920s. During the same period, United States of America began producing rigid airships. 'Shenandoah' was the earliest USA made rigid airship. It was deployed for passenger travel. It experienced disaster later (Keirns, 2004).

The interest in blimps was driven predominantly due to its ability to travel in air (sky)and carry personnel and/or cargo. However, in due course, blimps generated further interest among specialists and general public. Particularly, those who were interested in obtaining aerial images. Longer endurance meant continuous surveillance of ground features and better aerial images of them. Marzolff (2014) has noted that the first publication of aerial photo was done in 1888 by Arthur Batut (1888). A clear 100 years later, it has been recognized as a method of immense utility, to specialists in several topics and fields of activity. Aerial photos taken from small drones, blimps, aerostats, balloons and even kites have gained in popularity and usage. Most recent area where aerial photos, spectral analysis and collection of digital data is gaining in importance is the agrarian regions. The agronomic methods in future may hinge, to a great extent, on spectral data obtained by UAVs such as blimps. In the very near future, blimps, it seems could be a common feature above agrarian skyline.

PLATE 3.1   Goodyear Blimp.

Source: Cynthia Hess and Edward Markey, Goodyear Inc., Akron, Ohio, USA; Smithsonian National Air and Space Museum, Washington, USA; https://airandspace.si.edu/stories/editorial/blimpand https://www.goodyearblimp.com/behind-the-scenes/current-blimps.html.

3.1.2   RECENT HISTORY OF BLIMPS AND ZEPPELINS: ABOUT A RESURGENCE

During the past 2 decades, i.e., since 1990s, blimps and zeppelins seem to have staged a resurgence in different parts of the earth. So, let us now consider the more recent history of Blimps and Zeppelins. They seem to have made a return to sky because of several reasons such as ability to manufacture safer blimps, use of helium, need for large scale cargo and personnel transport, better economic efficiency and low carbon footprint. They say, zeppelins returned to the German skies in 1997. It was designed and developed by the German company, 'Zeppelin Luftschifftechnik Gmbh.' This zeppelin was meant mainly for recreational sight-seeing trips above German cities. Also, for short duration excursions to different places within Germany. In 2008, the prototype of this model of zeppelin from Germany was examined for regular use in California, in USA. This event indicated resurgence of interest in blimps within USA, particularly, after the crash of US Navy airships, USS Macon and USS Akron in 1930. In USA, it has taken a clear 50 years to introspect, analyze and opt again for blimps. The recently introduced Zeppelin NT from Germany are smaller at 247 ft., if compared with the erstwhile Hindenburg (845 ft long) that crashed out. Recent (i.e., 2000–2018) models of zeppelins from Germany are all inflated using helium and not hydrogen (Wiesenberger, 2017).

Since the turn of the century in 2000 AD, military establishments of developed nations, also aviation industries involved in civilian travel, cargo transport and in aerial photography (spectral analysis) have invested time and funds in greater proportion, on airships. Further, several modifications have occurred with regard to blimps and zeppelins, during past two decades. A few examples are as follows:

Lacroix et al. (2001) state that, there is a kind of resurgence of interest to design and produce blimps in Europe. Since 2000 AD, there is a spurt in designing and development of blimps, particularly, 'Hybrid Blimps,' i.e., 'Plimps.' A 'Plimp' combines a blimp, a helicopter and airplane (fixed-winged) in one vehicle. Such a combination adds to versatility of blimps. It was designed, developed, produced and marketed to clientele in 2017 (Borkhataria, 2018). This marks a major deviation from simple dirigibles and zeppelins that we have known for long.

There has also been a penchant to produce large hybrid blimps. This is to transport cargo. During this decade, i.e., 2010 and later, large hybrid blimps have taken to skies. For example, in 2016, Great Britain has manu-factured one of the largest hybrid blimps. It is known as 'Airlander 10.' It is a helium inflated blimp that measures 302 ft. in length. It floats at 20,000 ft above ground surface. 'Airlander 10' is a hybrid blimp that was released for public exhibition in 2016. It is actually a hybrid of blimp, helicopter and fixed-winged airplane. It can transit at a speed of 90 mph (148 kmph). It has particularly long endurance of 2 weeks in air. It keeps floating and obtaining spectral imagery (Lawless, 2018; Airlander 10, 2018).

Several industries have chosen to offer highly improvised blimps during past decades. Let us consider a few of them and their chronology. A few blimp models have also found special preference among agriculturists, particularly, those that offer high resolution aerial imagery, those that provide continuous surveillance to experimental farms and those able to spray the crop with agri-cultural chemicals (liquid fertilizers, fungicides, herbicides and pesticides).

In USA, blimp program for large scale transport of cargo and personnel was halted in 2013. Instead, a modified hybrid blimp was produced for use by military and civilian purposes (Lawless, 2018). One of the earliest efforts to use blimps to spray or dust crop with agricultural chemicals was reported from Russia by Salnikov et al. (2014). They have actually adopted a hybrid blimp that combines a blimp, a copter and airplane, to conduct aerial sprays. It is being manufactured by a company known as Bee Robotics Corporation.

During recent past, i.e., since a couple of decades, it seems Russian military is opting for a larger number of blimps to be used. This is mainly to conduct accurate surveillance, carry military personnel and cargo. They have even formed a separate hybrid-blimp force. Such a hybrid-blimp force constitutes blimps with high precision technology and computer-guided surveillance. For example, during the recent Syrian civilian war too blimps were used over combat zone and military camps. They say, blimps are being preferred because of their long endurance, relatively lower cost and accurate surveillance information (Winston, 2018).

Russian military has also opted for deploying a greater number of high precision technology blimps as a measure to thwart, enemy missiles. Blimps are to become part of missile detecting mechanism. Such hybrid blimps are said to become part of early warning systems. The blimps offer large surface space for radio equipment to be installed. Also, they can carry large payload of electronic equipment meant for surveillance of enemy missiles, aircrafts and submarine. (Sputnik News, 2018; Vaeth, 1992). These blimps are being designed and developed by Augur-RosAero Systems Inc., Moscow, Russia.

Russian military has utilized blimps since early 1900s. However, most recent report is about a hybrid blimp. It is a combination of blimp, copter and airplane. The blimp, 'Atlant' is a hi-tech 130 m long bodied blimp. There are actually two versions of 'Atlant' that is being produced for Russian Military. Larger model of the blimp travels at 86 kmph and the smaller one at 104 kmph. The smaller swifter one carries 16 tonnes cargo. The larger version carries 200 military personnel or 60 tonnes cargo. 'Atlant' uses sophisticated high precision computer technology, to control its lift-off, flight path and operate surveillance cameras. This blimp is to be released for use in 2018 (Stewart, 2018; Gertcyk and Lamble, 2016).

Reports clearly state that, at present, Russia is adopting many of Soviet era Airships (blimps), both in military and civilian tasks. However, recent models of blimps have also been inducted in good number. For example, Dozor-class blimps and DP-Unmanned blimps are being readied for use in many ways (Sputnik News, 2016). There are clear reports that Russian Airships are also operating in the hostile, cold conditions of Arctic zone. The aim is both, to further the military aspects and explore natural resources, like minerals, oil, etc. in the Arctic (Beckenhusen, 2016).

3.2 BLIMPS: THE AERIAL VEHICLES

3.2.1 WHAT ARE BLIMPS: DEFINITIONS AND EXPLANATIONS

Blimps are defined as a kind of airborne vehicles that levitate due to the pressure of 'lifting gas.' A blimp is generally filled with a lighter-than-air gas such as helium or coal gas. It is a non-rigid vehicle. Hence, it collapses, if it gets deflated. This is unlike zeppelins which have rigid internal framework (Gardner, 2015). Unlike semi-rigid and rigid airships (e.g., Zeppelins), blimps rely on the pressure of the lifting gas, usually helium rather than hydrogen, inside the envelope. Helium offers strength and shape to the blimp via internal pressure (Gerke et al., 2013; Wikipedia, 2017). Regarding zeppelins, they are also grouped as airships. The hull of zeppelins is provided with a rigid skeleton of aluminum or wooden framework. The envelope of a zeppelin is filled with a lighter-than-air gas such as helium. (Gerke et al., 2013; Figure 3.1; Plate 3.2).

There are several ways a blimp could be defined, described and classified. A blimp is generally grouped under lighter-than-air aerial vehicles. A blimp essentially comprises of a balloon like body made of tough fabric. It is called the 'Envelope.' It has a burner or a helium supply to keep it aloft. For field operations, blimps have a set of sensors for aerial photography (Krishna, 2019) and/or chemical detection probes (electrodes) to assess atmospheric composition (Gerke et al., 2013). The payload area in a blimp has essential computer processors and remote-control accessories (Carrivick et al., 2013). The sensors usually include cameras, namely visual bandwidth (R, G, B), infra-red (thermal) and red-edge cameras, to assess temperature variations. It has a gimbal that allows manipulation of angles and focusing of cameras. The cameras can be remotely controlled using ground control station (GCS) computers.

A blimp is usually an unmanned aerial vehicle (robotic). It could be entirely autonomous, if its flight path is pre-programmed or semi-autonomous if remote controlled, using radio signals or tethers. Blimps could also be manually piloted. A few types of blimps are tethered and moved across fields. Usually, 500 m rope is needed to tether the blimp considering the wind and air pressure on the blimp. This is to facilitate accurate placement of blimp to focus cameras and obtain stereoscopic images. Aerial photographs obtained using a blimp may cover 35m^2 to 20,000 m^2 depending on the altitude.

Adoption of blimps is dependent on general logistics, cost of the UAV, resolution of the imagery and purpose for which the photos or digital data are utilized. They say, there are blimp models that cost less than 1500 US$ per unit. Blimps are preferred when longer endurance in the air is needed, say 1–3 weeks. A recent google-based analysis by geoscientists at University of Leeds suggests that, blimps are among the cheapest methods to conduct aerial survey of terrain. Using blimps, we can obtain details regarding natural resources such as vegetation, crops, river water, surface minerals, etc. They have been adopted as UAVs since past 10 years. However, here, we should note that, small fixed-winged or copter drones are gaining in popularity and usage during past five years (Carrivick et al, 2013; see Krishna, 2017, 2019).

Overall, there are two kinds of floating lighter-than-air (or LTA) crafts: They are balloon and the airship. A balloon is an unpowered LTA craft that can lift. An airship is a powered LTA craft that can lift and then maneuver in any direction against the wind (Bellis, 2017). Unlike hot air balloons, airships have solid gondolas, engine-powered propellers and solid tailfins. There are three types of airships (see Table 3.2):

- **Rigid:** The rigid airships are usually long (greater than 360 ft./120 ft.). They are cigar-shaped with an internal frame.
- **Semi-Rigid:** The semi-rigid airships are a pressurized gas balloon or envelope attached to a lower metal keel.
- **Non-Rigid or Blimp:** The non-rigid airships (i.e., blimp) has large gas-filled envelopes, a good example being the 'Goodyear Blimp.'

Advantages and disadvantages of blimps need to be considered prior to opting an autonomous blimp (UAV) or a semi-autonomous one, for different purposes. A list of simple advantages attributable to blimps are as follows:

They are capable of carrying heavy loads up to 400 ton. They can reach cruising speed of 110–160 km/h. Blimps have very long autonomy, i.e., an endurance of over 15 days flight. They can transit long distances of several thousand kilometers. Blimps are functional in all-weather conditions, similar to cargo planes. They have 2–4 engines that are powerful and quiet in opera-tion. Blimps can lift-off or land from any rough surface on land. We can load and unload blimps anywhere, including at sea. Blimps are transformable into robotic drone for unmanned operation. They have detachable cargo-hold. amphibious and with crane fitting capability. Blimps need Low operating costs (50 to 500% lower cargo air service); Blimps cause reduced amounts of CO_2 emission. The environmental impact is 6 to 10 times less CO_2 than a plane of equal capacity (Seneport, 2018; EuroAirships, 2018a).

TABLE 3.2 A Few Useful Terminologies Relevant to Airship/Blimp Technology

Airship: Initially, the word 'Airship' meant any flying machine. Some called them 'Air Yatches.' Around 1930, large intercontinental blimps and zeppelins were called 'Airships.' At present, Airship denotes any lighter-than-air balloons or dirigibles. They could be rigid airships (zeppelin), semi-rigid or non-rigid (blimp).

Aerostat: Aerostat is an airship that stays afloat due to buoyancy or static lift. On the contrary, aerodyne is an airship that stays afloat due to dynamic movement in the air. Aerostats could be free floating or moored (i.e., tethered) balloon. Currently, aerostats manufactured in different nations can carry up to 1500 kg payload. They could be floating at high altitude of 3.5 km above ground surface. Aerostats can hold telecommunication equipment in the payload area. They are often used to measure weather parameters at different altitudes and relay them to meteorology stations.

Dirigible: The word 'Dirigible' was derived from French words Dirigeable or dirigible' balloons. It means that they are steerable. The word 'Dirigeable' was first used by the French inventor Henri Giffard, in 1852.

Blimp and Zeppelin: Blimp is a non-rigid airship. They are also referred as non-rigid aerostats. Whereas, the rigid version with aluminum frame is called the Zeppelin (Figures 3.1 and 3.2).

Hybrid Airship: Hybrid Airship is the term used to denote the airships that adopt a combination of airship technologies such as blimps, hover crafts, copter, airplanes, etc.

Tethered Blimps: These are tethered airships. They are of different sizes and the tether is usually more than 400 min length. A tethered blimp could be held fastened to a location or moved. Tethered blimps are also often mounted on a pick-up van and moved on to different locations.

Untethered Remote-Controlled Blimps: These are blimps with requisite telemetric connections via radio or internet connections. They possess a CPU that decodes the instructions sent from ground control station (handheld remote controller or iPad). They are *not* tethered. Instead, they are free-floating vehicles in the air. Their path could be pre-programmed. Then they are entirely autonomous and are called unmanned aerial blimps or vehicles, i.e., they are aerial robots.

Source: Wikipedia 2017, https://en.wikipedia.org/wiki/Airship; Lockheed Martin Inc., 2018; https://www.lockheedmartin.com/en-us/products/hybrid-airship.html; Egan Airships Inc., 2018; Airships. https://plimp.com/about-egan/; Boyle, 2018.

3.2.2 COMPONENTS OF A BLIMP

The major components or sections of a blimp are: the large balloon-like envelope; a gondola which is the equivalent of a pilot's cockpit in an airplane; and an engine that could be located either in the gondola or anywhere convenient and attached with envelope. The envelope is made of special texture. The envelope is filled with a lifting gas, usually helium. The envelop has fins at the rear-end. The fins help in stabilizing the blimp, in the sky (Figure 3.1).

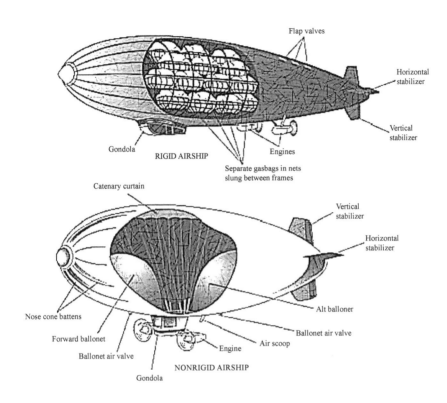

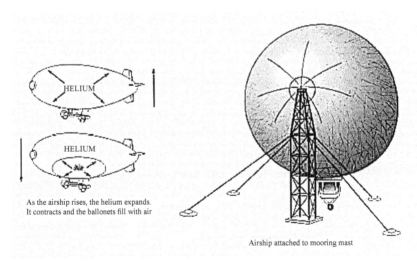

FIGURE 3.1   Parts of an Airship.

Source: Cynthia Hess and Edward Markey, Goodyear Inc., Akron, Ohio, USA and Advameg Inc. http://www.madehow.com/Volume-3/Airship.html.

- Lifting gas: Initially, hydrogen was used as the lifting gas. Hydrogen has been discontinued because it is inflammable. Instead, helium is being utilized as lifting gas. Helium is non-flammable and has 92% of the lifting properties of hydrogen.
- Gondola: It is the crew area in an airship. Gondola is usually located beneath the otherwise large envelope. It could beat the center or in front of the body envelope. Gondola could be tightly fitted to envelope or sometime slung a few feet below the blimp's body. Gondola often carries electronic payload, radiometry accessories, sensors to obtain visual, infra-red and lidar images. They may also carry equipment to record parameters relevant to atmospheric conditions and general weather. Blimps used for travel usually accommodate passengers in the gondola area. Gondola is converted aptly for luxury travel and excursions in the sky. It depends on the pay load permissible. Military blimps carry ammunition, supplies and personnel in the gondola.
- Blimps derive propulsion through IC engines that use petrol or diesel fuel. There are solar blimps too that derive energy from solar panels (Sonmez, 2015). Smaller blimps house the engines, usually 2–4 in number in the gondola. A large blimp that requires higher power houses its engines, in a separate chamber close to gondola and envelope. They are called 'engine car.' The engines are connected to propellers that provide the thrust. Propellers help in achieving suitable altitude and direction of movement in the air (Mowforth, 1991; Nitherclift, 1993; Sonmez, 2015)

3.2.3   AIRSHIPS (BLIMPS AND ZEPPELINS): THEIR PARTS IN DETAIL

A blimp's body is made of an inner layer, a bladder and an outer envelope. Bladders contain one or more cells or ballonets. The ballonets are filled with air, not helium. Usually, they are attached to the side of blimp. The ballonets can be controlled through air valves. The blimp has a nose cone. It helps as a point of attachment to the 'mooring pole.' It also adds to rigidity and strength of the nose region of the blimp. Inflated Blimps are held on the ground by tying it to a pole, i.e., 'mooring mast.' Blimps fastened to mooring mast may move a little, as and when wind lashes on the body.

The lighter-than-air helium gas is filled into bladder. There are catenary curtains and a cable system inside the envelope. This is to support the weight of the car (gondola). The shape of the envelope is dependent on pressure of helium gas.

The blimps have tails. The tails could be made in an any of the three different forms, namely cruciform, the 'X' or 'Y' form. Tails are made of thinner fabric. They weigh about 4.4 kg m². Tail fins help in controlling the direction of blimp's flight. It helps to maintain a correct flight path as determined by computer or controlled manually, using a remote controller. The tail fins are usually attached to the rear end via guide wires (strings). We should note that elevators and rudders also help in controlling the flight path and speed of the blimp.

The car or gondola of the blimp contains instrumentation, engines, computer unit and seats for pilot and passengers. Gondola is attached to the envelope. Gondola is similar to conventional aircraft in construction. Inflation of blimp's bladder is a key aspect of its preparation for flight. Inflation usually takes only few minutes. Helium gas is field into bladder from helium containing tanks situated on trucks. Pure helium (2100 psi) is filled into the bladder. Due care is taken to see that parts such as tail fins, nose cones, battens, air valves and helium valves are well fitted prior to inflation, i.e., erection (Advameg Inc., 2018a,b; Botting, 1980; Plate 3.2; Figure 3.1).

PLATE 3.2   A Blimp – it's typical exterior morphology.

Source: Cynthia Hess and Edward Markey, Goodyear Inc., Akron, Ohio.

Note: The above picture depicts external appearance of a simple blimp.

3.3   TYPES OF AIRSHIPS

During the past century, efforts to use blimps for different purposes such as in military, travel and cargo transport, etc. have fluctuated. There are periods when research to improvise blimps and use them has been conspicuous. Periods with reduced interest in blimps were also perceived. During this period really, wide range of blimp models have been designed. Pilot models have been tested stringently. Several of them have been even adopted to conduct different tasks. Specifications of blimps designed and manufactured vary enormously. Blimps can be classified based on following criteria. They are:

- Size: small, medium large or giga;
- Shape: variety of shapes such as oval, oblong, round, and fancy shapes too;
- Rigidity: non-rigid, semi-rigid and rigid (zeppelins);
- Altitude: low, medium and high;
- Filling gas or lighter-than-air gas (helium, coal gas or hydrogen, but note that hydrogen usage has been discontinued now);
- Power: steam-engine, petrol/diesel engine, electric engine, solar powered engine);
- Tethering: below the envelope, in front, center or rear end of envelope, above envelope (not common);
- Control of navigation: remote controlled (radio control, internet/GPS, autonomous (robotic);
- Autonomy: manual, using tethers and pick-up vans to move the blimp; semi-autonomous (i.e., piloted), entirely autonomous (flight path pre-programmed using computer programs such as *e-Motion*;
- Landing: water, rough ground surface anywhere, paved tarmac, sand, ice, snow mountains, etc.;
- Endurance: short (few hours), medium (1–3 weeks) and long (few weeks);
- Purpose: Most blimps could be easily classified based on purpose they serve. Broadly they serve the military in aerial reconnaissance, surveillance of installations, missiles, military cargo transport, etc. Regarding civilian uses, they serve wide range of civilian tasks such as travel, recreational travel, cargo transport, policing, advertisement, agriculture, etc.;
- Cost: low cost simple blimps, sophisticated high cost blimps for military, travel, and cargo.

Now, considering the focus of this book we should be concerned more about the blimps or models that are utilized for variety of agricultural operations and their immediate classification. First of all, we should realize that there are no blimps classified as 'agricultural blimp.' Perhaps we have not felt the impending need to group a certain set of blimps as agricultural blimps just because, they are more frequently adopted during farming. Farming, in fact, right now does not rely to any great extent on blimps and data derived by them or even for spraying or surveillance of crops, etc. However, it seems inevitable that, in near future, may be couple of years hence, we could be encountering terms such as agricultural blimps, sprayer blimps, aerial photography and field/soil mapping blimps, or even precision farming blimps. The precision farming blimps could collect detailed data showing variations in soil fertility, crop growth, pest/disease attacks, weeds and water status. 'Agricultural cargo transport blimps' that supply agricultural inputs at the field locations and those that transport fresh perishable vegetables, fruits, and flowers may become common in due course. Clay and Clement (1993) examined the feasibility of using blimps to carry the forest products such as logs from one location to another. It seems they are economically efficient compared to road-based methods. However, their evaluations suggested that blimp model has to be modified to lift a log of 5-ton weight efficiently, quickly and transport them safely.

There are several types and models of airships that have been designed, then, developed for testing and large-scale use. Many of them have made it to practical use. They have been utilized during travel of personnel, transport of cargo, and in aerial photographic assignments. A few have passed the test a bit better and got adopted during expeditions into remote regions such as Arctic, Antarctic or to float over volcanoes. A few other types of blimps have been utilized in long distance transport, reconnaissance and in networking of radio signals, etc.

Let us consider Airships, in general. Airships have been modified to improve their versatility and serve the purpose better and accurately. Broadly, 'Airship' is the terminology used for a group of larger balloon-like vehicles. They float in the sky. As stated earlier they are all called lighter-than-air vehicles, because, they are filled with lighter gases such as helium. Airships, in general, are classified as rigid type (e.g., Zeppelin NT); semi-rigid type (e.g., Wingfoot One by Goodyear Inc.); and non-rigid airships (Blimps).

Rigid type airships are filled with lighter-than-air gas such as hydrogen, helium or coal gas. The balloon body is encased in a rigid aluminum or wooden framework. The framework could be external or placed internally. The framework supports and offers a rigid body and shape to the airship. Zeppelins produced by Germans during World War II were rigid airships. Hindenburg that perished in 1937 in New Jersey, USA was a good example of rigid airship – a Zeppelin.

As stated earlier, non-rigid airships are commonly known as blimps. They possess large balloon like body made of strong cloth or poly-urethane material. the poly-urethane material does allow gases to leak, but negligibly. The blimps are usually filled with lighter-than-air gas such as helium. Helium is lighter than air. Plus, it is non-flammable. Hence, it is preferred. A few decades ago, they were filled with hydrogen. Hydrogen is also a lighter-than air gas, but it was prone to accidents. It is dangerous because hydrogen is highly inflammable gas.

Blimps stay rigid and do not loose shape because of the internal pressure of gas filled into balloon-like body. Blimps collapse, if deflated. There are indeed several brands, models and sizes of blimps being produced worldwide in different countries. Powered blimps have been in vogue since the development of steam engine supported blimp demonstrated, by Henri Giffard in France, in 1904. The 'Goodyear Blimp' is an example of non-rigid airship. Blimps operated currently differ with regard to traits such as material used to build the balloon like body, shape, size, weight, tethering if adopted, altitude of operation, speed of transit, autonomous or semi-autonomous nature of control of the blimps during transit, purpose and area of operation, endurance of flight, fuel used for powered blimps, etc.

Airships are also classified using different criteria. Airships were initially grouped based on size, power source, speed, altitude, sensors and purpose they serve. Airships were designated by American Military establishment as different 'Classes' of blimps. They were grouped into 'G' class, 'K' class, 'N' class, and Zeppelins.

- G-class Blimps: According to reports, G-Class blimps are a series of non-rigid airships that are in use by military in United States of America. The US Navy obtained Goodyear blimp to train blimp pilots and military personnel. They designated this class of Goodyear blimps as 'G-Class Blimps.' Earliest of G-Class blimps was named 'Defender.' It was a relatively larger blimp. The G-Class blimps used by US Navy has been generally large. They have served them in training personnel in blimp technology (Shock, 2001; Althoff, 1990). The G1 (Defender), first blimp in the series was in use by navy from 1935 till it was lost in the mid-air. To augment training needs, US Navy had bought a few more G-class blimps. They were named Goodyear ZNN-G (Z = lighter-than-air, N = nonrigid, N = trainer, G = blimp type or class). The series was designated as G-2 to G-5. The envelop size of G-class blimps is 390 m^3 Specifications of G class Blimps: Length: 57 m; Diameter:

13.1 m; Height: 19 m; Volume 18 m³· Pay load: 1867 kg; Power: 2 generators of 210 hp each; Crew 2 or 3; Maximum speed 92 kmph; Cruise speed: 77 kmph; Endurance: 16–17 hr; Total passenger capacity 7–8 (Wikipedia, 2018a);

- N-class Blimps: Again, N-class blimps were a series of airships manufactured by Goodyear Inc., Akron, Ohio, USA, for the US Navy. There were many versions of N-class blimps produced prior to World War II. Many were later re-designated in 1954. Several prototypes of N-class blimps were produced in 1940s. They were designated, initially, as ZPN-series. The N-class of blimps are predominantly used for aerial surveillance. The ZPN blimps are filled with helium gas. Specifications of N-class of Blimps: Length: 105 m; Diameter 23 m; Height: 33 m; Volume 29 m³, Power *Source:* 2 x Wright R-1300 engines that generate 800 hp each; Maximum speed: 128 kmph; Endurance: 200 hrs; crew-9–12.

- K-class Blimps: The K class blimps are a set of non-rigid airships. Again, many are built by Goodyear Inc of Akron, Ohio. These blimps are powered by petrol engines, to support propellers. It seems, about 134 K class blimps were manufactured prior to World War II. The K class blimps were first released for military activity, in 1938. Several more K-class blimps were added between 1938and 1942 (Stubblebine, 2018). Most of them were in opera-tion, mainly, to surveillance the submarines. The original designa-tion of K-class blimps has a suffix and a number. For example, ZNP-K-3, ZNP-K-18 etc. For daily use, the designations used is 'ZNP-K series.' The K-class blimps were improved periodically, regarding several aspects of material and operation. The sensors such as visual, infrared, lidar, magnetometers are vastly improved, if compared to early models.

At present, the K-class blimps are capable of flying low over ocean surface. So, they are good to detect submarine activity. Reports suggest that, K-class blimps were used effectively over Atlantic, Pacific, and Mediterranean oceans, to detect submarines (Vaeth, 1992; Althoff, 2009; McNally, 2013).

Specifications of K-class blimp: Length: 77 m; Diameter: 17.6 m; Volume: 12 m³; Maximum speed: 125 kmph; Cruise speed: 93 kmph; Range: 3537 km; Endurance: 38 hr 12 min; Payload: 3524 kg; Power Source: Prat and Whitney R 1340-AN-2 radial, 425 hp each; crew 9–10 (Wikipedia, 2018b)

Zeppelin Class: The zeppelins are a class of airships that are rigid due to a frame made of aluminum. This class of rigid airships was produced by Zeppelin Company in Germany. This company was founded by Ferdinand von Zeppelin. This class of airships were designated as LZ-class (L= Luftschiff, Z = Zeppelin). The civilian zeppelins were given specific names. The military zeppelins were just given series number and code numbers. During World War I, zeppelins from LZ-1 to LZ-25 were utilized. There were many unsuccessful zeppelins in the LZ series. Reasons were mostly lack of funding or sometimes fault in design. Also, vulnerability to enemy attacks the LZ series up to 114 were constructed during World War I. The later numbers were not done because of the treaty to stop end the war. The super-sized zeppelins were named 'Super Zeppelins' or 'S-class.' LZ-127 named by Graf Zeppelin was among the successful airships of Germany (Wikipedia, 2018c).

Worldwide, there are several models of blimps (airships) manufactured by companies and institutions. Each model comes with vastly improved electronics and capabilities for payload, cargo transport, and aerial survey possibilities. The sensors that are crucial components for spectral analysis of agrarian regions may also differ. No doubt, each model is vying for popularity with the farming community. The top blimps being produced and sought by agencies specializing in UAVs, aerial survey, and transport may change with time. The requirements of farmers regarding type of blimp, its cost, cost per flight, and aerial photos may vary. Accessories, repair, and maintenance also count. The top ten blimps, therefore, change depending on popularity with farming and urban agencies. For example, Newitz (2018) has listed the following companies, brands, and models of blimps as top ten accepted by users. They are shown in Table 3.3.

TABLE 3.3   Ten Top Blimps and Zeppelins Accepted by Clientele During Recent Years

Goodyear: It is one of the earliest blimp models sold since early 20th century. It is among the most popular blimps utilized in advertising. They are frequently seen in urban regions carrying banners and painted with different advertisements. They are also adopted to survey natural resources and rural regions. This blimp dates back to the early twentieth century. It has been used in USA for advertisement and to carry passengers.

Airship Ventures: This company produces 'Zeppelin NT Eureka.' The Zeppelin is helium filled for floating. It is made of rigid outer body that is tough. It is powered by 3 petrol/diesel engines. Yet, it is a quite (no noise) zeppelin. It could be used to carry passengers, cargo or to conduct aerial surveys.

SkyLifter: SkyLifter is an Australian company that produces heavy lifter zeppelin. It is apt when large sized and heavy cargo is to be lifted and transported, say, for short distances of 5–20 km.

TABLE 3.3   *(Continued)*

Nephelios: This zeppelin is frequently adopted by research personnel in institutions of France. It is a medium sized zeppelin at 72 x 18 ft. This zeppelin has been utilized to fly cargo across English Channel. In due course, it could be utilized to fly over agrarian regions, to conduct regular aerial survey of natural resources and asses crop growth and nutrient status. Also, to surveillance crops for disease/pest attack, drought, soil erosion, etc.

Blue Devil: It is a blimp produced by a company situated in North Carolina, USA, known as MAV6 LLC. It is being produced for use by US Air force. Particularly, for use in Afghanistan. It is a large sized blimp at 350 ft. length. It is used for aerial transport of large sized cargo.

Bullet 580: It is produced by a company known as E-Green Technology. This company is situated in Montgomery, Alabama, USA. It is a large sized zeppelin with a length of 230 ft. The zeppelin is said to serve military and private civilian purposes. Perhaps, it could find acceptance in agrarian regions too (The Christian Science Monitor, 2018; Hannaford, 2018).

Strato Cruiser: It is a rigid airship – i.e., a zeppelin made of carbon fiber and tough fabric. It has a viewing deck. It can carry several sensors for aerial photography.

Aeroscraft: It is produced by Aeros Inc., Montibello, California, USA. It is a rigid bodied zeppelin. It is utilized in advertisement, aerial surveillance and obtaining imagery of natural resources and terrain.

Firecat: It is a luxury blimp. It is produced by a British company – 'SkyCat Ltd.' It has capacity for large payload of 20 tons of water or any other cargo. It is useful to military and in answering disaster relief calls to quell fire by applying large volumes of water. It could be used for natural resource survey and spectral analysis of large areas of cropped field. Perhaps, it could be used to recover drought affected patches of cropped land. Blimps could sprinkle the life-saving irrigation water aerially on seedlings.

Source: Newitz, 2018; Schachtman, 2017.

Note: Most of the above popular blimps/zeppelins listed are large sized. They are suited well for general aerial survey, military cargo transport and other industrial purposes. They could be adapted to suit agricultural aerial survey and cargo transport. Perhaps, we should also consider smaller or very small blimps for aerial photography of farms. Agricultural cargo transport, if it involves small pay load then smaller blimps or zeppelins might be needed. Small blimps may have greater acceptance due to their easy maneuverability, easier remote control, low altitude floating and most importantly high-resolution imagery. They may also cost mush less than a sophisticated large blimp. We may note that a few hybrid blimps that are small are coupled with copters. Then, they are utilized for low altitude aerial photography and in spraying agrochemicals.

3.3.1   TETHERED AND UNTETHERED BLIMPS

A tethered balloon or blimp is probably the simplest UAV conceivable. It is easily controlled, especially its altitude. But it is quite unstable if the wind speeds are beyond threshold. The size of balloon can be selected to suit the volume and weight of the instruments that need to be carried. Untethered free-flying blimps are offered to the users as platform for monitoring and survey (Martinez Rubio, 2005; Vierling, 2006).

Tethered blimps are utilized to surveillance ground surface features from different altitudes. The sensors attached has to be appropriate to the desired level of resolution and clarity. Such blimps are usually small or medium in size. They possess a set of useful sensors such as visual (red, green and blue wavelength band), infra-red, rededge, lidar. Sometimes, tethered blimps may carry chemical sensors, particularly, if it is to be used, to assess gaseous composition of atmosphere above a geographic location. Such blimps could be tethered and tied to a location. Tethered blimps could also be moved using a pick-up van (Plate 3.3). In this case, the blimp is tied to a pick-up van. Blimps could be free without any tethering. Such blimps could be powered and controlled using remote control signals (radio network). Un-tethered blimps could be semi-autonomous, if it is piloted. Otherwise, they could be entirely autonomous, if its flight path and way points are pre-programmed, using a computer program (e.g., *e-Motion*) and GPS connectivity.

As stated earlier, tethered blimps have certain advantages when used in the realm of advertising (Editor 2018c). They are also useful in many ways in the agrarian regions. Farmers can rely on them to collect crucial data about crops. In fact, a tethered blimp could be floated at early seedling stage of the crop. It could then be allowed to last in the air until grain maturity/ senescence. During this period, the blimp could collect vast amounts of data about the crop, its growth pattern, and grain maturity trends. Tethered blimp can be fastened to a ground point and allowed to stay for weeks/months in sky with advertising panels showing up. They also act as good sentinels. Blimps could be allowed to float over large buildings, public places or above military installations, etc. as sentinels. A few of the models of tethered blimps are provided with remote control option so that, within limits we can move them. These tethered blimps could also be fastened to pick-up vans and moved across different locations (Plate 3.3). Now, what are the minimum specifications, if one wants to hoist a tethered blimp and get the requisite aerial images of say crop fields, a water storing reservoir, an agricultural experimental station or a grain storage zone. In order to use a tethered blimp efficiently in different settings, we should first have a good knowledge about all parts, methods of floating it and clear knowledge about specifications of different parts. Let us consider an example. A simple tethered blimp will have the following specifications and options that are extra. All standard tethered blimps packages include (see Editor, 2018c):

- High quality urethane balloon of different colors;
- Light weight carbon fiber and foam fins with hinged rudder;
- A remote-control system (iPad) or a radio control system;
- Nylon rope to tether the blimp;
- It is usually 135 ft. in length;
- Lithium polymer batteries and recharging system;
- A repair kit to seal the holes, if any;

An advertising blimp has a mechanism for dropping cards, envelops, and gift cards from air. Also, a set of spare parts.

There are several optional items that could be attached to tethered blimps. If used in agriculture, foremost, a set of sensors to get spectral data such as NDVI, leaf chlorophyll index, crop water stress index, i.e., IR data, etc. are included. If adopted in advertising, then, printed banners, digital slabs, LED lamps, electrical tether line and are provided. A pickup van to fasten tether and move the blimp to different locations is generally provided.

Now, let us consider possible utility of tethered blimps in agricultural crop production. This is a specific example. Recent report states that tethered blimp has been deployed exclusively for physiologists at the International Center for Maize and Wheat (CIMMYT), Mexico. Tethered blimps are utilized to observe and record crucial data about wheat and maize crops (Tattaris, 2014). The blimp stays at 700 ft. above the crop's canopy. It records data such as NDVI (crop growth), green index, i.e., leaf's chlorophyll content, crop's-N status, water stress index, maturity of grains, etc. They say, blimps enable physiologists to take data from large areas that previously had to be covered on foot (Tattaris, 2015; see Figure 3.17),

Tethered blimps have a great role to play in the surveillance of large fields, crops, irrigation set up and farm vehicles. Also, to note the progress of agronomic procedures in a farm. Tethered blimps could be used as sentinels above farms and structures, all through the day and night. Tethered blimps could be efficient in collecting spectral data and obtaining aerial photos (AeroDrum Inc., 2018; SkyDoc Aerostat Systems, 2018). As stated above, sensors on the tethered blimp can offer excellent spectral data about variations in crop growth, leaf chlorophyll (crop's N) status, water stress index, grain maturity and yield, etc. Tethered blimps could be used to monitor farm vehicles, farm animals and movement of trucks within and outside the large farm. Adoption of blimps by the International Center for Maize and

Wheat (CIMMYT) in Mexico and in Zimbabwe is a good example (Tattaris, 2014). They have adopted tethered blimp to make aerial spectral analysis of wheat/maize crop at various stages. They have also used blimps to assess the performance of several genotypes of wheat/maize for traits such as growth, N-uptake, water stress index, i.e., drought tolerance, disease/pest tolerance, grain productivity, etc. Blimps could eventually be helpful to plant breeders during genetic evaluation, selection and in assessing the results of crossing programs.

PLATE 3.3   A tethered blimp being transported using a pick-up van, for launch from a different site.

Source: Mijatovic, A., AeroDrum Ltd., Belgrade, Serbia.

Note: Such tethered blimps could be transported to different farms and hoisted, using a tether rope of appropriate length. Then, they could be utilized to draw useful data about crops.

Tethered blimps could be useful in many ways. For example, United States Department of Homeland Security adopts about 8 or even more of large tethered blimps at a place called Yuma, in Arizona. The tethered blimps are held lofted at 10,000 ft. above ground. They carry a series of radars and sensors. The entire set of installations is called 'Tethered Aerial Radar System (TARS).' These blimps supposedly surveillance the southern borders of USA, for trespassing, smuggling and other infringements. These tethered blimps are unmanned. They stay afloat but are moored to ground through strong nylon ropes. They conduct high altitude surveillance.

In 1978, the U.S. Air Force set up the first TARS site in Cudjoe Key, Florida. A second TARS went into service in 1983 at Fort Huachuca, Arizona. From 1988 to 1991, the U.S. Customs Service established more TARS sites at Yuma, Arizona, and three sites in Texas, including Marfa, Eagle Pass and Rio Grande City. By the end of 1994, additional TARS balloons were floating in Florida, Texas, Puerto Rico and even the Bahamas. The U.S. Air Force managed the TARS program until July 2013.

TARS is the only persistent wide-area air, maritime and land surveillance system. It is designed specifically for border security mission. Despite their effectiveness, unmanned aircraft systems are not designed for the same mission. "We cannot have enough UAS's carrying enough radar to duplicate the persistent ability of TARS to detect low-flying aircraft for 200 miles," (Neaves, 2014; US Department of Homeland Security-Science and Technology Directorate, 2018; Sutton, 2008). In 2013, TARS was responsible for detecting 586 suspicious flights, representing 42% of all the suspect flights along the Southwest border tracked that year, in USA.

They are also testing the potential of a family of smaller blimps, for other roles in border security. Recently, tactical blimps/aerostats have been transferred back following their use by U.S. forces in Afghanistan. The blimps include three models: The Persistent Threat Detection System; the Persistent Ground Surveillance System, and the smallest, the Rapid Aerostat Initial Deployment system. The smaller blimps operate at altitudes from 500 to 5,000 feet. They monitor ground activity with radars and electro-optical cameras. We may note that, data collected via TARS is usually shared by a network of other blimp stations.

SkyDoc Aerostat Systems offers tethered blimps with a minor modification that suits specific purposes during aerial survey. The tethered blimps resist wind disturbance better. However, it is interesting to note that this blimp system is sold along with two small fixed-winged drones, to conduct short duration surveillance. The blimp with its long endurance provides imagery continuously, for more than 2–3 weeks (SkyDoc Aerostat Systems, 2018; US Department of Homeland Security-Science and Technology Directorate, 2018). The blimp is provided with a slot to hold small drones. These tethered blimps could be mounted on trucks and flown to different sites. The blimp can be either networked or used singly to provide detailed aerial imagery (Plate 3.4).

PLATE 3.4   A tethered blimp fastened to a pick-up van.

Source: Dr. Kevin Hess, CTO, Drone Aviation Holding Company, Jacksonville, Florida, USA.

Tethered blimps can spy as efficiently, as rapidly and efficiently when compared with small drone aircrafts. The tethered blimps could easily capture every aspect of ground surface and happenings in the city below (Farivar, 2014). Patrolling the sky near major installations is another possibility using tethered blimps. They say, drone aircrafts are problematic. They could crash. There are many FAA regulations to follow. Also, license is needed for the drone flier. Such restrictions are low in cases of tethered blimp. If a tethered blimp is held below 500ft., then, FAA rules are less. This is because and they are deemed as 'moored aerial vehicle.' Tethered blimps have been adopted by police department in California and Tennessee (Farivar, 2014; Booth, 2018; Figures 3.3, 3.4, and 3.13). The initial cost of tethered blimps could be high. Right now, the market is at low level. However, there is every reason for tethered blimps to be most common, both, in urban locations and farming regions.

Let us consider yet another example of tethered blimp. This time, it is a tethered blimp that is predominantly used for aerial imagery of cities and adjoining area. The 'Photo blimp Inc.' is a company that is operative since 2007. It offers aerial photos of Vancouver suburbs in Canada. The blimps have long endurance of 8 hr., once in flight. It has a set of visual, infra-red and lidar sensors. It could be tethered to ground or on to a truck (Photo blimp Inc., 2018; Plate 3.5).

Untethered blimps are utilized for a wide range of purposes. They are usually powered using gasoline engine of different horsepower. Untethered blimps are used in recreation, travel, aerial observation and photography, spectral analysis, surveillance of mines, industries, public events, buildings, water bodies, irrigation channels, etc. Untethered blimps could be of different size, shape, material, and accessories depending on purpose. Larger blimps are used in travel and expedition. Untethered blimps could be entirely autonomous and programmed, using computer software such as Pixhawk, e-Motion, etc. AURORA I AS800 by Information Technology Institute, Campinas, Brazil; Anabatic Aero Blimp by Anabatic S.a.r.l., Switzerland; Solar Blimp by AeroDrum Ltd, Belgrade, Serbia; and Wingfoot 1 by Goodyear Inc., Ohio, USA are good examples of untethered blimps. Untethered blimps are highly versatile and possess long endurance. For example, they could be in the sky for 3–4 weeks and travel long distance. They were used for transatlantic routes long ago, in 1920s and 1930s.

PLATE 3.5   A tethered blimp being flown above a city, for Aerial Imagery.

Source: Dr. Kevin Hess, CTO, Drone Aviation Holding Company, Jacksonville, Florida, USA.

3.3.2   REMOTE CONTROLLED AND PILOTED BLIMPS

Blimps are aerial vehicles that are amenable for control using radio links. The ground control stations should possess necessary receivers and instruments, to direct and regulate the speed/direction of the blimp. The radio signals are effective only to a certain distance/radius from the ground control station (GCS). Blimps could also be regulated using satellite mediated communication and we can adopt GPS guided transit of blimps. The GCS, in this case, usually has a laptop as remote controller. Let us consider a few recent models of blimps that are either remote controlled (unmanned) or those that are piloted. If piloted, navigation instruments are usually found in the gondola.

3.3.2.1   WINGFOOT ONE (BY GOODYEAR INC., AKRON, OHIO, USA)

Goodyear Inc., located at Akron, Ohio, in USA is among the most popular companies that manufacture blimps. They offer several models of blimps to serve different requirements of the clientele. Let us consider a recent model designed, produced and sold by the Goodyear Inc., USA. Their recent models designated as Wingfoot 1, 2 and 3 are available for use by the clientele. The production of the model Wingfoot 1 got initiated in 2013. This blimp is a product designed by engineers from both Goodyear Inc., USA and ZLT Zeppelin Luftschifftechnik of Germany. Parts such as tail fins, gondola and a few others are manufactured in Germany and shipped to USA for assembly at the Goodyear workshop. The balloon-like body of the Wingfoot 1 is produced in USA. The body is made of polyester from Dupont Inc., and is known as Tedlar. The internal structure is semi-rigid in case of Wingfoot 1. This is unlike the other non-rigid blimps manufactured by the Goodyear company. Wingfoot 1 is a good example for piloted and/or remote-controlled blimps (Goodyear, 2018).

Now, let us consider a few salient specifications of Wingfoot One (Goodyear, 2018; Figure 3.6).

Wingfoot 1 was released for commercial use in 2014. Its overall length is 246.6 ft. Maximum width of the body is 64.8 ft. Overall height is 57.6 ft. It has an internal structure made of carbon fiber plus aluminum. The blimp's body is made of polyurethane, polyester and tedlar. The volume of the blimps body is 297 cubic ft. The poly-urethane body is said to last for 10 years withstanding all the vagaries of weather. The total weight of the blimp without the lifting-gas, i.e., helium is 19,780 lb (8954 kg). The blimp, Wingfoot One, has speed limit of 73 mph. The inner gondola accommodates about 14 passengers. The blimp has three engines. Each engine generates a 200 hp power. The endurance of the blimp in the sky is 24 to 40 hr. The long endurance period will allow transatlantic travel for 2–3 times. Wingfoot 1 is generally a quiet vehicle in the air. Its noise level is about 64 decibels. This is unlike the drone aircrafts. The Goodyear blimps such as Wingfoot series are also produced by several other companies, of course, under license. Goodyear blimps are among the common and easily visible in the sky with hoardings about variety of products. They are also used very efficiently during aerial photography of indoor public events, sports events and other functions (Plate 3.6). Such Goodyear blimps are also common in the outdoor surveillance, aerial photography, travel and in expeditions to different continents. Wingfoot one could be used to surveillance farms, large fields and cropping systems. Also, to note down spectral data of crops, assess water stress suffered by crops, and assess nutritional requirements of crops, particularly, crop's N demand (leaf chlorophyll estimation). Wingfoot 1 blimps could also survey disaster prone regions for soil erosion, loss of river embankments, floods, drought, etc. Remote control system helps to steer the blimp to vantage locations above crops, urban sites, disaster zones, rivers, mountains, buildings, etc. The remote-control system allows us to steer the blimp, to vantage altitude and geographic location (Goodyear, 2018).

PLATE 3.6   Wingfoot One- A Goodyear blimp.

Source: Cynthia Hess and Edward Markey, Goodyear Inc., Akron, Ohio, USA.

https://www.goodyearblimp.com/behind-the-scenes/current-blimps.html.

Specifications of RC outdoor blimp produced by AeroDrum Ltd., Belgrade, Serbia (AeroDrum Inc., 2018; Plate 3.7) are as follows: It has one or two motors to support the propellers; The main motor is powered by 12,000–20,000 mAh Lithium polymer battery. The back-up is powered using Lithium Polymer batteries of 3000–5000 mAh. Envelope material is 125 μ polyurethane. Flight time is 60 min endurance depending on flight path and payload. Double welding technology is used to build the blimp. Wind disturbance limit is 9 m s^{-1}. Helium permeability is 0.5 to 1 % on total volume. Expected lifetime of the blimp is several years. Electronic components such as remote controller and iPad, CPU in the blimp are tested for reliability. The blimp is available in ready to use condition, except that helium gas has to be filled. The blimp model is FAA (of USA) compliant.

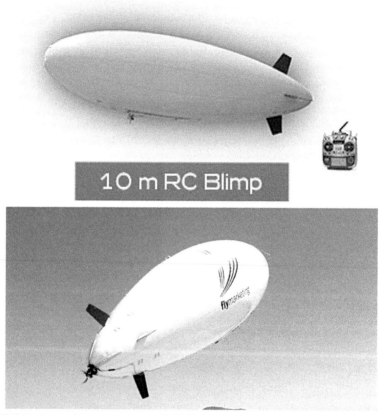

PLATE 3.7 An outdoor remote-controlled blimp.

Source: Mr. Mijatovic, A., AeroDrumLtd., Belgrade, Serbia.

Note: This is a 10 m long outdoor blimp produced by AeroDrum Ltd. of Serbia. It is one of the series of blimp models produced by the company. It is useful in obtaining aerial spectral data and photography. It carries a series of sensors such as the visual (R, G and B), infra-red, red edge, and lidar. The sensors are held in vibration-free, stable gimbal. It could be among blimps useful in assessing crops in different agrarian regions.

Now, let us consider a few other Remote-Controlled Blimps that are utilized in many ways. Remote controlled blimps are produced by several companies in China. They cater to a very large clientele. In that country, blimps serve several types of functions. They are: general aerial imagery, reconnaissance and monitoring activity above military and civilian zones. Blimps are also utilized in mapping natural resources, in monitoring rivers, vegetation and industrial sites, monitoring wildlife sanctuaries, etc. Specifications of individual blimp models may differ based on purpose it has to serve, the cost of material, durability, ease of operation, storage and economic gains anticipated, etc. Main markets for blimps turned out by the companies in Beijing or Shenzhen are actually located in North and South America, Eastern Europe and Southeast Asia. Each company may offer a range of blimp models, to suit specific purpose. A few of the models of blimps are exclusively utilized in agriculture and forestry. A blimp with long endurance and capability for continuous surveillance of land resources, crops, farm structures, etc. is useful in farming. In China, blimps are used to observe forest stand, logging, transport of forest products, etc. Following is an example of blimp model manufactured by a company in Beijing, in China (Beijing LONSAN United Aviation Technology Co Ltd. 2018).

RC AIRSHIP (LS-S1200): The RC Airship-LS 1200 is a portable blimp. It offers several advantages over others of similar size. It is a medium sized blimp. It cruises in the sky effortlessly, quietly (low noise) and safely. There are many safety features attached with this blimp. The manufacturer says its cost of operation is relatively low. 'RC Airship' is mainly utilized in obtaining aerial imagery of ground features, natural resources (vegetation, rivers, lakes, soil characteristics, cropping pattern, etc.). It is also used to monitor urban activities, public functions, parks, city traffic and in general policing of cities. These medium sized blimps are of great use in detecting disasters such as floods, drought, earthquake affected regions, large scale soil erosion, crop loss, etc. They could be used in monitoring oil pipelines, international borders, electric power lines, etc. Overall, such blimps could be very useful to city municipalities and state agencies, particularly, to draw aerial data rapidly.

Specifications of LS-S1200 (SP) RC Airship: Length: 12.9 m; Width: 4.2 m; Height: 4.7 m; Maximum diameter: 3.4 m; Ballonet volume: 80.0 m^3; Area for AD: 8.5 x 2.6 m^2 x 2; Maximum power: 7.5 x 2 hp; Gas Use Rate: 2.5 L h^{-1}; Speed: 0–80 kmph; Cruise speed: 20–30 kmph; Endurance: 2.0 h; Operating Wind speed: 12 M/s; Maximum payload: 28 kg;

Blimp manufacturing companies often turnout several different types of blimps. Mostly, they are specific to purposes such as sports and public event imagery, policing and law enforcement, military, cargo transport, personnel travel, etc. For example, a company in China caters blimps to agencies that conduct general aerial survey and offer imagery to clients. The same company also offers special types of blimps to suit policing and law enforcement, military surveillance, etc. A typical policing blimp would have specifications as follows:

Length: 11 m; Width: 2.6 m; Height: 3.8 m; Volume: 33 m³; Power plant: 62 cc, 5.4 hp; Fuel: gasoline; Maximum speed: 60 kmph; Cruise speed: 2–30 kmph; Maximum wind scale allowed 4–5 kmph; Pay load: 5 kg; Endurance: 4 hr.; Tail plane: cruciform; Remote control radius: 1.5 to 2 km from ground control center (Laptop computer) (Fei-Yu Aviation Science and Technology Co. Ltd., 2018). We may note that under each special type of blimp, these companies usually offer a series or sub-series of models with various modifications and sizes. This is to suit the particular function that blimp has to accomplish (EOMST, 2017; Fangzhou, 2018; Eblimp, 2018).

Fangzhou Ltd. is another UAV blimp related company situated in China. It classifies blimps based on the requirements of customers. They produce several different types of blimps (Fangzhou, 2018). Each type of blimp suits best for a specific purpose. They are:

1. A basic version that suits purposes like aerial advertisements. They are developed so that blimps could hover or stay still for general public to see the blimp, from ground locations. The specifications are: material: Polyethylene tetra-phthalate; Remote control range: 1.5 to 2 km radius; Operating altitude: 800 m above ground; Endurance: 2.5 to 4.0 hr; Engine: 7.6 hp; Pay load: 10–30; Remote control: Futuba 10; Maximum wind scale for operation: 6.
2. Aerial Survey Unmanned Blimps;
3. Ducted fan Powered Unmanned Blimps: These are blimp UAVs with ducted fan propulsion system. They could be classified further using the length as the criterion;
4. Electrically Powered Unmanned Blimps: Under this category the above Chinese company turns out four models with minor modifications. They are powered using electric Li-Po batteries.
5. Industrial Unmanned Blimps: These blimps are manufactured to suit missions for different industries and to monitor activity;
6. Research Unmanned Blimps: These are designed to accomplish tasks determined by researchers in different aspects of science. For

example, agricultural researchers would prefer blimps with ability for high resolution imagery of terrain, crops, irrigation channels, etc. No doubt, sensors to suit the purpose should be fitted (Tattaris, 2015)

7. Surveillance Blimps: These are blimps equipped with sensors to obtain sharp images and relay system with inbuilt central processing unit (computers). These surveillance blimps are used to survey and explore minerals and other geological features. Overall, we may note that types of blimps manufactured and made available depends on the need and demand for certain types.

Yet another Chinese blimp company produces only two types such as Unmanned blimps for general aerial survey and the other one for Police (EOMST, 2017). We may note that there are also blimp companies that specialize in a single brand of blimp model.

Hyperblimps are autonomous in terms of flight path, way points and aerial photography. The blimp utilizes solar energy to keep afloat. Hence, it costs less to operate a Hyperblimp. Just like other brands and models of blimps, there are many uses that can be attributed to Hyperblimps (Plate 3.8). Several different types of sensors and chemical detecting electrode (probes) could be placed in its payload area. It carries probes to detect mines, unexploded bombs and metallic objects. It also carries magnetometers to prepare maps. It carries instruments to detect water and water resources on surface. At the minimum, a Hyperblimp carries a set of visual, infra-red and red-edge cameras plus a lidar pod. These sensors are utilized to get accurate images and digital data about ground conditions. The same set of sensors could be of use to obtain useful data on agricultural fields and crops. They can also map disease/pest attacked zones in a cropped field. The infra-red imagery helps in detecting drought or water deficit regions in a cropped field. Hyperblimps have been adopted to study glaciers, ice bergs, snowfall and study their effects on terrain and vegetation (Plate 3.8). Hyperblimps could carry small payloads of 2–4 kg. Such payloads could be parcels of letters, food packages, and life-saving medicines for short distances of 30–40 km.

Let us consider a semi-rigid airship known commonly as Zeppelin. The model 'Zeppelin NT' is a unique airship that has a rigid internal structure. It was tested in 1997. It is produced by a German company, Zeppelin Luftschifftechnik Gmbh, situated at Friedrichshafen, Germany. Zeppelin NT has a rigid framework made of aluminum and carbon fiber. The low weight is said

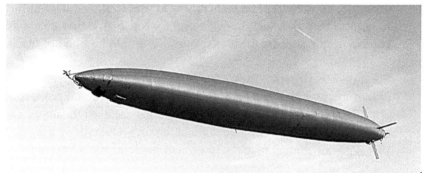

PLATE 3.8   A hyperblimp conducting aerial survey of natural vegetation and geological aspects of terrain in Utah State, USA. Top: A close up view of the blimp. Bottom: Blimp on a survey mission above natural vegetation.

Source: Daniel Geery, CEO, Hyperblimps LLC., Salt Lake City, Utah, USA; https://www.hyperblimp.com/#contact.

to allow better control during flight. The lift is provided by non-flammable helium. Helium is filled into the envelope made of tear-resistant material. The envelope with lighter-than-air gas helium provides the lift and allows the zeppelin to float. It has three propellers with a swivel angle of 120 degrees. The cockpit is a clearly structured high-tech workplace equipped with latest avionics. "Fly-by-wire" control systems with a joystick enable precise maneuvers and relieve the pilots. A mission display is available as an option. The cabin can accommodate 2 pilots and up to 12 passengers. The zeppelin is attached to the moore, at the nose. Zeppelin NT could be mounted with various sensors such as visual (R, G, and B), infra-red and red edge. It also carries a Lidar sensor. In addition, for geophysical measurements it has magnetometers. The Zeppelin NT has also been used to study atmospheric changes. It carries a set of electro-chemical electrodes for various greenhouse gases such as CO_2, SO_2, CH_4, NH_3, N_2O, particulate fraction, etc. A few reports from the manufacturer of Zeppelin NT suggests that it is well suited to conduct sampling and analysis of biological properties of atmosphere. The zeppelin can detect airborne propagules of infectious diseases above crops, in agrarian regions. The semi-rigid airship can detect contamination of atmosphere with harmful gas, particulates, etc. (Zeppelin Luftschiff-technik GmbH (ZLT), 2018; Zeppelin Science. de, 2018).

Specifications of Zeppelin NT: Volume: 8425 m³; Height: 19.4 m (64 ft.); Width: 15.5 m (46.5 ft.); Length: 75 m (246 ft.); Weight: 19,780 lbs (8980 kg); Endurance: 24 hr.; Payload: 2350 kg (5181 lbs); Maximum Flight Altitude: 10,000 ft. (3,048 m); Maximum speed: 125 kmph (78 mph); Engines: 3 Textron-Lycoming IO-360-C1G6 (197 hp); Fuel capacity: 306 gallons (5 tanks); Propellers: 3; Crew: 2 pilots; Passengers 15.

The Latin American lighter-than air technology is regaining its popularity enjoyed long ago during early 20th century. Let us consider a blimp model manufactured and utilized indigenously in Latin America. A new company located at Sao Carlos is supplying blimps that are based on lighter-than-air technology. It is a helium gas filled large blimp. Reports suggest that, Brazil aims to accentuate blimp technology for use in variety of professions including agriculture (Airship do Brazil Industrie, 2017). Here, we are concerned with a blimp model named ADB-3-X01 (Plate 3.9). Its inaugural flight was conducted in July 2016. Initially, the blimp was adopted by the Brazilian Armed Forces. However, eventually, it is to be used in civilian and agricultural situations too. Pilot training to maintain and utilize the blimp ADB-3-X01 is being imparted at the Lighter-than-Air School of Aviation (ESALTA) at Sao Carlos, Brazil. There is a list of uses possible with the ADB-3series blimps. They are in military surveillance missions; early warning systems to avert disasters such as floods, drought, etc.; recreational travel such as to obtain panoramic view of cities, geographical sites of interest, etc.; aerial survey and collection of digital data of land resources and crops; propaganda and general advertisements in the air; and monitoring public events. Initial reports suggest that ADB-3-X01 could perform efficiently compared to aircrafts or small drone aircrafts. The cargo transport done using blimps seems less costly. This blimp can carry 30 tons of cargo per flight at 125 kmph (Airship do Brazil Industrie, 2017; Eblimp, 2018).

PLATE 3.9   Airship ADB-3-X01.

Source: Airship do Brazil Industrie, Sao Carlos, Sao Paulo state, Brazil; http://www.adb.ind.br/contato.jsp.

Note: This blimp is versatile in its usage. It is useful in military, civilian and agricultural operations. In agriculture, it can help in aerial surveillance of agricultural experimental stations, cropped fields, water resource. It offers aerial photography and digital data useful during precision farming and in conducting several other agronomic procedures.

3.3.3   HYBRID BLIMPS OR 'PLIMPS'

Plimps are among newest aerial robots which are a combination of copter, a plane and a blimp. The net result is that, a plimp is not vulnerable to disastrous collapse and fall on buildings or people, if they are employed in urban areas. This is unlike small drone aircrafts that have failed in the mid-air due to mal-function of circuits, computer programming or exhaustion of battery power. Plimps float even if one of the aspects, say, a copter portion or airplane portion goes defective. Let us consider an example. Plimp is being produced by a company named Egan Airships, located at Seattle in Washington State, USA (Sigler, 2017; Boyle, 2018; Egan Airships Inc., 2018; Plate 3.10). Egan's plimp is 28 ft. long. and 20 ft. in diameter. It weighs 55lb. Its endurance is one hour using lithium polymer batteries. The lithium batteries actually support twin engines that generate 30 hp power, for its transit in the air. The plimp normally cruises at 20 kmph. Since it has copter portion, the plimp has ability for vertical take-off and landing. A plimp floats. It does not fall if copter or airplane aspect of it malfunctions. Borkhataria (2018) too has reported about an aerial robot called 'Plimp' (Plate 3.10). Its' ceiling altitude is 500 ft. above ground surface. The 'Plimp' is remote controlled using GCS computers. It is apt for applications in agriculture. Particularly, for aerial imagery of terrain and crop surveys, detecting disasters like soil erosion, floods, droughts, etc. It can be used to assess crop's growth rate. It can also be used to transport agricultural cargo. In the urban settings, it could be utilized to monitor public events, traffic, and in policing. Border patrolling is also a clear possibility. A few other uses are in patrolling railway lines and power lines. Using appropriate sensors, a plimp could be used to assess chemical pollution levels in the atmosphere. It can be used to monitor and direct forest fire control. It could be deployed to surveillance and monitor mines and mining activity, ore transport and dumping locations.

Hybrid airship-drone is a farm robot system for crop dusting, planting, fertilizing and other field jobs. They say, modern farming is currently being done by powerful ground equipment. Automated farming can also utilize small, agile, lightweight, energy-efficient automated robotic equipment that flies to do the same job. At the same time, such small drone planes can focus and analyze a single plant. We should note that, many of these functions are also accomplished by a blimp. A hybrid airship-drone has both 'passive lift' provided by a gas balloon and 'active lift' provided by propellers. A hybrid airship-drone may be cheaper and more stable in flight. It may require less maintenance than other aerial vehicles such as Quadro-copters. However, hybrid airship-drones may also be larger in size and have more inertia that needs to be overcome for starting, stopping and turning. These hybrid airships (plimps) conduct aerial photography, collect spectral data of crops, and transport cargo as payload. The hybrid blimp could be fitted with sprayer nozzles and pesticide/herbicide tanks (see Fima et al., 2018). Then, plimps spray or dust agricultural chemicals uniformly, all across fields. During spraying, they fly at low altitudes and keep buoyant due to lighter than-air gas helium (Salnikov et al., 2014; Bee Robotics Corporation, 2018). Here, we should note that hybrid airships could also be large versions. For example, Airlander 10 and Lockheed Martin's Hybrid Airship are a composite of blimp, copter and airplane.

PLATE 3.10   A 'plimp.'

Source: Dr. Daniel Nelson, Egan Airships Inc., Seattle, Washington State, USA. http://www.plimp.com.

Note: The PLIMP depicted above is a hybrid composed of a blimp, a copter and an aeroplane, all in one.

Hybrid Air Vehicles Inc., is not the only company making excessively large airships. Lockheed Martin's LMH1 airship is another example (Plate 3.11). The company touts to deliver cargo to remote areas. The aerospace firm has signed to deliver several units of hybrid airships. Several nations are backing the blimp-aided transport method. The French government recently backed a project to build an airship called the 'LCA60T.' A Chinese state-owned enterprise called AVIC General and a Moroccan private firm called Marita Group have also invested in the hybrid blimp project. Overall, the aim is to adopt hybrid airships for transport of cargo. It seems, the idea was born out of a problem in France's forestry industry. They found that despite a large amount of production, there was a problem in transporting wood to the mills (see Clay and Clement, 1993).

PLATE 3.11   A Lockheed Martin Hybrid Airship.

Source: Erica Tierney, Lockheed Martin Inc., USA; https://www.lockheedmartin.com/en-us/products/hybrid-airship.html.

Let us consider now a hybrid blimp designed, developed and that is being manufactured by a company named RosAero Systems, situated in Moscow, Russian Federation (see RosAero Systems, 2019). 'ATLANT' is the name of the blimp. It combines the characteristics of blimp, copter and airplane. It has been designed in three different models that can carry payloads (cargo) of 15, 60 and 170 tons. The hybrid blimp, Atlant, has been designed specifically to be efficient during large scale cargo transport in Russia. The efficiency reported for Atlant is 7 to 25 rubbles (Russian) per ton km. Atlant is supposedly economically efficient plus it is deemed environmentally safe. The blimp has a longer endurance and transit in-one-go from 150 to 5000 km. Reports state that, Atlant caters to special requirements of safety for precious military cargo. In addition, 'Atlant' specifically suits when heavy and over-sized cargo is to be transported. It helps in door-to-door delivery for various industries and military camps. It allows for crew to change routinely. It is useful for passenger travel and recreational tourism. Overall, the company believes that 'Atlant' and its various models will create a new mode of travel and transport of cargo in the Russian Federation, particularly, Far-North Russia, Siberia and Fareast Russia.

Specifications of 'ATLANT 30': Volume: 30,000 m³; Length: 75 m; Cruising speed: 140 kmph; Payload: 16,000 kg; Flight range with full load: 2000 km; Crew members: 3; passengers: up to 80 (RosAero Systems, 2019).

3.3.4   LARGE OR GIGA-BLIMPS

We have to note that blimps commonly in use are relatively large aerial vehicles. Particularly, when compared to other aerial robots and semi-autonomous drones such as fixed-winged aircrafts, copters, parafoils and aerostats. Therefore, among the options available for farmers, blimps are fairly large aerial vehicles that could offer them with aerial photographs and spectral data. Further, we have to note that, among blimps we have classes based on their size. Most blimps are large. Larger ones are 250–300 ft. in length (or even 400 ft.) and 15 ft. in width and height. However, in this section, we are concerned with blimps that are really very large. We may call them 'Giga-Blimps.' Some of them could easily reach a length of 350–400 ft.

At present, world's biggest aerial vehicle is a helium-inflated blimp known as 'Airlander-10.' It is designed and manufactured by the British company-'Hybrid Air Vehicles Ltd.' It incorporates aviation technology from helicopters and airplanes, to keep it aloft. It measures 302 feet (92 meters) in length. In comparison, largest passenger airplane, i.e., the Airbus A380 is only 232 feet (71 meters) in length. The Airlander-10 airship cruises at 20,000 feet (6,100 m) for up to two weeks. It is autonomous. Its flight path could be pre-programmed. Airlander 10 can take off and land from any surface (Airlander 10, 2018). It is useful in travel and cargo transport (Airlander 10, 2018; Daniels, 2016). Airlander-10 embarked on its maiden flight on August 17, 2016 It flew for about 19 minutes within a radius of 10 km from GCS in Bedfordshire, United Kingdom. It reached an altitude of 500 feet (152 m).

Lockheed Martin's Hybrid airship is another example of a large or giga-Airship (see Plates 3.11 and 3.12).

PLATE 3.12   Lockheed Martin's Hybrid Airship: A large or Giga blimp.

Source: Erica Martin, Lockheed Martin Inc., USA, https://www.lockheedmartin.com/en-us/products/hybrid-airship.html; Also see https://www.youtube.com/watch?v=aMRZ26cOxTc; https://www.youtube.com/watch?v=y6w5Iuh_reU.

Note: The hybrid airship manufactured by Lockheed Martin of USA is indeed large. Compare the sizes of African tribes, their living homes and food carrying truck with the size of envelope of the blimp.

3.3.5   TRAIN-LIKE BLIMP

It is clear that in very recent past, many of the forested regions on earth experienced rapid spread of fires and extensive damage to vegetation. Fires affect forest stand and the attached ecosystem. Sometimes, such a fire has repeatedly affected forest stands, in short intervals. For example, one of the reports from Massachusetts, USA. states that, fire related disasters could spread rapidly. The fire-fighting ground vehicles and helicopters do not carry with them sufficient fire extinguishing material, water and sand. Hence, containment of forest fire gets hampered. Further, the review mentions several advantages related to building along train of blimps, instead of singles that float slowly above, in the sky. A blimp may carry a lot more water and sand. Particularly, to quell the forest fire. However, a step further, it has been suggested that blimps too may fall short as far as transport of water and sand, if the fire spreads rapidly and the fire affected patch is large. Hence, one of the ideas touted, that is perhaps still on the drawing board is to construct a 'train of blimps,' almost a mile long. The mile-long blimps could be filled with water that suffices to extinguish the fire promptly.

3.4   USES OF BLIMPS

3.4.1   *USAGE OF BLIMPS IN MILITARY AND BORDER SECURITY*

3.4.1.1   *BLIMPS IN MILITARY ESTABLISHMENTS*

We may have to note that many aerial vehicles, either remote controlled or autonomous ones were first developed in military research institutions. They were, in fact, first tested then modified thoroughly and used by military establishments. For example, there are several models of aerial drones (UAVs) that are being successfully used during military operations. Blimps are no exception to this situation. Blimps too have initially attracted greater interest from military establishments. Otherwise, blimps were mostly opted during advertisement campaigns. They are being examined in agriculture only recently (CIMMYT, 2014; Krishna, 2019). So, promptly, let us first consider the salient uses of blimps in the military. Military engineers are constantly improving the performances of various models of blimps. So, each revised model may find a new or better usage in military establishments.

Recent reports emanating from United States Department of Defense suggest that there is a renewed interest in airships. There is a resurgence of funding to develop and adopt blimps for military surveillance. It seems, funding for blimps has increased by several millions of US$. Unmanned blimps are being utilized as sentinels that float at high altitudes (Editor, 2015). Let us consider a few newer brands/models of airships (blimps) being adopted for use in military:

'Blue Devil' is a large airship that hovers at over 20,000 ft. above ground surface. It has an on-board super-computer that stores and analyzes spectral data, rapidly. It relays data and allows ground control to react. The surveillance is done using a set of 12 sensors. The computers on board have ability for automatic analysis, without human intervention (Editor, 2015; Schachtman, 2017).

ISIS-DARPA is a 450 ft. long blimp. It hovers in the sky at 7000 ft. above ground surface. The blimp can stay afloat for 10 years, of course, using solar batteries. It can transit at a maximum speed of 230 kmph. It conducts surveillance using powerful high-resolution sensors that operate at different spectral band widths. It has an array of radars on board. The radar senses even slightest movement of a stray object (enemy planes, drones and missiles, etc.) in the sky. It tracks missiles moving at 375 miles from its location in the sky (Editor, 2015; Dahm, 2017).

Lockheed Martin Inc., produces tethered and untethered blimps that help in aerial surveillance and reconnaissance. They also offer lighter-than-air 'Persistent-Threat-Detection-System (PTDS).' The PTDS float between 800 to 1500 m above ground. This blimp is meant to provide a round-the-clock monitoring of military stations. They track vehicles that move in the vicinity. PTDS offer visual and infrared- videos to ground control stations. It was meant to help US military located in the far-off Afghanistan's hilly tracts (Brinton, 2010; Bowley, 2012). The PTDS system is also used by the US Department of Homeland Security.

Hale-D is a blimp produced by Lockheed Martin Inc., They say, Hale-D is an autonomous blimp that could continue to float in the sky at 20–30,000 ft. altitude. It can hover for long periods covering millions of miles of ground surface and sky. It traces other flying objects. It has been adopted as part of missile shield by the United States Air force (Editor, 2015; Brinton, 2010).

Hisentinel-80 is an autonomous blimp. It transits at high altitude of over 20,000 ft. from ground surface. It has long endurance. It is a blimp developed for US Army. It is also adopted by the missile defense command. It draws power from solar panels. It can survey up to 600 km radius, using powerful radar system. This blimp has a series of sensors for spectral imagery and reconnaissance (Editor, 2015).

LEMV (Long Endurance Multi-Intelligence) airship is a new addition to US military intelligence system. The blimp surveillances continuously for 3 weeks. It uses special set of sensors. It has hybrid propulsion system. It combines a lighter-than-air blimp filled with helium plus an airplane with propellers (Editor, 2015). It is produced by Northrup Grumman Inc., Virginia, USA.

Sanswire-TAO STS-111 is among blimps selected for use by the military establishments of many nations. It is unusually shaped. It is 111 ft. long and 11 ft. in diameter. It is held aloft by helium gas. It was tested in 2009 for long duration surveillance. It can be used to carry a payload of military cargo. It adopts parachute-aided drop system (Editor, 2015).

3.4.1.2   BLIMPS AS EARLY WARNING SENTINELS ABOVE MISSILE SITES

As stated earlier, blimps have been adopted to surveillance ground activity in and around locations where cruise missiles are held. Recently, the US Military has developed and tested the 'all seeing' surveillance blimps. These blimps carry very high-resolution cameras and radars. These recent models can spot even a small sized object, say, a man moving on the ground, from 340 miles away. Such blimps have been in service now above missile, locations in Maryland, USA. They conduct surveillance all through the day (24 (h) and all days of week (7 days). The aim of radars on blimps is to conduct aerial survey and forewarn incursions of planes and missiles, into east coast of USA (Priggs, 2014; Koebler, 2014). This blimp-aided military surveillance system is called JLENS (Joint Land Attack Cruise Missile Defense Elevated Netted Sensor System). It is composed of blimps that fly at 10,000 ft. above ground surface. The blimps have radar system. It scans for cruise missiles, manned and unmanned aircrafts, surface vehicles, boats and missile launchers (Priggs, 2014; Froomkin, 2014; Bach, 2015).

There is no doubt that the surveillance of military installations, vehicles and most importantly the missile centers is important. It seems, US military is now organizing a set of at least two blimps to hover above each 'cruise missile center.' The blimps selected keep watch over a 150 km radius on the ground. They track vehicles and personnel movement in and around the missile sites. Similar blimps endowed with radars have been adopted to surveillance US bases in Afghanistan and in other regions.

Blimps could serve as excellent aerial spy vehicles. For example, reports suggest that US Airforce has a project to develop a blimp that lasts longer in the sky. It supposedly stays hovering at very high altitude of 60,000 ft. above ground surface. This dirigible is long at 450 ft. length. They say, it can stay afloat for 10 years. Its radar can therefore provide information continuously. The blimp, when fully operational is destined to provide superior intelligence, surveillance and reconnaissance for the US military, particularly, above missile sites.

Let us quote another example. A new unmanned airship named 'Dread Zeppelin' has been designed to carry out surveillance mission in Afghanistan. It has been developed under a project named 'Long Endurance Multi-INT Vehicle (LEMV).' It is also to be used efficiently as a cargo transport blimp. It has a long endurance of 3 weeks in air transit. It carries a payload of 2500 lb (1200 kg) and covers a distance of 2500 km per take-off. It floats at an altitude of 2000 ft. above ground surface. Its transit speed with cargo ranges from 20–80 knots (Smith, 2009).

Blimp units are transportable and re-deployable. They can be put to use in any of the prevailing weather conditions. An interesting report emanating from border patrol experts in southern United States of America tells us that, blimps were used in good number in Afghanistan. The aim was to surveillance US military posts and camps. Blimps' sizes and capabilities for aerial photography and transmission of images varied. In Afghanistan, they were used to detect drug smuggling, ammunition dumps of insurgents, illegal border crossing and vehicular movement, in general (Richmond, 2014; Bowley, 2012). Since, the end of occupation of Afghanistan by US military, they have opted to dismantle blimp-aided surveillance system. They have now organized the same set of blimps under Homeland Security System to patrol and guard Mexican border posts in South Texas, New Mexico and California (Rees, 2012; Sherman, 2012). Such a blimp system costs millions of US dollars, if they have to be newly manufactured. Blimps were also relocated to observe Florida coastline and restrict smuggling (Sutton, 2008).

Recent reports from the Russian Federation websites indicate that, a large blimp network is being contemplated, to protect missile sites. During this year, they are set to induct a series of surveillance blimp. These blimps are meant to track enemy and forewarn against attacks (Lighter Than Air Society, 2017).

3.4.1.3 BLIMPS TO SURVEILLANCE INTERNATIONAL BORDERS

Airships provide long-term surveillance capabilities over extended regions or borders. Such surveillance is possible even when the blimps are entirely autonomous. Blimps are economically efficient compared to traditional aircraft and drones. Further, airships also offer the possibility of landing or loading/unloading without ground-based infrastructure, in the remote border areas (Euro Airships, 2018c).

Blimp aided surveillance system, particularly, at the international border is in vogue, since decades. For example, in USA, huge blimps either tethered or freely floating ones have been adopted, to obtain aerial photos. They help in tracing movement of vehicles, personnel and cargo transit across the border between USA and Mexico. In fact, there is a network of blimps that float above the 'hot spots' between USA and Mexican border. The network is called 'Tethered Aerial Radar System (TARS).' They help in aerial surveillance and exchange of information between Customs and Police check posts. It seems, such blimps have also been adopted to police the unauthorized airplane traffic across the borders (Figure 3.2; Neaves, 2014; Rees, 2012).

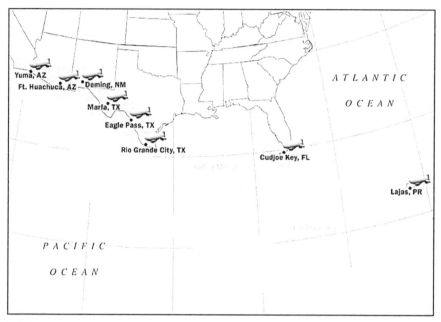

FIGURE 3.2 A network of 'tethered blimps' stationed on USA–Mexico border, to surveillance movement of vehicles and human beings.

Source: United States Department of Homeland Security, Yuma, Arizona, USA. https://www. cbp.gov/frontline/frontline-november-aerostats.

Blimps could be playing a vital role in tracking smugglers and smuggling routes across coastal lines. For example, in 1950s and 60s, it seems, US Airforce used tethered balloons to monitor coastal activity in Florida. Sutton (2008) reports that, along with several other methods of surveillance to detect illegal transport of goods (smuggling), blimps stationed above coastal locations are finding acceptance. US Navy and Coast Guard, for example, has tested a helium-filled blimp in the coastal regions of Florida. A blimp known as 'Skyship 600' belonging to US Navy has been stationed above coastline. This is to cover parts of Key West of Florida and Cuba. The blimp is equivalent to Boeing 747 jet airplane in size. The blimp has two Porche engines that utilize about 10–12 gallons of patrol per hour. The endurance of this blimp is about 52 hours at a stretch.

Reports by border patrol suggest that they are assessing the usefulness of two models of blimps produced by a company in North Carolina. They are being deployed to hover in the sky above border post between Texas and Mexico, and along the river Rio Grande. They were earlier adopted in Iraq and Afghanistan to surveillance troopers, military installations and vehicular traffic. So far, it is clear that blimps with sensors and detection/alarm system (radars) are good bets as sentinels. These are examples drawn predominantly from USA. We ought to realize that many other nations do adopt similar blimp-aided surveillance of international borders.

3.4.1.4   BLIMPS IN POLICING AND INTERNAL SECURITY

Homeland Security of USA has introduced blimps to conduct aerial surveillance of traffic, public events and interstate borders. At Ogden, in Utah, the state police have introduced blimps to surveillance neighborhoods and vehicular traffic (Homeland Security News Wire, 2011; Rees, 2012). They have assessed the cost of using a police car versus a blimp that floats and hovers above. They report that a police car costs 40,000 US$ plus fuel charges. Whereas, a blimp costs around 15,000 US$ and US 100$ per week, to keep helium gas replenished. This blimp is 54 ft. in length. Its ceiling altitude is 400 ft. above ground. It has several cameras to record the proceedings on the ground. It has an endurance of 7–10 hr. at a time (Plate 3.13).

3.4.2   BLIMPS IN SPACE SCIENCE

Aerial vehicles floating in the atmosphere above the planet's/moon's surface and navigating based on remote directions or autonomously (pre-programmed) seems essential. Extended periods of aerial photography and sampling is needed, if one wants to explore the heavenly bodies, in great detail. The endurance to stay afloat is indeed crucial. Autonomous Blimps seem to possess such abilities. Robotic blimps and parafoils may answer space researcher's requirements. On the contrary, parachutes or balloons allow us very short duration during which they can relay information. The time lapse between release into heavenly body's atmosphere and until it touches the surface is only the period, when parachute works usefully for us.

It seems NASA's Cassini Spacecraft beamed detailed images about the surface, but only during its brief descent into the atmosphere and until it crashed into surface. However, the next mission is supposedly contemplated to send detailed images of surface, using a 'floating blimp.' Among several types of aerial robots considered, blimps were said to be feasible to make close aerial survey of Titan's (Saturn's moon) surface. Titans atmosphere is 21 km thick and blimp is to localize at around 6 m above the surface and provide imagery. The endurance of the blimp is expected to be long. Say, six months at the least. So, it might relay photographs for a very long time to earth's stations. The European Space Agency has contemplated to send an

'Orbiter' and expects to lower down a balloon or blimp to survey the Titan surface and atmosphere (Wall, 2010; Heun et al., 1998).

PLATE 3.13   Tethered blimp mounted and fastened to pick-up trucks.

Source: Dr. Charlie Steffan, SkyDoc Aerostat Inc., http://skydoc.com/index.html.

Note: These tethered blimps could be used during surveillance of ground vehicles and general activity in public places.

Space science experts have adopted several different types of UAVs. This is to discover and understand earth's exterior space as well as those of other heavenly bodies. About 15 years ago, a few scientists from Jet Propulsion Laboratory in California had thought of several different types of uses for blimps. Blimps, it seems offer certain clear advantages while exploring other planetary bodies (Elfes et al., 1998, 2003; Jet Propulsion Lab., 2000). Blimps could have a major role in exploring the atmosphere. They have to carry instrumentation that analyzes the composition of planet's atmosphere. Robotic airships, balloons and copters allow accurate flight path above the surface of the planet. Blimps with lighter-than-air gas could be used to make long range observations of the surface and atmosphere of planets. This will be possible because of the longer endurance, if they are solar powered. Further, Elfes et al. (2000, 2003) and Ramos et al. (2000) state that autonomous blimps could be modeled for specific atmosphere that prevails over a planet or its moon. Target locations for sampling surface or atmosphere could be pre-programmed. We may note that, research program- AURORA (Autonomous Unmanned Monitoring Robotic Airship) is an effort to use robotic airships, to explore heavenly bodies and their atmosphere. AURORA may even help us to get vast details about our own space, i.e., surrounding earth. Also, about atmosphere and space surrounding other planets (Ramos et al., 2000; Bueno et al., 2002).

Robotic unmanned aerial vehicles have great potential as surveying and instrument deployment platforms, in the exploration of planets and moons with an atmosphere. Among the various types of planetary aero vehicles proposed, lighter-than-atmosphere (LTA) systems are of particular interest. It is because of their extended mission duration and long endurance. A few other unique characteristics of robotic airships also make them ideal candidates for exploration of atmosphere of planetary bodies with an atmosphere. Blimps offer certain advantages like, (1) precise flight path execution for surveying purposes; (2) long-range as well as close-up ground observations; (3) long-term monitoring of high science value sites; (4) transportation and deployment of scientific instruments and *in situ* laboratory facilities across vast distances (Elfes et al., 1998, 2000, 2003).

3.4.3 CIVILIAN USES OF BLIMPS

Aerial photography is an important aspect for which, not only blimps, several other types of UAVs, balloons and kites are utilized. Aber et al. (1995) have reviewed and discussed about usage of aerial vehicles (UAV aircrafts, copters, airplanes, blimps, kites) during aerial survey and photography. Blimps, in particular, allow us certain special advantages while obtaining aerial photographs of ground features. They can stay afloat for longer duration. Their endurance runs for weeks. Also, they fly at low altitude. Therefore, blimps offer high resolution images. Although, interest to adopt blimps in a big way is confined to advertising, during recent past, say last 5 years, there has been consistent effort to adopt blimps. Most recently, blimps are being examined for their utility in obtaining aerial images of high resolution and thermal images about terrain and natural resources (CIMMYT, 2014; see Krishna, 2019). They are trying to adopt blimps in agriculture, rather extensively.

Vericat et al. (2008) have stated that lighter than-air blimps are useful in monitoring natural resources such as mining zones, natural vegetation, rivers and river valleys. Blimps provide aerial photography periodically. They could also be programmed to provide aerial imagery of rivers and their banks continuously. They say, aerial photography derived using blimps are highly useful. Particularly, when images are dispatched continuously. Blimps, it seems are among low-cost platforms to obtain aerial photos of disaster affected regions. We should note that imagery obtained using blimps may differ in terms of accuracy and resolution. High quality aerial images are possible. Aerial photos of river basins could be obtained rapidly, using blimps attached with CPU and connectivity to ground control computers (Vericat et al., 2008; Jo et al., 2014).

There are also aerial photography-based consulting agencies in many regions of North America and Europe. They offer images drawn using blimps stationed at vantage locations. They offer images of requisite resolution depending on the purpose. Some of the purposes include study of natural resources, agricultural terrain, cropping systems, irrigation channels, industries, mining centers, traffic, etc. Such agencies offer training involving remote control of blimps, image processing and storage of digital data. Commercial consulting with blimp agencies costs around 250 US$ per day for governmental agencies in USA (Aber and Aber, 2013).

Blimps could be utilized to offer continuous around-the-clock surveillance and warning systems above important government buildings and installations relevant to civil administration of cities/states. For example, a recent report by Kaye (2015) points out that there are intrusions of important buildings. Hence, the USA administration is considering placing a few blimps above the administration area. This is to surveillance buildings, including the Capitol Hill, Washington D.C. (Kaye, 2015).

Let us consider an example wherein a blimp is used to monitor city's traffic. Procter (2015) reports that transport departments in several cities of USA are either inclined to or they have already adopted lighter-than-air blimps. A blimp called 'SkySentry' has been flown above Denver city in Colorado. It is meant to monitor city's traffic. These blimps offer images and lots of data about traffic patterns and density. We may note that this blimp has offered images of national highway 1–25 in USA. The highway is actually surveillanced all through the day/night (24 h), using appropriate sensors. This blimp is produced by a company SkySentry LLC located in Colorado Springs, USA.

Blimps could serve as excellent sentinels above mines and mine dump areas. They could be used to monitor the entire set of activities of open mining areas, track the vehicles, help in sending information to minors, monitor mine dumping area and its vegetational areas, if any. Since blimps are often larger than a small fixed-winged or copter drone, they cannot go too close to mining area. The miners will have to get satisfied with high-resolution imagery offered by blimp's cameras. This is a constraint with using blimps. However, long endurance, continuous monitoring and low costs are advantages attributable to blimps. Such advantages make blimps feasible for surveillance and aerial imagery of mining zones. Let us consider an example involving diamond mines of South Africa. It seems famous companies such as DeBeers Inc., and others have been mining diamonds in this region for long. It is not uncommon to strike a low or nil during excavations. Exhaustion of diamonds in an area leads them to survey for long hours. They use Kimberlite, an ore, as the indicator of existence of diamonds. Recently, DeBeers Inc., and Bell Geospace Inc., have pooled their research ability. They are surveying for Kimberlite using geophysical methods-based on gravity of the ore. It is opined that geophysical methods are more efficient in detecting diamond reserves. This method of blimps and geophysical techniques has now been adopted in other diamond rich regions in Zambia, Zimbabwe and South Africa. A few reports by blimp producing companies suggest that Remote controlled large hybrid blimps are a boon to African mining industries. For example, a 'Hybrid blimp' of the type produced by Lockheed Martin Inc., could be a major asset to mining companies in Africa. They can be highly useful in mineral prospecting. In fact, they say, with a hybrid blimp it is 'no road no problem' the blimps can transit and reach the exact spot in air. The hybrid blimp can land on water, rough land, sand, and even ice, if required (Piquepaille, 2005; CNN Staff, 2016; Lockheed Martin Inc., 2018; Boaz, 2018; Geere, 2013).

Zeppelins fitted with appropriate visual high-resolution multispectral cameras, infra-red cameras and other accessories such as magnetometers, gravity change detectors are also being used to conduct aerial surveys. They have been used to detect natural reserves. The zeppelins fly low over the geological sites and offer information for miners, to act upon. The German Zeppelin producer says that the airships are useful and cost efficient. Particularly, while prospecting for ores in mining regions (Zeppelin Lufschifftechnik Gmbh., 2010).

Blimps stationed above disaster prone geographic locations is a good idea. They can detect and inform the ground control about the magnitude of disaster, immediately. Fire fighting in forests is an important task that requires efficient and quick transport of fire extinguishing material. Blimps may have a role in overcoming such a disaster. Blimps could be very efficient aerial vehicles in distributing water and fire extinguishing material above forests, buildings and other infrastructure, industries, etc. Blimps carry relatively more water than helicopters or ground trucks. One idea professed to overcome the situation is to build a long train-like airship, a long inter-connected row of blimps. The train-like blimps can carry lots of water and sand to quell the fire quickly (Nadis, 1997).

3.4.3.1 BLIMPS ARE UTILIZED TO ADVERTISE PRODUCTS AND PUBLIC EVENTS

During indoor activities, blimps that form part of lighter-than-air technology driven UAVs are being increasingly preferred. This is to obtain aerial photography (stills and movies) and to surveillance the event itself. Remote controlled blimps or those with pre-programmed flight path and way points in indoor conditions are possible. They are mostly in use during public events such as games (e.g., Olympics, Football or Tennis matches, Horse Racing event, etc.). Such blimps are usually made of thick fabric, fitted with sets of cameras for aerial photography from different angles and distances. They are also simultaneously used for advertisements, plus surveillance of games or similar events (Raven Aerostar Inc., 2018; Plate 3.14).

Blimps are conspicuous vehicles when they float in the sky. They attract human attention instantly. Hence, they are preferred by advertising agencies. Blimps are affordable and have low recurring costs. Blimps purchased once lasts for longer duration. Usually, the wrinkle-less poly-urethane material is preferred for advertisements. They are durable and re-usable frequently, to revise the advertisements. 'Goodyear' is perhaps the most frequently seen advertisement on blimps. Blimps have the ability for hovering or moving very slowly over urban and populated locations. So, we can maneuver them to correct spots above, in the sky. They can be stationed at vantage locations or moved frequently, if required, during an advertisement spree (See Inflatable, 2000, 2018).

PLATE 3.14   An autonomous blimp floats and hangs still, to obtain images and surveillance the crowd.

Source: Cynthia Hess, Goodyear Inc., Akron, Ohio, USA and Raven Aerostar LLC., Sioux Falls, South Dakota, USA., https://ravenaerostar.com/contact.

Note: These blimps can simultaneously advertise a product plus surveillance the entire stadium.

3.4.3.2   BLIMPS IN TRAVEL AND CARGO TRANSPORT

Airships perhaps ruled the long-distance travel during early 1920s. They were popular until the series of mishaps that halted their use. It seems blimps and zeppelins received greater attention than the aircrafts, from passengers. Aircraft travel was still a fledgling enterprise in early 1930s. Airships, both blimps and zeppelins were utilized to carry a group of passengers. Long distance travel was possible because of extended flight endurance of 130 to 150 hr. Transatlantic journeys too were accomplished frequently, using airships. A report by Sims (2017) says that, there was a big dip and rejection of blimps/zeppelins during past few decades. However, at present there has been a tendency to adopt the airships for short or even long-distance travel. It seems, in Europe, commercial travel agencies run about 12 routes from different capitals. There seems to be a certain degree of interest even in North America, to adopt blimps for passenger travel. Further, it is interesting to note that, European airship companies are now targeting to exploit the Chinese travel routes. A few airship enthusiasts believe that 'Hybrid Blimps' with greater safety features attached are apt to be utilized, for long-distance travel of passengers. Therefore, in due course, Hybrid blimps may get popular support with travel agencies.

One set of opinions suggest that it would be nice, if this entire planet could use zeppelin routes instead of tarmac roads. This could be possible, especially, for countryside transport away from cities. A zeppelin for one person is about the size of a lorry. In future, we can have added safety features like sensors, weather prediction, solar conversion, etc. A recent report states that production and use of blimps is getting renewed interest even in Latin America. In Brazil, for example, blimps are being manufactured and tested for use in transport of large cargo. Blimps to suit transport of cargo will highly useful when road transport may not be easily feasible (Airship do Brazil Industrie, 2017; Grabish, 2018; Boaz, 2018; EuroAirships, 2018b).

Jowit (2010) has stated that, helium powered ships (blimps) could eventually get a greater degree of acceptance with regard to transport of cargo. In Europe, companies have been evaluating blimps to carry fresh fruit, vegetables, flowers and other perishable items. Blimps could carry them rapidly from one place to another within the nation. In United Kingdom, there are forecasts that blimps could revolutionize transport of large sized cargos. They could be replacing aeroplanes to a certain extent. Particularly, when the distance to be covered is relatively small. A few other forecasts state that, replacement of aircrafts by large blimps could help in reducing the cost, for transport of goods. Further, blimps may cause less greenhouse gas emission compared to cargo transport via aircrafts.

Some of the chain stores in USA could be employing blimps in relatively larger scale. It is mainly to supply food items and other goods, to all outlets in a region. They could be owning their flotilla or blimp network attending different outlets. So, sooner, we may come across airships serving people as a flying delivery vehicle with their daily requirements. Blimps may be visiting Walmart-like outlets periodically (Sigler, 2017). Therefore, a good portion of goods transport trucks may get relieved, due to airships. So far, we have no idea about economics of trucks versus blimp mediated transport of cargo such as food items, etc. Of course, there are many forecasts. The new airship technologies may reduce the cost of moving things per ton-mile. They say, it depends on the size of the airship. A larger airship can reduce costs a lot more than a smaller ship. It seems designing of a plimp that can lift up to 500 tons could be more fuel-efficient than even a truck (Sigler, 2017).

Blimps have made visits to Arctic circle. It is said that in Canada, people throng the southern borders. Cargo movement to and from is also localized to south of the nation. People in the Arctic circle may have to buy goods in one lump before the onset of winter and store them. The cost of food items and raw material gets exorbitant. It is because of high air freight charges during winter. The Arctic depends on air freight, seasonal sea shipments and ice roads. Living costs become exorbitant in arctic. To overcome it, government planners have touted idea such as rapid pipeline transit or railways that operate under most frigid conditions (Prentice and Turiff, 2002; Prentice et al., 2005). These seem not feasible. Instead, a new idea propagated is use of a flock of airships that are made resistant to below freeze conditions. Blimps may be able to operate smoothly, daily from southern borders to arctic. They could transport cargo faster and at all seasons throughout the year (Editor, 2018b; Prentice et al., 2005). Ultimately, airship flotillas are forecasted to play vital role in the economy of countries within the arctic region.

Reed (2015) points out that blimps were to become predominant mode of air travel, much before the era of aircrafts, during early 1900s. The dramatic disasters and blimps being slower than aircrafts made them loose out in the sky travel. They faded from the sky around mid-20th century. However, recently, some of the blimp models produced by companies in North America and Europe could engineer a come-back of blimp travel. For example, Aeroscrafts blimp offers rapid transit at 120 knots and carries up to 66 tons. It has a range of 5000 miles. So, these Aeroscraft's blimps or models with similar specification turned out by other companies could get preferred (Plate 3.15). They say, blimps may mainly out compete aircrafts during cargo transport.

A report about Hybrid blimps suggests that blimp producing companies such as Lockheed Martin Inc., and PRL Logistics of USA, the United Kingdom-based Hybrid Airships Ltd. and Straight-line Aviation Ltd. are combining their effort, to produce a 'Hybrid blimp- LMH-1' (PRNewswire, 2018). The hybrid blimp is said to be resistant to cold conditions and blizzards that occur in Alaska or locations with similar weather conditions. The idea is to develop hybrids that can operate into Alaska and transport goods and personnel, safely. The Hybrid Airship-LMH-1' operation is forecasted to begin in 2019. It is to start from PRL Logistics Operation's Center in Kenai, Alaska. The hybrid blimp carries 22 tons of cargo, and 18 passengers plus crew per each flight. The airship enables access to Alaska's most isolated regions. It is designed to protect the sensitive ecological environment. LMH-1 lands on any type of surface including snow, ice, gravel and even water. 'Hybrid Airship LMH-1'provides low cost and environmentally friendly solutions, for moving freight and personnel to the most isolated regions of Northern Canada (PRNewswire, 2018). Overall, we may note that, like aircrafts, airships capable of negotiating difficult weather and harsh terrains would be eventually available to us. It needs research effort to develop suitable materials and techniques to withstand such harsh conditions.

PLATE 3.15   A travel and cargo transport blimp.

Source: Irene Guerrero, Aeroscraft Inc., Montebello. California, USA.

Note: The above blimp is one of the series of ultra-modern blimps produced by different aviation companies. They could help in popularizing both travel and cargo transport via blimps. Blimps may regain a portion of lost sky, in due course.

A recent report from North America indicates that blimps could be modified to add a few useful characteristics of planes and helicopters. This way, the new vehicle called 'PLIMP' could be more effective. It could safely transport personnel. The plimp has ability for vertical take-off and landing. It adopts a hybrid electric/gas- based power source. The plimp floats at low altitudes, say 500 ft. over any type of terrain. It is kept afloat by helium. It can maintain speed of 65 km. It tolerates wind turbulence better than a traditional blimp. The plimp carries at least 10 people or 900 kg per flight. They say, cost of travel and maintenance of plimp is lesser compared to blimp. It has good endurance and can transit a distance of 1300 km without re-fueling (WSBuzz, 2018).

Blimps in tourism is a good idea. Blimps are utilized for transport of personnel, although not very frequently like an aircraft. We may realize that, blimps offer excellent opportunity for small to medium distance travel. They offer excellent aerial view of the terrain and geological features. Since they move relatively slowly, a good panoramic view is possible, all around 360 degrees. Blimps have been used to serve as luxury aerial cruise vehicles, by groups of people. Blimps can travel and hover above sights that are otherwise rugged, if we attempt to reach via surface (road). Blimps can stay still or hover around historical or geographical sites of interest, say above a mountain peak or a geyser. Blimps can offer a good view of a game sanctuary, etc. Blimps are excellent for people carrying out ecological expeditions. There could be many more possibilities where blimps are used in the tourism industry (EuroAirships, 2018b; Tucker, 2012).

3.4.3.3   BLIMPS IN ECOLOGICAL STUDIES

Blimps, like other UAVs such as fixed-winged or copter drone aircrafts, parafoils, microlights, etc. are useful in observing natural vegetation and wildlife sanctuaries from above (Krishna, 2016, 2017, 2019; Cumberford, 2016). They offer excellent overview of large expanses of vegetation. The spectral analysis possible allows us to study the species diversity of trees and ground flora and fauna. Blimps may be used to monitor wildlife, rather continuously, because of the long endurance. High resolution close-up photos are possible, using the sensors. Blimps do fly over regions that are difficult to transit by road. They also make sorties over large bodies of water, rivers, seacoast and even high seas. So, it is possible to monitor animals such as wales, dolphins, sharks, etc. For example, here in this case, blimps have been adopted to study the number and ecological aspects of dolphins that inhabit seas. Oliviera et al. (2017) conducted an experiment by surveying Araguaian river dolphin (*Inia araguaiaensis*) during low water season, in the Parque Estadual do Cantão. This is a seasonally flooded lake system in the Amazonian forest of Brazil. Visual counts of dolphins are possible. However, human error may creep in. So, spectral imagery using blimps could be adopted, to study the dolphin movement. On an average, visual counts detected only <75% of the dolphins recorded by the blimp's camera. Compared to the visual survey, the aerial method was able to detect more individuals, also, more dolphin groups. It was possible to record more dolphin calves, which are usually difficult to detect. Blimps avoid human drudgery. A blimp also avoids errors related to human fatigue and oversight. Blimps offer excellent storage of digital data in their CPU.

Reports suggest that SkySentry LLC has produced blimps that were flown above Lake Erie, in North America. The main aim was to monitor algal blooms and vegetation on the coasts. Simultaneously, the blimp helped in monitoring proceedings of institutions/roads in the vicinity of the Lake Erie (Proctor, 2015).

Navy-MZ-3A is an airship that patrolled the sea off the coast of New Orleans, during the aftermath of large oil spill. The blimp helped in identifying the location of oil spills and directing the clean-up operations by other vehicles, ships, helicopters, etc. It helped ecologists to trace out the endangered species of turtles, dolphins, seabirds, whales and fishes (Editor, 2015). Rescue of seaborne animals was faster due to blimp-aided surveillance of the sea.

As stated earlier, blimps possess long endurance. They are relatively slow in movement. They offer better aerial images. Such features make them preferred aerial vehicles to conduct ecological studies. We can observe ecological changes above a location, say a national park, a water body, an eco region, etc. Let us now consider a report from Biscayne National Park, in Florida (Burnett, 2011). Here, blimps offer excellent aerial imagery of the national park, its terrain, vegetation and species diversity. The staff at the park state that blimps are quite or low-noise machines. They float at 250 ft. above the canopy of trees or ground surface in Everglades, Florida, USA. This is unlike helicopter sorties that are costly and noisy. Helicopters' movement close to surface/canopy of trees affects wildlife. It distorts recordings about the natural reactions of wildlife to environment. Recently, the Biscayne National Park replaced helicopters with a blimp called 'Snoopy-1.' 'Snoopy-1' is a small blimp.

It keeps track of tree canopies and the happenings in the park. It seems, the blimp provided excellent data about nesting of squirrels in the national park. Birds, it seems, first flew away from blimps but soon they got accustomed to the large bodied airship in the vicinity (Burnett, 2011). Also, movements of other animals in the park could be monitored. The Snoopy-1, it seems is also used in wildlife parks of other states such as in Texas, Nevada, Arizona and California. They are used to study vegetation, rivulets, the land mammals and their reactions to seasonal changes in environment. No doubt, blimps offer specific advantages while monitoring the natural reserves, wildlife sanctuaries, and other geographical regions on earth. Perhaps, blimps are good alternatives in many geographical regions. Blimps could monitor natural regions such as an arid desert or semi-arid region, a large river or lake and its surroundings, fringes of dam sites, hills and mountain regions, tropical expanses, cropping zones, etc. Blimps could eventually become sentinels that keep an exclusive watch over geographical/ecological sites of interest.

3.5   BLIMPS IN AGRARIAN REGIONS

This section on blimps and their role in agricultural expanses forms the major thrust of the chapter. Blimps or non-rigid airships are relatively a recent introduction into agrarian regions. Particularly, when compared with ground robots or small UAV aircrafts of utility to farmers. Blimps have not attracted great attention like the small fixed-winged or copter drones. These drone aircrafts are being adopted briskly by agricultural researchers and farmers world-wide. Perhaps, it is a matter time before agricultural agencies and farmers alike start realizing specific benefits of a blimp hovering over their farms. Blimps too could beproviding farmers with explicit aerial photography and digital data. They say, the long endurance, low cost per hour of vigilance and aerial survey possible with blimps should make them more competitive, in agrarian regions. Blimps may look large and cumbersome to maintain. However, they may not turn out to be 'white elephants' in any case.' Their utility to governmental agencies and individual farmers seems more assured and useful. Once, agricultural cargo transport via blimps picks up, perhaps, they will be sought frequently by the rural community. Blimps have a range of applications in the realm of agriculture, in general. During crop production, blimps offer variety of services to farmers. They make it easier for farmers to carry various agronomic procedures, by offering excellent images of their farm, from time to time. Also, they offer digital data that could be introduced into ground robots that are utilized during precision farming. Blimps can detect disease/pest incidence via the sensors that they carry, in the payload area. They can also help farmers by providing data about water stress index (CWSI) of crop. Indirectly, they help in scheduling the irrigation events in a crop field more accurately. Now, in the following paragraphs, let us consider the knowledge that we have accumulated so far, specifically, about blimps and their futility in farming. We should note that, there are not many evaluations of blimps made in farms. Reports too are feeble. There are only few short reports. Research effort to adopt them in a big way in farming, no doubt, is essential.

Blimps could be a good idea, if farmers intend to make a round-the-clock surveillance of field soils and their effect on crop productivity. Blimps have been already used to study the landform changes. They have offered excellent aerial photographs about detrimental effects of gully erosion. Since blimps have longer endurance, the assessment of loss of topsoil and gully formation could be done continuously. Such information could be relayed to ground computers. Early detection of soil erosion helps in timely remedial measures (Ries and Marzolff, 2003; Wells et al., 2017).

Heneghan (2018) forecasts that, in future, aerial surveillance of crops using Unmanned Aerial Vehicles may become too common in the agrarian belts. There are indeed several types of platforms from which the farmers could choose. Blimps with their specific traits of long endurance, slow transit, ability to hover and fly low may get preferred by farmers. At present, a stiff competition is in progress among the various types of UAVs that could eventually rule the sky above farmland. The popularity of a particular type of UAV could of course be transitory. Popularity of a blimp or a drone aircraft may fluctuate. The best UAV or most chosen one may also change depending on agrarian location, farm size, cropping systems and farmer's economic status. In agrarian regions the competition is actually among fixed-winged small drones, small helicopters and multi-copters, blimps, tethered aerostats, balloons, parafoils and kites. It is quite a competition among engineering gadgets. It decides the aerial robots that would eventually dominate the agrarian belts. If we consider blimps exclusively, then, there are many aspects of agricultural crop production that could be done swiftly, accurately and at low cost to farmers. Blimps could remove a certain degree of drudgery in farms, by obtaining spectral data from above the crop fields. Farmers may avoid excessive field work. Instead, they may opt to use blimps with sensors and computer-based data capture and analysis. Prescription and timing of agricultural inputs could become more accurate. Now, let us first consider a few major uses of blimps in the farm world. No doubt, it will begin with advertisement about brands of farm vehicles, inputs and related products. Blimps have been used predominantly to advertise products and events. However, now, blimps are also excellent aerial vehicles to obtain data about crops.

A range of different types of UAVs have been utilized to obtain aerial images of ground surface and characteristics of crops in agrarian belts. Blimps fitted with visual and infra-red band width sensors (i.e., thermal sensors) and LIDAR can provide us with excellent images of crop land. Maps with accurate depiction of physical features of terrain and crops could be prepared, in short time. In a farming zone, we can utilize aerial imagery from blimps to demarcate the boundaries of farms, crops and irrigation channels with greater accuracy. Maps of land holding of farms could be very useful to agencies that cater farm inputs. Also, farm insurance companies may benefit from such accurate maps. Let us consider an example. In Zimbabwe, Ali (2017) has examined blimp derived aerial imagery for use in demarcating land holdings, classifying types of land holdings and their boundaries, cropping systems and their fringes, etc. It is said that, we can study in greater detail, the land characteristics, map them and match the features with cropping zones and their boundaries. No doubt, blimps could provide images useful to assess land holdings and soil fertility pattern. It ultimately helps to plan proper cropping systems.

Airships (blimps) can transport fresh produce directly from the farm to ports, local distribution centers or processing plants. Blimps can transport agricultural harvests even when the distances are great and transport infra-structure is of poor quality or not available. Faster, fresher, cheaper delivery, as well as the opportunity to support economic development in remote regions is possible, if we adopt blimps (EuroAirships, 2018a,b).

There are many companies that manufacture blimps. They offer models that suit the requirements of agricultural researcher and farmers. Many of them seem apt to be adopted by national agricultural agencies and large farming companies. For example, SkySentry LLC has produced blimps that promise a big role in agricultural regions. Their blimps are useful in monitoring the crops throughout the season. They help farmers to obtain rapid and regular information about crops. These blimps are flown at 200 to 700 m above the crop while obtaining spectral data. The aerial photography shows up areas that are affected by drought, floods, insects/diseases. Crop's vital data like NDVI, plant height, leaf chlorophyll index, crop's-N status and water stress index could be measured. Then, relayed to ground station computers (Proctor, 2015).

Petzer (2013) has reported that in South Africa, there are clear plans to station blimps across different locations in the nation. These blimps help in telemetry and radio signal transmission. These blimps help farming community to get better connectivity via broadband mobile connections. They help in disseminating information about farming, to rural South Africa. This is in addition to general telecommunication. It seems companies such as Google and Microsoft are also opting to use a network of blimps, to cater mobile usage requirements of rural(agricultural) South Africa. A different report suggests that in South Africa blimps are returning after a gap of 80 years. They expect the hybrid airships to revolutionize cargo transport, in general. Airships are being touted as an environmentally greener options to farmland (Airships Africa- South Africa, 2009; Geere, 2013). They are meant to transport over 200 tons of general or agricultural cargo for a distance of 250 km, at a stretch.

Jowit (2010) has opined that helium-filled powered ships (blimps) could eventually get greater degree of acceptance with regard to transport of cargo. Helium-powered ships could be carrying freight and even passengers in as little as a decade's time. However, we may note that we are yet to notice routine use of blimps in agricultural or other civilian transport functions. In Europe, companies have been evaluating blimps to carry fresh fruit, vegetables, flowers and other perishable items. These harvests are being transshipped rapidly from one place to another, within the nation. In the United Kingdom, there are forecasts that blimps could revolutionize transport of large sized agricultural cargos. They could be replacing aeroplanes to a certain extent, particularly, when distance to be covered is relatively short. A few other forecasts state that, replacement of aircrafts by large blimps could help in reducing cost for transport of goods. Further, blimps may cause less greenhouse gas emission compared to cargo transport aircrafts. Autonomous blimps or those that could be controlled using radio signals are available in many countries. In India, reports suggest that blimps have been utilized in monitoring ground activities, particularly, in locations dealing with agriculture and military (Blimps India Pvt Ltd. 2015).

3.5.1 BLIMPS TO SURVEY NATURAL VEGETATION, AGRICULTURAL TERRAIN, AND AGROECOSYSTEMS

Several different types of platforms have been utilized to obtain aerial imagery of natural vegetation and record changes, if any, periodically. Among them, most frequently utilized platforms are satellites, aircrafts, small drones (fixed-winged and copters). Blimps, tethered balloons and kites have also been utilized to serve the same purpose. Even static platforms located at vantage points have been adopted. Each type has its advantages and lacunae. Blimps possess relatively longer endurance in flight. They float for longer durations, say, a few days at a stretch. Hence, they can keep continuous watch over an agricultural terrain and specific cropped fields. Blimps could be kept floating at low altitudes. Therefore, they offer highly accurate aerial imagery. They can easily accommodate sensors and CPU in the payload area. Salami et al. (2014) have reviewed UAVs with regard to their several different characteristics such as size, take-off, altitude during flight, speed, endurance, fuel needs, sensors, etc. and their acceptability for different purposes. They say a blimp with long endurance is apt, if one aims at continuous surveillance and aerial photography of natural vegetation and crops. A blimp experiences relatively less disturbance and can even stand still. Particularly, while accumulating data via aerial photography and detailed spectral analysis. Blimps could be economically efficient, if adopted for long-term surveillance and spectral analysis of natural vegetation (Dunford et al., 2009; Martinez Rubio et al., 2005; Salami et al., 2014; Table 3.4).

Vericat et al. (2008) have stated that lighter-than-air blimps are useful in monitoring natural resources such as rivers and river valleys. Blimps provide aerial photography of such geographic features, periodically. They could also be programmed to provide aerial imagery of rivers and their banks, continuously. They say, aerial photography derived using blimps are highly useful in planning, establishing and managing irrigation projects. Particularly, when images are dispatched continuously. Blimps are among low-cost platforms that could be adopted, to obtain aerial photos of disaster affected regions. Particularly, if the crop production zones are located in flood prone regions. We should note that imagery obtained using blimps may differ in terms of accuracy and resolution. High quality aerial images are possible. Aerial photos of river basins could be obtained rapidly using blimps attached with CPU and connectivity to ground control computers (Vericat et al., 2008). This way, blimps could be of utility while managing irrigation projects.

Blimps could be utilized to monitor any drastic changes in terrain. We can note extent of soil erosion, loss of embankments and floods, if any, in a river basin. In other words, blimps could help in monitoring natural resources and their deterioration, if any. The aerial photography obtained periodically could help agricultural planners, to judge the damage due to floods/gully erosion. Then, they may apply remedial measures such as contour bunding, accordingly (Ries and Marzolff, 2003).

TABLE 3.4 Adoption of Blimps in Different Agrarian Regions

Country/Blimp Model/Manufacturer	Remarks on Use of Blimps During Various Aspects of Crop Production	References
Canada/Model-NA/Manufacturer-NA	Utilized to assess wheat crop growth and biomass accumulation. Aerial spectral analysis of wheat for growth (NDVI), leaf chlorophyll content, and disease/pest attack. Also, to develop digital surface models (DSMs)of wheat	Cao, 2018
USA/B256PRT2.2-5 Eblimps/Microlight Inc. Lake Forest, California, USA	Popular advertisement for agricultural inputs. Use of LED to popularize agricultural products	Eblimps, 2018
USA/ Hybrid Blimp-PLIMP/ Egan Airships Inc., Seattle Washington State, USA	Aerial survey of cropped land, detecting, soil fertility, crop growth variations, crop' water status and drought-affected patches.	Egan Airships Inc.,2018; Jorge Lois Alonso, 2015
USA/ Model- NA Seattle, Washington State, USA	Crop phenotyping using aerial data from blimps/drones	Sankaran et al., 2015
USA/ Model-Tethered Blimp Southern Balloon Works Inc., Deland, Florida, USA	Aerial imagery, collection of spectral data and values for vegetation indices such as NDVI, CWSI, LAI, leaf chlorophyll index (crop's N status)	Grace, 2004; Haire, 2004; Ritchie et al. 2010
USA/Hyperblimp/ Hyperblimp Inc., Utah, USA	Hyperblimp used to spray agricultural fields with herbicides to control weeds	Geery, 2018

TABLE 3.4 (Continued)

Country/Blimp Model/Manufacturer	Remarks on Use of Blimps During Various Aspects of Crop Production	References
Mexico/ AB 1100/ Cameron Balloons Ltd. Bristol, United Kingdom	Rapid detection of NDVI values of wheat crop. Detection and mapping of wheat rust disease and aphid attack. Evaluation of wheat genotypes for drought tolerance and disease/aphid resistance	Listman, 2018; CIMMYT, 2012;
Mexico/ AB 1100/ Cameron Balloons Ltd. Bristol, United Kingdom	Aerial survey of soil conditions. Collection of physiological data of utility to crop breeders (e.g. maize and wheat). Crop phenotyping and development Digital Surface Models.	CIMMYT, 2014
Brazil/ ADB 3-X01/ Airship do Brazil Industrie, Brasilia, Brazil	Photogrammetry of natural vegetation, spectral analysis and digital data collection about cropped fields. Mapping of natural resources, soil fertility variations and cropping systems.	Airship do Brazil Industrie, 2017
Argentina/ Model- NA Manufacturer-NA	Preparation of farm inventory, land and soil maps. Aerial survey and digital data procurement from above the forest stands and cropping stretches, national parks and rivers.	Cervantes, 2010
The United Kingdom/ Helium Airships/ Hybrid Airships Ltd. Bradford, United Kingdom	Transport of large agricultural cargo.	Jowit, 2010
Germany/ Zeppelin-NT/ Deutsche Zeppelin, Friederchshafen, Germany	Aerial survey of natural vegetation and cropped land. Measurement of atmospheric quality above agrarian regions. Investigation related to atmospheric biology above crop fields. Collection of spectral data about crops, to assess nitrogen and water status.	Zeppelin Science.de, 2018

TABLE 3.4 *(Continued)*

Country/Blimp Model/Manufacturer	Remarks on Use of Blimps During Various Aspects of Crop Production	References
Switzerland/Helium Blimp/ Blimp Works Inc. Statesville, N.C., USA	Blimp used to collect atmospheric gaseous and particulate samples above cropped fields. Assessment of methane emissions	Stieger et al., 2015
Netherlands/ Blimp-type Airship/ Fern University, Hagen, Netherlands	Aerial survey and monitoring of crops in farms. Assessment of greenhouse gas emissions, atmospheric dust and visibility above crop fields.	Gerke et al., 2013; Bouche et al.,2016
Russia / Hybrid blimp/ Bee Robotic Corporation Millbrae, California	Pilot testing of blimps with spray bar and nozzles to distribute pesticides and fungicide uniformly on crops.	Salnikov et al., 2014
Zimbabwe/ AB-1100 Cameron Balloons Ltd. Bristol, United Kingdom	Aerial survey of maize. Collection of data for phenotyping crops and preparing. 'Digital Surface Models'.	CIMMYT, 2012; 2014
Zimbabwe/ Blimp model: NA Manufacturer: NA	Aerial survey of farmland. Demarcating landholdings and cropped fields. Mapping boundaries of different cropping systems.	Ali, 2017
South Africa/ Google Blimp/ Blimp network, Natal, South Africa	Aerial survey of large stretches of farming zones. Blimp network is utilized to transmit information using mobile phone technology.' A concept called as 'Phenomobiles' is being developed in farming belts of South Africa. Phenomobiles help in rapid relay of data on crop growth rate.	Geere, 2013
South Africa/ DynaLifter/ DynaLifter Inc. Durban South Africa	Transport of heavy cargo pertaining to agricultural inputs or harvests.	Airships Africa; South Africa, 2007

TABLE 3.4 *(Continued)*

Country/Blimp Model/Manufacturer	Remarks on Use of Blimps During Various Aspects of Crop Production	References
China/ Autonomous Airships	Aerial survey, photogrammetry and observations on agricultural terrain.	Wei, 2010
China/Blimps Manufacture -NA	Crop phenotyping and collection of data periodically about plant height, leaf area index, leaf chlorophyll and NDVI (Biomass),	Yang et al., 2017
Australia/ Tethered Blimp/ Univ. of South Queensland/ Toowoomba, Australia	Aerial imagery of cereals and cotton using optical and thermal sensors. Thermal imagery utilized to prescribe irrigation. The digital data showing variation in water stress index used in precision farming.	Smith et al. 2010

Note: Each of the examples in the above list depicts the use of a blimp in agricultural farms or related activities. It is indicative of the relevance of blimps to farmland. However, it is not an exhaustive list of usage of blimps in Agriculture. One of the examples shown above is a small zeppelin (Zeppelin NT), a rigid airship, and not a blimp. NA = Data Not Available.

Cervantes (2010) has stated that, airship technology is an essential aspect of monitoring the natural reserves, forests, agroforestry projects, cropping belts and national park areas in many of the Latin American nations. Blimp technology brings together and trains personnel in aerial imagery. Blimp manufacture helps in imparting training to agriculturists, engineers, airship enthusiasts, forest technicians, experts in renewable energy resources and university students. Blimps project trains them regarding uses of aerial surveillance, survey and spectral analysis. There are several UAV companies in Mexico, Equador, Argentina, and Brazil. Blimp project is disseminating knowledge about light-fabric technology and thin film solar cell technology. Most importantly, they are imparting training in production of unmanned aerial vehicles and blimps. According to Cervantes (2010), initially, the Latin American Blimp program started operations by conducting aerial survey from 1 to 9000 ft. above the ground surface. They adopted 3 small airships ranging in size from 7.8 m to 14 m long. They were all equipped with electric motors for thrust. Blimps could travel at 25 to 40 kmph. Flight endurance was 1–2 hr. The autonomous blimps allowed a maximum payload of 8kg. The blimps were filled with helium. The autopilot capability extended for up to 5 km from ground control station.

Again, regarding use of small unmanned blimps in Latin America, Cervantes (2010) states, they are mainly adopted to monitor the greenhouse gas mitigating projects; community level agriculture, agroforestry and forestry projects; monitoring forest vegetation at risk; developing soil inventory, studying plant diversity and mapping plant species. We may note that, satellite imagery and manned aircrafts have been the usual methods adopted to develop maps. However, with the advent of airship technology, particularly, long endurance blimps could take over mapping of vegetation. Species diversity could be studied in great detail because blimps can float low above the canopy. Blimps are slow during transit. Therefore, high resolution imagery of a particular spot is possible.

3.5.2 BLIMPS IN AERIAL SPECTRAL ANALYSIS OF CROPS

Inoue et al. (2010) have clearly shown that low altitude blimps have a role in assessing crops' performance. They have put forth an idea that blimps with their ability for continuous spectral analysis of crops and nearby vegetation are well suited, to study crop productivity. Blimps may also help us in obtaining spectral data about agroecosystems. Several parameters such as NDVI, biomass accumulation, drought and its implications to a patch of vegetation could all be assessed, using spectral data from blimps. They have further emphasized that longer endurance of 5–10 days possible with blimps is a great advantage. Particularly, during continuous collection of data. Blimps could be entirely autonomous. Their flight path, way points and camera shutter activity could all be pre-programmed. Also, they could be remotely controlled using radio signals. Blimps move slowly at < 40 kmph. Blimps can hover at a single point above crops. Hence, they do offer high resolution spectral data and visual imagery of crops. It is said that, spatial variability of soils, their fertility and crops could be studied accurately, using such blimps (Plate 3.16).

PLATE 3.16   A PLIMP- a hybrid blimp, copter and airplane.

Source: Dr. Daniel Nelson, Egan Airships Inc., Seattle, Washington State, USA.

Note: The blimp helps in assessing crop growth pattern. It identifies water stressed patches, disease/pest attacked regions and grain maturity.

Unmanned aerial vehicles include various types of autonomous blimps. Autonomous blimps are currently getting popular, both, in the urban and rural agricultural regions of mainland China. Wei (2010) describes efforts by Chinese Academy of Autonomous Airships to utilize them in aerial survey and photogrammetry above the general terrain and agricultural regions. It is said that, in China, an autonomous blimp may offer greater number of advantages and cost efficiency, to operators/farmers. Satellite-based methods, as we know have constraints regarding re-visit time, lower resolution, haziness due to clouds and distance, etc. Low flying aircrafts too has many constraints. They are expensive to use repeatedly. Small UAVs, (i.e., fixed-winged or copter drone aircrafts) too have constraints related to short endurance and cost. Hence, the Chinese Airship Academy is trying to tout the use of autonomous blimps that transit slowly or stay stand-still, while obtaining the aerial imagery of crops. Autonomous blimps do offer images with high resolution. It is because they float at low altitudes of just 100 m above cropped fields. Further, Wei (2010) states that such autonomous blimps adopted to float above agrarian regions or urban locations, often have a set of 4 cameras (sensors). They are less costly to purchase and utilize. The commonly used computer software regulates flight path, speed during floating, way points and sensor shutters. The image processing is done carefully using software such as Pix4D Mapper, MAP-DEM to obtain digital elevation models. MAP-DSM is used to obtain digital surface models (DSMs) of crops in fields. MAP-DOM is adopted to store and process digital ortho-images. (Wei, 2010; Jian et al., 2007; Tuan et al., 2010).

Let us consider an example, wherein, cotton crop (*Gossypium hirsutum*) has been studied using tethered blimp-mounted sensors. The tethered blimp (Southern Balloon Works, Deland, Florida, USA) could carry a pay load of 4 kg. The pay load cameras included a Nikon 4300 mega-pixel digital cameras. One of them was a near-infrared camera. Ritchie et al. (2010) state that, a combination of green, red, red-edge and NIR reflectance data derived from blimp-mounted cameras were most useful, in assessing ground cover fraction (GCF). They further report that, data derived using sensors on blimps and those recorded, using ground-based sensors correlated significantly. They have concluded that green/red ratio index derived using blimps could be quick, simple and accurate. It allows us to assess NDVI, biomass and groundcover fraction of cotton, at different stages of the crop. The blimp-derived data depicts crop growth trends and stress effects, if we account and separate out the effects of soil and atmosphere on spectral readings. No doubt, in future, crops' growth rate could be measured frequently, during the season. Measurements could be done accurately using sensors mounted on blimps. We are yet to assess several crop species regarding feasibility of blimps and blimp-mounted sensors. Particularly, to collect spectral data throughout the season and arrive at appropriate prescriptions.

3.5.3   BLIMPS IN CROP PHENOTYPING

Crop phenotyping is relatively a new methodology that is adopted by agricultural experts dealing with crop genetic selection, crossing programs, agronomic evaluation and yield formation. At present, many of the farming companies and farmers with large holdings and even small farmers are adopting crop phenotyping as a method. Crop phenotyping helps farmers in developing input schedules in relation to their growth pattern and yield expectations. Crop phenotyping is a tedious task, if conducted using farm scouts. It might take longer hours of drudgery in open fields. It involves travelling from one spot to another in a field, assessing crop parameters and recording data as accurately as possible. Avoiding human related errors and fatigue is equally important, if crop phenotyping is done using skilled farm technicians. At present, the trend is to opt for small drone aircrafts (fixed-winged, helicopters, or multi-copters) with sensors, to collect data useful in crop phenotyping. These aerial robots finish the same task of getting digital data, processing and developing prescription, in a matter few minutes or at best a couple of hours. Blimps are also adaptable to obtain data useful in crop phenotyping and developing digital surface models (DSM). Such data is useful to develop fertilizer and irrigation schedules. Of course, after due consideration of yield goals. Blimps are only a recent introduction in crop production areas *per se*. However, already, there are many instances, wherein, farm experts have tried adopting blimps to procure parameters relevant to crop phenotyping. Let us consider few facts and examples.

Researchers in China have made considerable progress in adopting different types of UAVs and tethered aerial vehicles, during crop production. They have examined the blimps that are of utility in monitoring crop growth and general health. Also, in providing useful digital data to judge crop's need for different inputs. Yang et al. (2017) too state that, in China, blimps have net advantages during aerial survey and phenotyping of crops. Blimps can hover at a single spot. They have ability for vertical take-off and landing. They are slow and so high-resolution imagery is possible. Blimps offer extensive data that is useful in crop's phenomics studies. In fact, phenotyping of crops is done utilizing traits such as plant height, leaf number, leaf area index, leaf chlorophyll content, etc. Data derived using sensors on blimps could easily become the basis for scheduling inputs. Also, in selection of crop genotypes with better traits. Forecasts suggest that blimps could play a major role in collecting relevant data for phenotyping and phenomics in agricultural experimental farms. Yang et al. (2017) and Li et al. (2017) report that blimps could be used most efficiently to assess crop growth. In particular, they could help in developing digital surface models.

Let us consider an example. In China, a blimp-based remote sensing system was developed. It aimed at monitoring land surface variables from low altitudes. The blimp had a helium-filled envelope. The blimp was 23 m long with a maximum diameter of 7 m. The blimp cruised at a speed of 40 kmph. It could hover at 30 to 400 m altitude, under radio control, with a payload of 100 kg. The blimp carried four monochromatic video cameras. The ability of the blimp to hover allowed the collection of a continuous set of high-quality images. The spatial variability between and within agricultural fields was mapped at a fine spatial resolution. The blimp-based system proved to be suitable for low-altitude remote sensing of plant variables (Yang et al., 2017).

The International Maize and Wheat Center in Mexico has been evaluating the usefulness of blimps in obtaining aerial images and data sets, to phenotype both maize and wheat, all through their growth period. Blimps specifically meant for phenotyping crops are being adopted by wheat physiologists, at the Mexico center (CIMMYT, 2014; Reynolds et al., 2012; Plate 3.17). Further, reports suggest that blimps with sensors (infra-red) have been applied to select crop genotypes that tolerate drought and water stress better. Crop phenotyping using blimps offers certain clear advantages to physiologists. Ground analysis of crops using hand-held sensors and ground vehicles could be tedious. Also, human fatigue related and other types of errors creep in. Blimps offer accurate spectral images and data to prepare digital surface models of wheat/ maize crops. Farmers could compare such DS to channel appropriately sized inputs, to standing crops.

Aerial photography-based phenotyping platforms are being considered as an alternative option, to overcome limitations associated with ground-based phenotyping platforms. Aerial-based phenotyping platforms enable the rapid characterization of many plots within minutes. Initial phenotyping platforms used were small airplanes (e.g., crop-dusting airplanes). However, this is costly, and it is difficult to safely achieve the low speeds required for high-definition images at low altitude. The current generation of phenotyping platforms, vary significantly, in terms of the payload, initial costs, maintenance costs, and control. Recently developed alternatives include 'phenotowers.' However, these types of phenotyping platforms (phenotowers) have a maximum height of 50 m. Blimps can be held at an appropriate altitude. They have sensors mounted underneath the envelope. Their advantages are their ability to carry a heavy payload (several kilograms). They can make many sensors work concurrently. Therefore, blimps may become popular with farmers needing information crop's phenotypic progress.

PLATE 3.17   A blimp scouting the field at the International Maize and Wheat Center's (CIMMYT) Experimental Station, at Ciudad Obregon, Mexico, scouting the fields.

Source: Dr. Clyde Beaver, International Maize and Wheat Center (CIMMYT) Archives, Mexico, USA

Note: Blimps could reduce the need for skilled farm technicians during collection of crop's data.

Li et al. (2017) too have pointed out that, at present, there are several ground-based methods of assessing crops using sensors. We can collect useful data for rapid phenotyping of crops. Field vehicles mounted with sensors are often termed 'Phenomobiles.' They could be moved from one plot to another. However, it could be time consuming. The data is obtained at plot level. Therefore, it needs no further processing. Aerial robots, such as robotic drone aircrafts or blimps, in comparison, cover large areas and obtain spectral data in a matter of minutes. However, such data needs processing using appropriate image processing software. Prescriptions for nutrient application, irrigation or pesticide application could be based on DSM, but they need specific software. Overall, since blimps are swift in collecting data about crops, they're being touted as possible alternatives to ground-based phenotyping methods.

Sankaran et al. (2015) have reviewed various low altitude, high resolution aerial survey methods to study the crop's phenology. During recent years, plant breeders/physiologists rely on crop phenotyping while selecting genotypes with characters for better performance. Crop phenotyping is also done to assess the progress of a crop in field, at different intervals. They say, among various aerial vehicles, blimps could be a cost-effective method for phenotyping crops. Blimps have the advantage of long endurance that allows crop breeders to study and obtain data continuously.

Virlet et al. (2017) point out that ground-based methods of phenotyping of crops is tedious and only a few sensors could be adopted at a time. Human error creeps in often, if the fields are very large. Therefore, phenotyping using fixed-winged or copter drones are getting popular. Blimps too could aid in rapid collection of data. Blimps help in phenotyping crops at short intervals. Blimps can carry a larger payload of sensors. They have longer endurance and hovering ability. Data from blimps can help in developing 'digital surface models (DSM).' As stated earlier, DSMs are highly helpful to farmers. DSMs provide an idea about the performance of crops, at a given instance.

Blimps are among the most suited aerial platforms to phenotype a crop, all through the season. We have to re-state that its ability for long endurance, hovering and slow movement in the sky are added advantages, particularly, while phenotyping crops. In Canada, blimps could be used to assess canola crop. We can get data about canola's growth rate, leaf chlorophyll content (i.e., crop's-N status), water stress conditions, if any, disease/pest attack and weed infestation in fields. The spectral data from blimps can help farmers, to decide the quantum and timing of fertilizer-N and irrigation. In Saskatch-ewan, researchers have shown that it is possible to use different platforms such as blimps, to phenotype a crop of canola. They say, we can even arrive at tangible yield forecast, using aerial survey data (Cao, 2018; Jose et al., 2014). Digital Surface Models (DSMs) have been useful in assessing crop's performance, periodically.

3.5.4 BLIMPS/ZEPPELINS IN PRECISION FARMING

Blimps/Zeppelins, like any other aerial platform that supports sensors have a role to play, if farmers want to apply precision farming techniques. We may note that during past two decades, precision farming methods have gained acceptance with farmers and farming companies with large holdings, particularly in developed nations. Precision farming, in the general course, requires tedious collection of data about spatial variations in soil types, soil fertility factors, disease/pest attack patterns, drought and crop water stress index patterns and finally grain yield patterns (see Krishna, 2012, 2016, 2017; Zhang, 2015; Sharma, 2007; Stafford et al, 2000). Many of the inputs are channeled at variable rates based on digital data obtained. The soil/crop related data could be obtained using skilled farm technician (scouts). It involves intense application of skilled labor. Further, human related errors and fatigue could reduce accuracy of the data. Also, if farm human labor is involved, then, we may not be able to repeat data collection too often. Above all, use of skilled farm scouts may be economically costly and turn out to be inefficient, if we consider the net gains attributable to practice of precision farming principles. It is this background that induced a great surge in use of small fixed-winged, helicopter and multi-copter drones that are less costly, efficient and easy to manage. These drones could be flown swiftly over the field/crop as many times in short intervals (SenseFly Inc., 2015a; 2015b; Precision Hawk Inc., 2014; Trimble Inc, 2015; DJI 2019; RMAX, 2015). The data collected into computer chips could be used in autonomous seed planters and pesticide/fungicide applicators. Also, in fertilizer inocula-tors. Precision farming, in other words, variable-rate application of inputs may reduce use of fertilizers, pesticides and other harmful chemicals and irrigation water. So, there is no doubt that aerial platforms (e.g., blimps) that offer digital data about variations in field topography, soil types, soil fertility and crop yield are sought. First, the data showing variations helps in demar-cating 'management block.' Management blocks make adoption of precision techniques less cumbersome and more accurate. Aerial imagery collected by blimps and transmitted to ground computers can be highly useful in marking 'management block.'

Now, consider the possibility of using blimps/zeppelins during precision farming. Blimps/zeppelins are relatively slow to move above the agrarian regions, farms, land resources and cropped fields. Therefore, blimps offer aerial imagery and digital data that is more accurate and of higher spatial resolution than other platforms. Blimps allow relatively large payload volume and weight. The payload can include a series of electro-optical sensors (R, G, B, Infra-red, Rededge, etc.), electro-chemical probes for greenhouse gas emissions, lidar, magnetometers, etc. Blimps have longer endurance say, a few weeks in the air. They can collect and relay data to ground vehicles for longer periods. There are now examples, where in low flying blimps could also be adopted, to spray fertilizers and pesticides on to crop's canopy (Salnikov, 2014; Geere, 2018; Jorge Luis Alonso; 2015). We need to conduct several experimental trials to evaluate agronomic and economic efficiency of blimps, during precision farming. Overall, studies reported about use of blimps in precision agriculture is feeble. Clearly, use of blimps in precision farming needs greater attention from agricultural experts and farmers.

3.5.5   BLIMPS TO MEASURE CROP WATER STRESS INDICES

Crop specialists at the University of Georgia (Georgia, USA) have recently adopted a 6.5 ft. blimp. The blimp was manufactured by a company situated in Florida, USA. They have attached a series of cameras to obtain high resolution spectral data about crops such as cotton, maize, melons, etc. (Grace, 2004; Haire, 2004; See Table 3.4; Plate 3.18). The blimp floats at a height of 300 ft. above crop's canopy. Researchers use blimps to obtain a series of images of crops. The primary aim is to asses growth (vegetation indices) and water status (CWSI). Details about crop's water stress index (CWSI) forms the basis for irrigation. Accurate timing of irrigation is essential. This is because, farmers in Georgia, USA are known to incur loss due to improper irrigation of cotton. State-wide, about 80 million US$, it seems, are lost due to water stress. Drought affects cotton crop, at various stages. A blimp above cotton fields may offer information to farmers, so that, they could correct the situation. Further, it is said that, spectral data about crops that blimps collect may help farmers to apply insecticides, herbicides and fungicides, at proper times. Blimp derived details also help in channeling correct quantity of water. Irrigation is applied only at the spots affected by drought. Researchers at the University of Georgia hope that blimp-based technology will be available to farmers, sooner.

Blimps could play a vital role in forewarning drought effects on crops such as cereals, pulses, cotton, etc. grown in semi-arid belts. A report from Researchers from National Center for Engineering in Agriculture, Australia states that, tethered blimps are being adopted to help cotton farmers in Queensland, Australia (Land and Food Resources, 2008). The blimp was stationed at 90 m above the crop's canopy. The 6 m long blimp was evaluated for usefulness and efficiency in obtaining, aerial optical and thermal imagery of cotton crop. The thermal imagery was utilized in assessing the cotton crop's water stress index. The digital data was processed and utilized in channeling irrigation water, at variable rates. Such blimps may find use in drought prone areas and in farms that depend on aerial survey, imagery and precision irrigation (Smith et al., 2010).

PLATE 3.18   A farm research blimp.

Source: The University of Georgia, Camilla, Georgia, USA.

Note: Researchers at University of Georgia Experimental Station are using a blimp (airship) stationed above crops. This is done to get detailed data about topography, cropping pattern, nutrient and water needs, etc. Sensors such as visual, multi-spectral and infra-red bandwidth are used to obtain spectral data. The airship could also be used to collect atmospheric samples, for further analysis. At present, these blimps are used to collect data about crop's water stress index. In due course, blimps may relay data about disease/pest attack on cotton crop in Georgia, USA. Spots affected with excessive weeds could also be traced using blimps.

3.5.6   BLIMPS TO STUDY CROP DISEASE AND PEST ATTACK

Historical records show that, after World War II, farmers did utilize blimps to spray pesticide and control boll weevil on cotton. They also sprayed a million pounds of carbon arsenate dust over corn, using blimps (McWilliams, 2014). So, blimps were adopted to conduct aerial spraying long time ago, during 1940s and 1950s.

A blimp integrated quad-copter aerial automated pesticide sprayer (AAPS) was developed for pesticide spraying. The blimp operates based on the GPS coordinates and at very low altitude above the canopy of the crops (Mogili and Deepak, 2018; Varadhan et al., 2018). A report by Yogita et al. (2017) states that blimps attached to copter drones are highly useful, during spraying crops with pesticides. Such a 'blimp + copter' hybrid requires relatively lower levels of energy to complete spraying the same area, if compared to a human scout or aircraft. Blimps assure floating of the hybrid craft. They point out that such a hybrid spraying blimp is cost efficient plus accurate. Blimps avoid farmer from getting into contact with harmful agricultural chemicals.

There are reports that hybrid airships with facility for spraying are available for use on cropped fields (Salnikov, 2014). It is said that using such blimps we can conduct aerial application of fertilizers. Pesticide sprays and regular crop dusting of crops is also done, using hybrid blimps. Usually, such hybrid blimps have pesticide or liquid fertilizer reservoirs (tanks) and spray bars. They transit low over crops to avoid wind disturbance.

Hyperblimps can quietly watch wildlife or float into wilderness areas that are otherwise inaccessible. With certain modifications, they can identify and selectively spray the invasive weeds with herbicides or help in devising methods to eradicate pine beetles (Geery, 2018).

In due course, may be less than a couple of years from now, blimps may get adopted more frequently in farm settings. This may include spraying activity using blimps rather commonly (Fima et al., 2018). Regarding other types of aerial robots and semi-autonomous vehicles, we have fairly well stated rules and regulations. Most of these drone aircrafts are low flying and they may avoid drift of chemical sprays, efficiently. However, blimps may have to stay afloat at a slightly higher altitude during spraying. So, either spray bars have to be extended and lowered to reach the crop canopy or suitable guidelines for blimp-aided spray of agricultural chemicals need to be formulated. There are well publicized guidelines and precautions that should be adopted during aerial spray of agricultural chemicals (Lavers, 2001). There are also FAA guidelines decided by the USA. Similarly, rules have been formulated by other nations such as Canada, Australia, etc. These regulations pertain to drone aircrafts that could be used as robotic aerial spray machines, during crop production. Regarding blimps, we may need well defined and apt rules and regulations for adoption in agricultural farms. Particularly, for purposes such as aerial photography from different heights, surveillance, and spray of agricultural chemicals.

Agricultural blimps (smaller versions) seem to be less costly when they are made as hybrids attached to small robotic multi-copters. These multi-copters have already established a niche for themselves in rapid, accurate and easier handling. As such, sprayer copters are gaining acceptance while spraying of pesticides, fungicides, herbicides, and even liquid fertilizers. Small blimps add to safety, if the copter fails. Hence, we could forecast that in due course, the 'small hybrid blimps' that stay afloat or loiter all through the season, above the crop fields would be a common site. Blimps with sprayer bars and ability to float at low altitudes above the crop is the need of the hour. Further, If, blimps are energy and cost efficient, then, they may get preferred over other aerial robots, during application of plant protection chemicals.

3.5.7 BLIMPS TO STUDY ATMOSPHERIC QUALITY AND GREENHOUSE GAS EMISSIONS ABOVE CROPPING BELTS

Several different types of unmanned aerial vehicles (UAVs) have been utilized, to assess weather conditions prevailing in the atmosphere above the crops. For example, small copters with chemical sensors and probes are utilized to detect particulate faction in the atmosphere. The copter dronestoo supply useful data about atmospheric quality above crops (Scenteroid Inc., 2017; Krishna, 2019). Weather balloons are also utilized to assess atmospheric conditions above crops. Balloons relay data from different heights in the atmosphere above crops (Bouche et al., 2016). More recent suggestion is to adopt blimps with a series of optical and chemical sensors. Blimps could be utilized to monitor visibility, particulate fraction, gaseous composition (e.g., CO_2, NO, N_2O, CH_4 SO_2 and H_2S) and moisture above crop fields. Blimps have also been utilized to collect data from sensors and sensor networks that assess soil moisture, soil organic matter content and soil moisture (Gerke et al., 2013).

Agricultural crop production causes greenhouse gas emissions. The extent of loss of CO_2, CH_4, N_2O, NO_2 and NH_3 may vary depending on the soil type, cropping systems and fertilizer supply trends. Emissions from cattle too are generally high, especially, as NH_3 and CH_4. These greenhouse gas emissions need to be monitored constantly and accurately though out the year. Greenhouse gas emissions per crop season too may vary. Emissions vary a lot based on season, soil, agronomic procedures and crops grown in a location. Farm experts and skilled technicians usually need long ours of sample collection. Sample collection above the crop's canopy is not easy. Chemical analysis of samples at the laboratory are also tedious. Hence, there is an urgent need to standardize air sample collection procedures, using one or the other UAVs. We know that copters that fly low over crop's canopy have been utilized to collect atmospheric gas sample. Blimps collect samples until an altitude of 300 ft. above ground surface (Stieger et al., 2015). Let us consider an example, wherein, air samples were collected from farms in Switzerland. Greenhouse gas emission (e.g., CH_4 and CO_2) has been authenticated using samples collected by the blimp. The collection of gas sample was done at two different stages of the crops, during 2011 and 2012. They have reported that 77% of farms with crops and livestock do emit CH_4 into atmosphere.

At this stage we may note that, it is possible to place as many chemical probes and detectors on blimps. For example, probes that detect, say, NH_3, N_2O, SO_2, CH_4, CO_2, and particulate fraction are feasible to be placed on blimps. Probes could be placed in the blimp's gondola itself. With an efficient CPU also placed in the gondola, it should be possible to relay the readings for atmospheric composition. Blimps with ability for long endurance and continuous relay of atmospheric quality would be a boon, to national weather agencies and farmers.

3.5.8   BLIMPS TO SURVEILLANCE AND MONITOR CROPS IN AN AGRICULTURAL EXPERIMENTAL STATION

Blimps that float (move) or stay static over an agricultural experimental field offer certain advantages. The sensors could record data about crop's progress, continuously. Later, we may transmit the same data to ground computers for instant analysis (see reports by CIMMYT, 2012; Krishna, 2019). In fact, globally, a network of experimental stations having blimps hovering over each location is a boon to crop researchers. Particularly, while evaluating their genotypes under multi-locational trials worldwide. For example, there is a collaboration across continents, wherein wheat genotypes are evaluated for aphid and virus attack. Such a network operates in USA, Australia, and New Zealand. Blimps based network can have a major role in rapidly collecting data at several different experimental stations and exchanging it. Next, Listman (2018) states that, there is a need for evaluating wheat genotypes against the rapidly mutating rust fungus (*Puccinia striiformis f. sp. tritici*). We have to select resistant genotypes of wheat based on multi location trials. In this case, blimps equipped with sensors and floated at multi-locations can collect relevant data, simultaneously. Using blimps, researchers may find out resistant genotypes, rapidly. They say, the current need is to find genotypes that are tolerant to several strains of pathogen, in given locations. It is opined that, the data exchange and evaluation could be managed more efficiently, using a network of blimps across 'hot spots.' Incidentally, there are reports that blimps have been adopted to collect physiological data of large number of wheat/maize genotypes, in-one-go (see CIMMYT, 2012; Krishna, 2019).

Blimps can play a vital role as sentinels when stationed above agricultural experimental stations. They can keep a watchful eye over the installations within farms. They can trace movement of farm vehicles and keep an account. Also, store data about happenings in the farm. They can relay information about the progress of agronomic procedures within the experimental station, to ground control computers.

3.6   FUTURE OF AIRSHIPS: BLIMPS AND ZEPPELINS IN AGRICULTURE

So far, it is clear from the above narration that, airships in general have played their role in military, cargo transport, travel, advertisement industry, in policing urban locations and public events. Blimps and Zeppelins, in particular, went through a period of disaster followed by rejection or neglect for a long period of few decades. They say, during recent years, with better safety ensured for their use, blimps/zeppelins are making a kind of resurgence, in many of the activities on earth. Aerial imagery has become the basis for a wide range of professions and activities. Aerial imaging, *per se*, is done using sensors placed on platforms. Airships (Blimps and Zeppelins) are among the few platforms that are experiencing a renewed interest and popularity. During past 5 years, there is actually an underling competition going on among the different types of aerial robots, utilized, to acquire aerial images. Right now, platforms most accepted and popular are the small drones (fixed-winged and copters). They offer aerial images quickly that could eventually be processed, analyzed and utilized by various agencies. In agriculture too, the small aircraft drones are currently the most sought-after aerial robots. However, during past few years blimps/zeppelins have also been examined for their utility in farmland. A blimps/zeppelin successful in aerial reconnaissance, cargo transport and advertisement could also be a successful aerial vehicle in crop production zones. The sensor technology adopted, which is the crucial portion for spectral analysis and collection of digital data is the same. At present, blimps have been examined for their use on major crops such as wheat, maize, lentils, oil seeds and forest plantations. Blimps act as sentinels above large farms, plantations and experimental stations. There are great many advantages listed for blimps. Hence, the forecast is that blimps may garner a fair portion of market that relates to aerial robotics in farmland. There are instances, when researchers/ farmers have opted for both blimps on a long a term basis and small drone for quick aerial surveys and digital data collection. In that case, each type of platform may be serving the clients, to satisfy a particular aspect of analysis.

Heneghan (2018) believes that drones, i.e., small fixed-winged and copter airplanes may eventually take over the task of survey, monitoring and collecting data about crops in the field. The above statement may also apply to other types of UAVs such as blimps. In this regard, we may note that blimps with long endurance and ability for continuous stretch of surveillance may garner sizeable share of activity, in the sky above crops. Agrarian regions may encounter intrusion of a large number of floating blimps fitted with a set of sensors. Blimps may offer images and spectral data about crop growth, water requirements, disease/pest attack, etc. to the farmer. So, blimps could be replacing a sizeable number of field scouts trained specifically, to collect data about crops in large fields.

Further, Heneghan (2018) states that blimps could be garnering a portion aerial robotic activity above agricultural sky. Blimps could be cost effective on a long range and if we consider community ownership of a blimp unit. Also, when agricultural agencies utilize them to serve farmers with requisite aerial photos, digital data and apt prescription. Whatever is the basis of guess or authentic forecast, we ought to realize that future of blimps in many aspects of human activities, particularly, in agriculture is actually indicated by the market for them. The number of units of blimps/zeppelins manufactured and sold to different agencies is a good indicator. Blimps were used originally for military, travel and advertisements. Demand for blimps to cater to these very activities continues as such. It is the new demand created by agricultural agencies, companies with large land holdings, agricultural service agencies, national agricultural departments, and farmers' societies which can induce greater exchequer, to airship industries.

Recent reports suggest that usage and renewed demand for blimps and related aerial survey activity was highest in North America, followed by Europe, Southeast Asia, and rest of world in that order. Latin American agriculture that constitutes a major portion of global agriculture is just poised to use blimps. Latin American farmers could adopt them in large numbers. During 2016, global market size for airships (blimps/zeppelins) was about 174 million US$. It is expected to increase by 7.2% for the next few years. It has been forecasted that, by the year 2024, airship market may rise to 304 million US$ per year (Variant Market Research, 2018; PR Newswire, 2017). Yet another report states that, global demand for airships at present is around 180 million US$. It is expected to grow at 6.9% per year. So, the market may reach 260 million US$ in 2023 (Sambit, 2019).

The above reports further suggest that, right now companies such as Lockheed Martin Inc., Airborne Industries Inc., Hybrid Airships Ltd., Lindstrand Technologies, GEFA-FLuGGmbh, Shangai Vantage Airship manufacturing Co., Skyship Services Inc., RosAeroSystems, s.r.a.l and few others are major manufacturers (Variant Market Research, 2018). So, these companies and others that may appear on the scene newly may affect the investment and desire, to develop blimps as useful and popular platforms in farm world. There could be many more new models and new improvements in drone aircrafts and blimps. Such modifications may attract farmers and lead to greater adoption of these aerial vehicles. The hybrid airships are one example. Hybrid copters with helium balloon to float freely and a copter-sprayer attached to it is being touted, as a better aerial vehicle in farmland (see Fima et al., 2018; Vardhan et al., 2018; Salnikov et al., 2014; Minor, 2004). In future, they may garner more space in the agricultural sky. At the bottom line, reason for demand for aerial robots is of course the human drudgery that it reduces and cost effectiveness. Also, the ease with which a particular farm procedure can be conducted and successfully completed, using blimps. There is absolutely no doubt that, in about a decade, the agricultural skyline is going to transform. It will show up large number of different kinds of aerial robots. Airships (i.e., blimps/zeppelins) are going to be a conspicuous part of the skyline above all agrarian regions of the world.

KEYWORDS

- automated pesticide sprayer
- Autonomous Unmanned Monitoring Robotic Airship
- crop water stress index
- digital surface models
- ground cover fraction
- ground control station
- long endurance multi-intelligence
- lighter-than-atmosphere
- persistent-threat-detection-system
- unmanned aerial vehicle

REFERENCES

Abbot, P., & Walmsley, N., (1998). *British Airships in Pictures: An Illustrated History* (pp. 59–69). House of Lochar, Great Britain.

Aber, J. S., & Aber, S. W., (2013). *Aerial Photography Consulting and Sales* (pp. 1–7). http://geospectra.net/kite/consult/consult.htm. (accessed on 30 July 2020).

Aber, J., Marzolff, I., & Ries, J., (1995). *Small Format Aerial Photography and UAS Imagery* (p. 268). Principles, Techniques and Uses. Elsevier Science, New York, USA.

Advameg Inc., (2018a). *Airship* (pp. 1–4). http://www.madehow.com/Volume-3/Airship.html (accessed on 30 July 2020).

Advameg Inc., (2018b). *How Products are Made: Airship* (pp. 1–12). http://www.madehow.com/Volume-3/Airship.html#ixzz5WkKXoeG9 (accessed on 30 July 2020).

AeroDrum Inc., (2018). *10 m Long Blimps* (p. 161). AeroDrum Inc., Belgrade, Serbia https://www.rc-zeppelin.com/outdoor-rc-blimps.html (accessed on 30 July 2020).

Airlander 10, (2018). *Airlander: Rethink the Skies* (pp. 1–12). Hybrid Air Vehicles, Beddfordshire, United Kingdom. https://www.hybridairvehicles.com/ (accessed on 30 July 2020).

Airship do Brazil Industrie, (2017). *Airship do Brazil Conducts Test Flight of Latin America's First Manned Airship* (pp. 1–6). http://www.adb.ind.br/noticiaDetalhada?id=77 (accessed on 30 July 2020).

Airships Africa- South Africa, (2007). *Dyna Lifter* (p. 1). A revolution in transport system for Africa and its trading partners and the whole world. https://www.airshipsafrica.com/News.html/ (accessed on August 8th, 2020).

Ali, F., (2017). *FIT-for-Purpose Boundary Mapping and Valuation of Agricultural Lands Using UAVs: The Case Study of A1 Farms in Zimbabwe* (p. 289). Department of Geo-Information Sciences. University of Twente, Netherlands. Ph D. Theses Dissertation.

Althoff, W. F., (1990). *Sky Ships* (p. 229). Orion Books, New York, USA.

Althoff, W. F., (2009). *Forgotten weapon: U.S. Navy Airships and the U-Boat War* (p. 419). Annapolis, MD: Naval Institute Press.

Arthur Batut (1888). In: Earth from above. *International Conference on Kite Aerial Photography* (pp. 1–3). https://www.wokipi.com/Kapined/art-va-html (accessed on 30 July 2020).

Bach, J., (2015). *A Raytheon Blimp Broke free From its Tethered* (pp. 1–3). https://www.bizjournals.com/washington/blog/fedbiz_daily/2015/10/a-raytheon-blimp-broke-free-from-its-tether-and.html (accessed on 30 July 2020).

Beckenhusen, R., (2016). *The Kremlin Dreams of an Arctic Airship Fleet* (pp. 1–3). https://warisboring.com/the-kremlin-dreams-of-an-arctic-airship-fleet/ (accessed on 30 July 2020).

Bee Robotics Corporation, (2018). *Hybrid Airship Drone Farming Robot System for Crop Dusting, Planting, Fertilizing and Other Field Jobs* (pp. 1, 2). https://patents.justia.com/inventor/vladimir-salnikov (accessed on 30 July 2020).

Beijing LONSAN United Aviation Technology Co. Ltd., (2018). *Remote control Airships and Blimps* (pp. 1–4). https://liutaocat.fm.alibaba.com/?spm=a2700.icbuShop.88.2.281b7a2bH2i8fC (accessed on 30 July 2020).

Bellis, M., (2017). *Background and Definitions: Airships and Balloons* (pp. 1–8). https://www.thoughtco.com/history-of-airships-and-balloons-1991241 (accessed on 30 July 2020).

Blimps India Pvt Ltd., (2015). *Blimps in Agriculture and Military* (pp. 1–7). http://www.blimpsindia.com/ (accessed on 30 July 2020).

Boaz, D., (2018). *The Hybrid Airship: Lockheed Martin's Latest Advance in Air Freight Technology* (pp. 1–3). Airfreight.com. https://www.airfreight.com/blog/hybrid-airship-lockheed-martin-air-freight-technology/ ((accessed on 30 July 2020).

Booth, R., (2018). *Tethered Blimps* (pp. 1–3). http://www.blimpguys.com/index.html (accessed on 30 July 2020).

Borkhataria, C., (2018). *Meet the Plimp: A Bizarre Hybrid Craft Combines a Plane, a Helicopter and Blimp* (pp. 1–4). Mail Online. https://www.dailymail.co.uk/sciencetech/article-4950070/Meet-PLIMP-hybrid-craft-combines-plane-blimp.html (accessed on 30 July 2020).

Botting, D., (1980). *The Giant Airships* (p. 180). Time-Life Books, Alexandria, Virginia, USA.

Bouche, A., Beck-Winchatz, B., & Potochak, M. J., (2016). A high altitude balloon platform for determining carbon-di-oxide over agricultural landscapes. *Atmospheric Measurement Techniques, 9,*3707–3717 doi: 10.5174/atm-9–5707–2016/ (accessed on 30 July 2020).

Bowley, G., (2012). *Spy Balloons Become Part of the Afghanistan Landscape, Stirring Unease* (pp. 1–5). https://www.nytimes.com/2012/05/13/world/asia/in-afghanistan-spy-balloons-now-part-of-landscape.html/ (accessed on 30 July 2020).

Boyle, A., (2018). *Egan Airships Floats a Sales Campaign to Get its Plimp Aircraft off the Ground* (pp. 1–6). https://www.geekwire.com/2018/egan-airships-floats-sales-campaign-get-plimp-hybrid-aircraft-off-ground/ (accessed on 30 July 2020).

Brinton, T., (2010). *Lockheed Martin Ramps up Surveillance Blimp Production* (pp. 1–4). Space News, https://spacenews.com/lockheed-martin-ramps-surveillance-blimp-production/ (accessed on 30 July 2020).

Brooks, P. W., (1992). Zeppelin: Rigid Airships 1893–1940 (p. 221). Smithsonian Institution Press, Washington D.C. USA, https://catalog.hathitrust.org/Record/002817510/ (accessed on 30 July 2020).

Bueno, S. S., Azinheira, J. R., Ramos, J. J. G., Paiva, E. C., Rives, P., Elfes, A., Carvalho, J. R. H., & Silveira, G. F., (2002). Project aurora: Towards an autonomous robotic airship. In: *Proceedings of the 2002 IEEE/RSJ International Conference on Intelligent Robots and Systems (IROS 2002)* (pp. 1–11). Lausanne, Switzerland, IEEE/RSJ http://docshare04.docshare.tips/files/25249/252490420.pdf (accessed on 30 July 2020).

Burnett, J., (2011). *Blimps used for Natural Resource Monitoring at Biscayne National Park* (pp. 1–7). https://www.nationalparkstraveler.org/2011/01/blimp-used-natural-resource-monitoring-biscayne-national-park7539 (accessed on 30 July 2020).

Buzz, W. S., (2018). *Meet the PLIMP: Craft Combines Planes, Blimps and Helicopters* (pp. 1–12) http://wsbuzz.com/science/meet-the-plimp-craft-combines-planes-blimps-and-helicopters/ (accessed on 30 July 2020).

Cao, H. T., (2018). A low-cost depth imaging platform for canola phenotyping. Department of Engineering, University of Saskatchewan, Canada. *Master's Thesis Dissertation,* 125. https://harvest.usask.ca/bitstream/handle/10388/8526/CAO-THESIS-2018.pdf?sequence=1&isAllowed=y/ (accessed on 30 July 2020).

Carrivick, J. L., Smith, M. W., & Quincey, D. J., (2013). Developments in budget remote sensing for the geosciences. *Geology Today, 29*(4). 138–143. https://doi.org/10.1111/gto.12015 (accessed on 30 July 2020).

Cervantes, A. P., (2010). Small unmanned helium airships with electric power plant as low cost remote sensing platforms. *Proceedings of the International Airship Convention* (pp. 1–13). Bedford, UK. Paper 20, https://www.slideshare.net/airship701/small-unmanned-airships-for-remote-perception-2010.

CIMMYT, (2012). *Obregon Blimp Airborne and Eyeing Plots* (pp. 1–4). http://www.cimmyt. org/obregon-blimp-irborne-and-eyeing-plots/ (accessed on 30 July 2020).

CIMMYT, (2014). *An Aerial Remote Sensing Platform for High Throughput Phenotyping of Genetic Resources* (pp. 1–24). International center for Maize and Wheat Mexico. www.slideshare.net/CIMMYT/an-aerial-remote-sensing-platform-for-high-throughput-phenotyping-of-genetic-resources/ (accessed on 30 July 2020).

Clay, J. W., & Clement, C. R., (1993). *Selection of Species and Strategies to Enhance Income Generation from Amazonian Forest* (pp. 1–270). Food and Agricultural Organization of the United Nations, Rome, Italy, FAO/93/6 Working Paper http://www.fao.org/3/v0784e/v0784e.pdf (accessed on 30 July 2020).

CNNStaff, (2016). *Can Super Blimp Unlock Hidden Riches of Africa* (pp. 1–4). https://edition.cnn.com/style/article/superblimp-africa/index.html/ (accessed on 30 July 2020).

Colvin, J. D., (2017). *History of Airship* (pp. 1–8). Airship research labs. Houston, Texas, USA. http://airship-research-lab.com/History_of_Airships.html (accessed on 30 July 2020).

Crouch, T., (2017). *Blimps* (pp. 1–17). Smithsonian National Air and Space Museum. https://airandspace.si.edu/stories/editorial/blimp (accessed on 30 July 2020).

Cumberford, R., (2016). *Blimps for Ecological Observations* (p. 1). Whole Earth Catalogue. http://www.wholeearth.com/issue/2074/article/339/blimps.for.ecological.observation (accessed on 30 July 2020).

Dahm, W. J. A., (2017). *Air Force and DARPA Develop Spy Blimp* (pp. 1–9). The Future of Things. https://thefutureofthings.com/4015-air-force-and-darpa-develop-spy-blimp// (accessed on 30 July 2020).

Daniels, C., (2016). *The Airlander 10 Airship Gets Ready for Flight Head* (pp. 1–7). Hybrid Air Vehicles. https://www.youtube.com/watch?v=c-0aYicv26M/ (accessed on 30 July 2020).

Dick, H. G., & Robinson, D. H., (1992). *The Golden Age of the Great Passenger Airships: Graf Zeppelin and Hindenburg* (p. 193) Smithsonian Press Inc., Washington D.C.

DJI, (2019). *DJI AGRAS MG-1* (pp. 1–8) DJI Company Ltd. Shenzhen, China, https://www.dji.com/mg-1/info (accessed on 30 July 2020).

Dunford, R., Michel, K., Gagnage, M., Piégay, H., & Trémelo, M. L., (2009). Potential and constraints of Unmanned Aerial Vehicle technology for the characterization of Mediterranean riparian forest. *Int. J. Remote Sens., 30,*4915–4935.

Eblimp, (2018). *Remote Controlled Blimps* (pp. 1–3). Eblimp.com http://www.eblimp.com/best-international-rc-airship-company/index.html/ (accessed on 30 July 2020).

Editor, (2015). *Airships are Back in Fashion* (pp. 1, 2). https://www.uasvision.com/2011/08/05/the-blimps-of-war/ (accessed on 30 July 2020).

Editor, (2018a). *Airship* (p. 2). Encyclopedia Britannica. https://www.britannica.com/technology/airship (accessed on 30 July 2020).

Editor, (2018b). *Airships in the Arctic: Dirigibles are Being Floated as the Future Mules of the Great White North* (pp. 1–8). The Americas. https://www.economist.com/the-americas/2016/06/09/airships-in-the-arctic/ (accessed on 30 July 2020).

Editor, (2018c). *All Standard Tethered Remote Control eBlimp Complete Packages* (pp. 1–12). eBlimp.com (accessed on 30 July 2020).

Egan Airships Inc., (2018). *Plimp-Airship Revolutionizing Flight* (pp. 1–4). https://plimp.com/ (accessed on 30 July 2020).

Elfes, A., Bueno, S. S., Bergerman, M., & Ramos, J. J. G., (1998). A semi-autonomous robotic airship for environmental monitoring missions. In: *Proceedings of the 1998 IEEE International Conference on Robotics and Automation* (p. 1) Leuven, Belgium, IEEE. https://ieeexplore.ieee.org/abstract/document/680971 (accessed on 30 July 2020).

Elfes, A., Bueno, S. S., Bergerman, M., Paiva, E. C. D., Ramos, J. G., & Azinheira, J. R., (2003). Robotic airships for exploration of planetary bodies with an atmosphere: Autonomy challenges. *Autonomous Robots, 14,*147–164.

Elfes, A., Campos, M. F. M., Bergerman, M., Bueno, S. S., & Podnar, G. W., (2000). A robotic unmanned aerial vehicle for environmental research and monitoring. In: *Proceedings of the First Scientific Conference on the Large-Scale Biosphere-Atmosphere Experiment in Amazonia (LBA)* (pp. 220–229) Belém, Pará, Brazil, LBA Central Office, CPTEC/INPE, Rod. Presidente Dutra, km 40,12630–000 CachoeiraPaulista, SP, Brazil.

EOMST, (2017). *Unmanned Blimps* (pp. 1–4). Easy Operation Models Science and technology co. Ltd. https://wikivisually.com/wiki/EOMST_unmanned_blimps/ (accessed on 30 July 2020).

EuroAirships, (2018a). *Agriculture* (p. 1). http://www.euroairship.eu/index.php/agriculture/ (accessed on 30 July 2020).

EuroAirships, (2018b). *Unparalleled Touristic Experience* (p. 1). http://www.euroairship.eu/index.php/luxury-tourism/ (accessed on 30 July 2020).

EuroAirships, (2018c). Extended regional and border surveillance. http://www.euroairship.eu/index.php/military/ (accessed on 30 July 2020).

Fangzhou, (2018). *Fangzhou UAV* (pp. 1–4). https://en.wikipedia.org/wiki/Fangzhou_UAV (accessed on 30 July 2020).

Farivar, C., (2014). *Forget Drones: These Tethered Blimps Can Spy on Cities Below* (pp. 1–6). https://arstechnica.com/tech-policy/2014/08/forget-drones-how-about-tethered-blimps-to-spy-on-cities-below/ (accessed on 30 July 2020).

Fei-Yu Aviation Science and Technology Company Ltd. (2018) Unmanned Blimps. https://www.wikiwand.com/en/FYFT_S-series_unmanned_blimp/ (pp 1–3), (accessed on August 8th, 2020).

Fima, P. G., Gagliano, T., & Pope, R. E., (2018). *Liquid Dispensing Lighter-than-Air Airship System* (pp. 1–8). https://patents.google.com/patent/US6769493B1/en/ (accessed on 30 July 2020).

Froomkin, M., (2014). *US Army to Launch Two Massive Blimps Over Maryland* (pp. 1–5). https://theintercept.com/2014/12/17/billion-dollar-surveillance-blimp-launch-maryland/ (accessed on 30 July 2020).

Gardner, G., (2015). *What is the Etymology of the Word 'Blimp'* (p. 1). https://www.etymonline.com/word/blimp. The Lawrence Hall of Science, Berkeley, California, USA (accessed on 30 July 2020).

Geere, D., (2013). *Google Blimps Will Carry Wireless Signal Across Africa* (pp. 1–4). WIRED https://www.wired.co.uk/article/google-blimps/ (accessed on 30 July 2020).)

Geery, D., (2018). *Mission Statement: Uses and Potential Uses of the Hyperblimp* (pp. 1–8). Hyperblimp Inc., Utah, USA, https://hyperblimp.com/ (January 5th, 2019).

Gerke, M., Masar, I., Borgolte, U., & Rohrig, C., (2013). Farmland Monitoring by Sensor Networks and Airships. *Fourth IFAC Conference on Modeling and Control in Agriculture, Horticulture and Post-harvest Industry* (pp. 321–326) Finland. doi: 10.3182/20130828-2-SF-3019.00024.

Gertcyk, O., & Lamble, D., (2016). *Hi-Tech Airships Heralded as the Future of Transport in the Region* (p. 17). The Siberian News. https://siberiantimes.com/business/others/features/f0126-hi-tech-airships-heralded-as-the-future-of-transport-in-the-region// (accessed on 30 July 2020).

Goodyear Inc., (2018). *Wingfoot One* (pp. 1–5). https://www.goodyearblimp.com/behind-the-scenes/current-blimps.html (accessed on 30 July 2020).

Grabish, A., (2018). *Winnipeg Airship Business Signs MOU with Brazilian Company* (pp. 1–3). https://www.cbc.ca/news/canada/manitoba/airship-manitoba-brazil-mou-1.4571955/ (accessed on 30 July 2020).

Grace, F., (2004). *Farmers Buddy Up to Blimps* (pp. 1–5). The University of South Georgia, USA. https://www.cbsnews.com/news/farmers-buddy-up-to-blimps/ (accessed on 30 July 2020).

Greens, A., (2013). *A Brief History of Airship* (pp. 1–4). Popular Mechanics. https://www.popularmechanics.com/flight/g1281/a-brief-history-of-the-airship/ (accessed on 30 July 2020).

Grossnick, R. A., (1987). *Kite Balloons to Airships: The Navy's Lighter-Than-Air Experience* (p. 266). U.S. Government Printing Office, Washington D.C.

Haire, B., (2004). *Blimp Helps Fine-Tune Irrigation* (pp. 1–6). University of Georgia Extension Service, Georgia, USA. http://grains.caes.uga.edu/news/story.html?storyid=2045&story=Infrared-cotton/ (accessed on 30 July 2020).

Hannaford, K., (2018). *Bullet 580 Blimp is the World's Largest Airship—Which You Can Hire* (pp. 1, 2). https://gizmodo.com/5543413/bullet-580-blimp-is-the-worlds-largest-airshipwhich-you-can-hire. (accessed on 30 July 2020).

Heneghan, C., (2018). *Drones: The Future of Crop Surveillance* (p. 88). https://www.fooddive.com/news/drones-the-future-of-crop-surveillance/356234/ (accessed on 30 July 2020).

Heun, M. K., Jones, J. A., & Neck, K., (1998). Solar/infrared aerobots for exploring several planets. JPL new technology report: Npo-20264, Jet Propulsion Laboratory (JPL), Pasadena, CA, USA. *Autonomous Robots, 14,*147–164.

History Forum, (2018). *Jules Henri Giffard's Steam Airship* (p. 1). http://www.thehistoryforum.com/airships/henri_giffard/ (accessed on 30 July 2020).

Historylines, (2012). *Navy Opens a Blimp Base in New Jersey* (pp. 1–3). https://www.history.com/this-day-in-history/navy-opens-a-blimp-base-in-new-jersey/ (accessed on 30 July 2020).

Homeland Security News Wire, (2011). Ogde Utah Police-First in Nation to Use Surveillance Blimp (pp. 1–5). US Department of Homeland Security. http://www.homelandsecuritynewswire.com/ogden-utah-police-first-nation-use-surveillance-blimp/ (accessed on 30 July 2020).

Inflatable-2000, (2018). *Advertising Blimps: Custom Advertising Blimps* (pp. 1–3). https://i2kplay.com/custom-inflatable-products/helium-blimps-and-balloons/ (accessed on 30 July 2020).

Inoue, Y., Moringa, S., & Tomita, A., (2010). *A blimp Based Remote Sensing System for Low-Altitude Monitoring of Plant Variable: A Preliminary Experiment for Agricultural and Ecological Applications* (pp. 1–8). https://doi.org/10.1080/014311600210894 (accessed on 30 July 2020).

Jet Propulsion Lab, (2000). *Planetary Aerovehicles: Balloons and Ballutes* (pp. 1–3) https://www2.jpl.nasa.gov/adv_tech/balloons/summary_overview.htm (accessed on 30 July 2020).

Jian, Z., Jie, S., Zhao-Qin, L., & Li, Y. R., (2007). *Unmanned Low-Altitude Remote Sensing.* Beijing: The second session of the China (Beijing) Urban public safety audio-visual information seminar.

Jo, H. S., Tsun, M. S., & Young, C. M. L., (2014). *Development of Blimp platform for Aerial Photography* (pp. 1–3). https://www.scientific.net/AMM.629.170 (accessed on 30 July 2020).

Jorge, L. A. G., (2015). *Everything Potato Growers Should Know About Remote Sensing* (pp. 1–5). RedePapa. https://medium.com/@redepapa/everything-potato-farmers-should-know-about-remote-sensing-2a5d47e2e2e0/ (accessed on 30 July 2020).

Jose, P., Araus, L., & Cairns, J. E., (2014). Field high-throughput phenotyping: The new crop breeding frontier. *Trends Journal, 19,* 52–61 https://doi.org/10.1016/j.tplants.2013.09.008 (accessed on 30 July 2020).

Jowit, J., (2010). *Blimps Could Replace Aircraft in Freight Transport, Say Scientists* (p. 1). The Guardian, https://www.theguardian.com/environment/2010/jun/30/blimps-aircraft-freight/ (accessed on 30 July 2020).

Kaye, K., (2015). *Coming Soon to Capitol Hill* (pp. 1–3). Giant surveillance blimps? Legal Insurrection https://legalinsurrection.com/2015/05/coming-soon-to-capitol-hill-giant-surveillance-blimps/ (accessed on 30 July 2020).

Keirns, A. J., (2004). *America's Forgotten Airship Disaster: The Crash of the USS Shenandoah* (p. 122). Howard Little River Press, Ohio, USA.

Koebler, J., (2014). Giant military surveillance blimps to constantly monitor east coast. *The Sleuth Journal* (pp. 1–4). https://www.thesleuthjournal.com/giant-military-surveillance-blimps-constantly-monitor-east-coast/ (accessed on 30 July 2020).

Krishna, K. R., (2012). *Precision Farming: Soil Fertility and Productivity Aspects* (p. 189). Apple Academic Press Inc., Waretown, New Jersey, USA.

Krishna, K. R., (2016). *Push Button Agriculture: Robotics, Drones and Satellite guided Soil and Crop Management* (p. 448). Apple Academic Press Inc., Waretown, New Jersey, USA.

Krishna, K. R., (2017). *Agricultural Drones: A Peaceful Pursuit* (p. 396). Apple Academic Press Inc., Waretown, New Jersey, USA.

Krishna, K. R., (2019). *Unmanned aerial Vehicle Systems in Crop Production: A Compendium* (p. 675). Apple Academic Press Inc., Waretown, New Jersey, USA.

Lacroix, S., Jung, I., Sourres, P., Hygonenc, E., & Berry, J. P., (2001). *The Autonomous Blimp Project of LAAS/CNRS Current Status and Research Challenges* (p. 11). http://www.prisma.unina.it/iser02/papers/53.pdf (accessed on 30 July 2020).

Land and Food Resources, (2008). *Thermal Imagery to Improve Irrigation Efficiency and Productivity* (pp. 1–22). University of Melbourne, Australia. http://www.dookie.unimelb.edu.au/research/thermal.html./ (accessed on 30 July 2020).

Laskas, J. M., (2016). *Helium Dreams: A New Generation of Airships is Born* (pp. 1–9). The New Yorker. https://www.newyorker.com/magazine/2016/02/29/a-new-generation-of-airships-is-born/ (accessed on 30 July 2020).

Lavers, A., (2001). *Guidelines for Good Practices for Aerial Application of Pesticides* (p. 178). Food and Agricultural Organization of the United Nations. Rome.

Lawless, J., (2018). *Giant Helium-Filled Airship 'Airlander' taKes off for First Time* (pp. 1–6). https://phys.org/news/2016–08-giant-helium-filled-airship-airlander.html (accessed on 30 July 2020).

Li, L., Zhang, Q., & Huang, D., (2017). A review of imaging technique for Plant phenotyping. *Sensors (Basel), 14,*20078–20111 doi: 10.3390/s141120078.

Lighter Than Air Society, (2017). *Russians Engineers Develop Surveillance Missile Defense Airships* (pp. 1–8). Augur RosAero Systems, Moscow.

Listman, M., (2018). *Research Program on Wheat CGIAR* (pp. 1–3). Consultative Group on International Agricultural Research, FAO HQ, Rome, Italy.

Lockheed Martin Inc., (2018). *Hybrid Airship* (pp. 1–5). https://www.lockheedmartin.com/en-us/products/hybrid-airship.html (accessed on 30 July 2020).

Lord, V., & Kolesnik, E. M., (1982). *Airship Saga: The History of Airships Seen Through the Eyes of the Men Who Designed, Built and Flew Them* (p. 189). Blandford Press, Poole, England.

Martinez, R. J., Lahoz, J., Aguilera, D., Codes, J., (2005). IMAP3D: Low-cost photogrammetry for cultural heritage. In: *Proceedings of the 20ᵗʰCIPA International Symposium* (pp. 447–451) Torino, Italy.

Marzolff, I., (2014). The sky is the limit? 20 years of small aerial photography taken from UAS for monitoring geomorphological processes. Proceedings of European geological union general assembly, Vienna, Austria. *Geophysical Research Abstracts, 16,* 7005.

McNally, B., (2013). *A Very Short History of Post-War Military Airships* (pp. 1–4). Defense Media Network https://www.defensemedianetwork.com/stories/a-very-short-history-of-postwar-military-airships/ (accessed on 30 July 2020).

McWilliams, J., (2014). *A Brief History of Insect Control* (pp. 1–7). The Paris Review. https://www.theparisreview.org/blog/2014/11/17/a-brief-history-of-insect-control/ (accessed on 30 July 2020).

Minor, F., (2004). *Researchers Turn to Blimps to Fine Tune Irrigation,* (pp 1–2), https://lmtribune.com/northwest/researchers-turn-to-blimps-to-fine-tune-irrigation/article_97f5632b-1436-5919-abde-e9285e22ae4e.html/ (accessed on August 8th, 2020)

Mogili, U. R., & Deepak, B. B. V. L., (2018). Review on application of drone system in precision agriculture. International conference on Robotics and Smart manufacturing. *Procedia., 133,*502–509.

Mowforth, E., (1991). *An Introduction to the Airship* (p. 157) The Airship Association Ltd. Folkstone, Kent, London, United Kingdom.

Nadis, S., (1997). *The Zeppelins also Rises* (pp. 1–5). Massachusetts Institute of Technology Review. https://www.technologyreview.com/s/400115/the-zepplin-also-rises/ (accessed on 30 July 2020).

Neaves, A., (2014). *Balloon Surveillance System Back in Action Between Marfa and Valentine* (pp. 1–4). News West 9. http://www.newswest9.com/story/26750918/balloon-surveillance-system-back-in-action-between-marfa-and-valentine/ (accessed on 30 July 2020).

Newitz, A., (2018). *10 Airships of the Present Day-and Near* (pp. 1–12) future. https://io9.gizmodo.com/5815576/zeppelins-of-the-present-day--and-near-future/ (accessed on 30 July 2020).

Nitherclift, O. J., (1993). *Airships Today and Tomorrow* (p. 167). The Airship Association Ltd. Folkestone, Kent, United Kingdom.

Oliviera, J. S. F., Campello, S., Brandao, R. A., & Ciuti, S., (2017). *Improving River Dolphins Monitoring Using Arial Surveys* (pp. 1–7). Ecosphere https://doi:.org/10.1002/ecs2.1912 (accessed on 30 July 2020).

Petzer, B., (2013). *Blimps Across the Savannah: How to Side-Step the Broad Band Gap by Thinking Big* (pp. 1–5). The South African. https://www.thesouthafrican.com/news/blimps-across-the-savannah-how-to-sidestep-the-broadband-gap-by-thinking-big/ (accessed on 30 July 2020).

Photo blimp Inc., (2018). *Photo blimp Aerial Photography is the Silent and Safe Solution that you have been Looking for.* https://Photo blimp.com/about-Photo blimp/ (accessed on 30 July 2020).

Piquepaille, R., (2005). *Using Blimps to Find Diamonds* (pp. 1–5). https://www.zdnet.com/article/using-blimps-to-find-diamonds/ (accessed on 30 July 2020).

Precision Hawk LLC, (2014). *Lancaster Platforms in Agriculture* (pp. 1–4). https://www.precisionhawk.com/blog/media/topic/lancaster-5 (accessed on 30 July 2020).

Prentice, B. E., & Turiff, S., (2002). *Application for Northern Transportation: Airships to the Arctic* (p. 187). Transport Institute, University of Manitoba, Canada.

Prentice, B. E., Cohen, S., & Duncan, D. B., (2005). *Airships to the Arctic-Iii: Sustainable Northern Transportation: Proceedings* (p. 223). Transportation Institute, the University of Manitoba, Manitoba, Canada.

Priggs, M., (2014). *The US Army's Controversial 'All Seeing' Surveillance Blimps Lift off Above MARYLAND-and can Spot Objects as Small as a Person 340 Miles Away* (pp. 1–5). https://www.dailymail.co.uk/sciencetech/article-2890570/ (accessed on 30 July 2020).

PRNewswire, (2017). *Airships Market: Global Industry Analysis, Size, Share, Growth, Trends and Forecasts* (pp. 1–6). CISTON PRNewswire. https://www.prnewswire.com/news-releases/airships-market---global-industry-analysis-size-share-growth-trends-and-forecast-2016---2024-300422828.html (accessed on 30 July 2020).

PRNewswire, (2018). *Airships are Coming to Alaska* (pp. 1, 2). The lighter-than-airships-Society   http://www.blimpinfo.com/airships/hybrid-airships-are-coming-to-alaska/ (accessed on 30 July 2020).

Proctor, C., (2015). *CDOT to Test Fly Blimp Over 1–25 to Monitor Denver Area* (pp. 1–8). Lighter-than-air   Society,   http://www.blimpinfo.com/airships/aerostats-airships/cdot-to-test-fly-blimp-over-i-25-to-monitor-denver-area-traffic/ (November 30th, 2018)

Ramos, J. J. G., Paiva, E. C., Maeta, S. M., Mirisola, L. G. B., Azinheira, J. R., Faria, B. G., et al. (2000). Project aurora: A status report. In: *Proceedings of the 3rd International Airship Convention and Exhibition (IACE 2000)* (pp. 174–194) Friedrichshafen, Germany, The Airship Association: UK.

Raven Aerostar Inc., (2018). *Remote Controlled Blimps* (pp. 1–5). https://ravenaerostar.com/products/balloons-airships (accessed on 30 July 2020).

Reed, D., (2015). *Why Blimps and Airships Died out and how they Might Make a Comeback* (pp. 1–2). New Statesman. https://www.citymetric.com/transport/why-blimps-and-airships-died-out-and-how-they-might-make-comeback-722?page=7/ (accessed on 30 July 2020).

Rees, M., (2012). *US to use Unmanned Surveillance Blimps at the Mexico Border* (pp. 1–4). Unmanned News Systems WSJ. https://www.unmannedsystemstechnology.com/2012/08/us-to-use-unmanned-surveillance-blimps-at-mexico-border/ (accessed on 30 July 2020).

Reynolds, M., Brau, & Quilligan, E., (2012). Development and delivery of breeding lines encompassing yield potential. *Proceedings of the 2nd International Workshop of the Wheat Yield Consortium* (pp. 1–64). International Maize and Wheat Center (CIMMYT), Mexico.

Ric, A., (1994). *Hindenburg: An Illustrated History* (p. 287). Warner Books Inc., New York, USA.

Richmond, B., (2014). *From Afghanistan to Texas, Surveillance Blimps are on the Rise* (pp. 1–8). https://motherboard.vice.com/en_us/article/8qxmj4/from-afghanistan-to-texas-surveillance-blimps-are-on-the-rise/ (accessed on 30 July 2020).

Ries, J. B., & Marzolff, I., (2003). Monitoring of gully erosion in the central Ebro basin by large-scale aerial photography taken from a remotely controlled blimp. *Catena, 50,*309–328.

Ritchie, G. L., Sullivan, D. G., Vencill, W. K., Bednarz, C. W., & Hook, J. E., (2010). Sensitivities of normalized differences vegetation index and green/red ratio index to cotton ground cover fraction. *Crop Science, 50,*1000–1010.

RMAX, (2015). *RMAX Specifications* (pp. 1–4). Yamaha Motor Company, Japan. https://barnardmicrosystems.com/UAV/uav_list/yamaha_rmax.html/ (accessed on August 8th, 2020).

RosAero Systems, (2019). *ATLANT 30-Hybrid Aircraft* (pp. 1–12). Ros Aero Systems, Moscow, Russia. http://www.aerall.org/projet_RosAreosystems-Atlant30.htm (accessed on 30 July 2020).

Salami, E., Barrado, C., & Pastor, E., (2014). Review UAV flight experiments applied to the remote sensing of vegetated areas. *Remote Sensing, 6,*11051–11081; doi: 10.3390/rs61111051; www.mdpi.com/journal/remotesensing (accessed on 30 July 2020).

Salnikov, V., Filin, A., & Bureuma, H., (2014). *Hybrid Airship-Drone Farm Robot System for Crop Dusting, Planting, Fertilizing and other Field Jobs* (pp. 1–12). https://patents.google.com/patent/US20160307448A1/en/ (accessed on 30 July 2020).

Sambit, K., (2019). *Airships Market Report, Include Product Scope, Overview, Opportunities and Risk, Driving Force Analysis with Global Forecast* (pp. 1–14). https://bulletinline.com/2020/07/29/airships-market-2020-by-new-tools-technology-advancement-opportunities-risk-driving-force-and-forecast-to-2024/ (accessed on 30 July 2020).

Sankaran, S., Khot, L. R., Espinoza, C. Z., Jarolmasjed, S., Sathuvalli, S., Vandemark, G. J., Miklas, P. N., et al. (2015). Low altitude, high resolution aerial imagery system for row and field crop phenotyping. *European Journal of Agronomy, 70,*112–123.

Scentroid Inc., (2017). *Scentroid: The Future of Sensory Technology* (pp. 1–12). http://scentroid.com/scentroid-sampling-drone/ (accessed on 30 July 2020).

Schachtman, N., (2017). *Giant Spy Blimp Battle could Decide Surveillance Future* (pp. 1–3). https://www.wired.com/2011/07/spy-blimp-battle/ (accessed on 30 July 2020).

Seneport, M., (2018). *Euro Airships SAS* (pp. 1–8). http://www.euroairship.eu (accessed on 30 July 2020).

SenseFly Inc., (2015a). *Drones for Agriculture* (pp. 1–9). http://www.sensefly.com/applications/agriculture.html (accessed on 30 July 2020).

SenseFly Inc., (2015b). *eBee by Sensefly* (pp. 1–4). http://www.sensefly.com (accessed on 30 July 2020).

Sharma, P., (2007). *Precision Farming* (p. 276). Gene-Tech Books, New Delhi, India.

Sharp, T., (2012). *The First Powered Airship: The Greatest Moments in Flight* (pp. 1, 2). https://www.space.com/16623-first-powered-airship.html (accessed on 30 July 2020).

Sherman, C., (2012). *US Testing Balloon on Mexican Border* (pp. 1–3). Border Patrol. Associated press in. https://www.mprnews.org/story/2012/08/22/news/us-testing-surveillance-balloons-on-mexico-border/ (accessed on 30 July 2020).

Shock, J. R., (2001). *US Navy Airships, 1915–1962* (pp. 1–24). Edgewater, Florida. Atlantis Productions.   https://www.libraries.wright.edu/special/collectionguides/files/ms388.pdf (accessed on 30 July 2020).

Sigler, D., (2017). *Zeppelins, Blimps and Plimps* (pp. 1–8). Sustainable Skies. http:// sustainableskies.org/zeppelins-blimps-plimps/ (accessed on 30 July 2020).

Sims, J., (2017). *Airships Revolutionize Luxury Travel and Companies Target the Chinese Market* (pp. 1–3). https://www.scmp.com/magazines/style/tech-design/article/2097307/ airships-revolutionise-luxury-transport-companies (accessed on 30 July 2020).

SkyDoc Aerostat Systems Inc., (2018). *Home of the Category-2 Hurricane Proof Aerostat* (pp. 1–7). http://skydoc.com/index.html (accessed on 30 July 2020).

Smith, D., (2009). *Dread Zeppelin: The Army's New Surveillance Blimp* (pp. 1–3). Popular Science   https://www.popsci.com/military-aviation-amp-space/article/2009-06/dread-zeppelin-armys-new-surveillance-blimp/ (accessed on 30 July 2020).

Smith, R. J., Baillie, J. N., McCarthy, A. C., Raine, S. R., & Ballie, S. P., (2010). *Review of Precision Irrigation Technology and their Application* (pp. 1–48). National Center for Engineering in Agriculture, University of South Queensland, Toowoomba, Queensland, Australia. http://www.insidecotton.com/jspui/bitstream/1/4130/1/USQ5024%20Final%20 Report.pdf (accessed on 30 July 2020).

Sonmez, O. S., (2015). *Design of a Solar Powered Unmanned Airship* (p. 124). Department of Mechanical Engineering, Middle east Technological University, Master's Thesis, Middle East Technology University, Cankaya, Turkey.

Sputnik News, (2016). *Battle Blimps: Russian Military Expands the use of Blimps* (pp. 1–3).   https://sputniknews.com/science/201606031040716862-russia-military-blimps-surveillance/ (accessed on 30 July 2020).

Sputnik News, (2018). *Russian Engineers Develop New Surveillance, Missile Defense Airships* (pp. 1–3). https://sputniknews.com/science/201705061053345054-early-warning-blimp-development/ (accessed on 30 July 2020).

Stafford, J. V., (2000). Implementing precision agriculture in the 21st century. *Journal of Agricultural Engineering Research, 76,*267–275.

Stewart, W., (2018). *Red Zeppelin: Russia Set to Unveil Military Airships Capable of Carrying 200 Personnel and Traveling up to 105mph Without the Need for a Runway* (pp. 1–5). MailOnline. https://www.dailymail.co.uk/news/article-3144854/Red-zeppelin-Russia-set-unveil-military-airships-capable-carrying-200-personnel-traveling-105mph-without-need-runway.html (accessed on 30 July 2020).

Stieger, J., Bamberger, I., Buchmann, N., & Eurgser, W. W., (2015). Validation of farm-scale methane emissions using nocturnal boundary layer budgets. *Atmospheric Chemistry and Physics, 15,*14055–14069. doi: 10.5194/acp-15–14055-2015.

Stubblebine, D., (2018). *K-Class Airship*. World War II data base. https://ww2db.com/ aircraft_spec.php?aircraft_model_id=493 (accessed on 30 July 2020).

Sutton, J., (2008). *Blimp Joins Anti-Smuggling Patrols off Florida* (pp. 1–3). https://www. reuters.com/article/us-usa-security-blimp-idusn6o43930320080624 (accessed on 30 July 2020).

Tattaris, M., (2014). *Researcher Helps Remote Sensing Soar* (pp. 1–5). Lighter than Air Society. http://www.blimpinfo.com/airships/blimps/researcher-helps-remote-sensing-soar/ (accessed on 30 July 2020).

Tattaris, M., (2015). *Obregon Blimp Airborne and Eyeing Plots* (pp. 1–3). CIMMYT, Mexico, https://www.cimmyt.org/news/obregon-blimp-airborne-and-eyeing-plots/ (accessed on 30 July 2020).

The Christian Science Monitor, (2018). *World's Largest Inflatable vehicle-Bullet 58 Airship-Inflated and Ready for Action* (pp. 1–6). https://www.csmonitor.com/About/Contact (accessed on 30 July 2020).

Thiesen, W. H., (2010). LCDR Edward 'Iceberg' smith 13 and artic expedition of 1931of the German airship Graf zeppelin. *The Foundation for Coast Guard History Bulletin, 23*, 58–62. www.ICGH.org (accessed on 30 July 2020).

Trimble Inc., (2015). *Trimble UX5 Aerial Imaging Solution for Agriculture* (pp. 1–3). http://www.trimble.com/Agriculture/UX5.aspx (accessed on 30 July 2020).

Tuan, Z., Xiao-Ping, W., & Jian, T., (2010). *UAV Digital Photogrammetry System Design and Application of Xian: China Monitoring Network* (pp. 1–7). http://www.chinamca.com/. (accessed on 30 July 2020).

Tucker, A., (2012). *Blimp with a View: A Special Ride in the Hood Blimp* (pp. 1–3). The New England Today. https://newengland.com/today/living/new-england-nostalgia/blimp-with-a-view-a-special-ride-in-the-hood-blimp/ (accessed on 30 July 2020).

US Department of Homeland Security-Science and Technology Directorate, (2018). *Tethered Aerostat System Application Note* (pp. 1–18). https://www.dhs.gov/sites/default/files/publications/TetheredAerostat_AppN_0913-508.pdf (accessed on 30 July 2020).

Vaeth, J. G., (1992). *Blimps & U-Boats*. US Naval Institute Press, Annapolis, Maryland, (pp 205).

Valiulis, A. V., (2014). Dirigibles. *A History of Materials and Technologies Development* (p. 444). Vilnius: Technika.

Vardhan, P. H., Dheepak, S., Aditya, P. T., & Arul, S., (2018). Development of automated aerial pesticide sprayer. *International Journal of Engineering Science and Research Technology, 3*(4), 88–84.

Variant Market Research, (2018). *Airships Market Overview* (pp. 1–4). https://www.variantmarketresearch.com/report-categories/defense-aerospace/airships-market/ (accessed on 30 July 2020).

Vericat, D., Brasington, J., Weaton, J., & Cowie, M., (2008). *Accuracy Assessment of Aerial Photographs Acquired Using Lighter-Than-Air Blimps: Low-Cost Tools for Mapping River Corridors* (pp. 1–22, 985–1000). https://doi.org/10.1002/rra.1198 (accessed on 30 July 2020).

Vierling, L. A., (2006). The Short-Wave Aerostat-Mounted Imager (SWAMI): A novel platform for acquiring remotely sensed data from a tethered balloon. *Remote Sensing of Environment, 103,*255–264.

Virlet, N., Sabermanesh, K., Sadeghi-Tehran, P., & Hawkesford, M. J., (2017). Field scanalyzer: An automated robotic field phenotyping for detailed crop monitoring. *Functional Biology, 44,*143–153. http://dx.doi.org/10.1071/FP16163 (accessed on 30 July 2020).

Wall, M. D., (2010). *Titan from Above: Blimp Survey Options to Study Saturn's Moon* (pp. 1–8). National Aeronautics and Space Agency. https://www.space.com/10544-titan-blimp-survey-options-study-saturn-moon.html (accessed on 30 July 2020).

Wei, L., (2010). *Low Unmanned Airship Photogrammetry* (pp. 165–168). Semantic Scholar. org https://pdfs.semanticscholar.org/5bfb/f9d42536ba9e2bbcf33e16de7930728c648d.pdf (accessed on 30 July 2020).

Wells, M. H., & Castillo, C., (2017). Quantifying uncertainty in high-resolution remotely sensed topographic surveys for ephemeral gully channel monitoring. *Earth Surface Dynamics, 5*,347–367. https://doi.org/10.5194/esurf-5-347-2017 (accessed on 30 July 2020).

Wiesenberger, (2017). *Flying High: 7 Post-Hindenburg Airships* (pp. 1–5). https://www. livescience.com/58988-post-hindenburg-airships.html (accessed on 30 July 2020).

Wikipedia, (2017). *Airships* (p. 24). https://en.wikipedia.org/wiki/Airship (accessed on 30 July 2020).

Wikipedia, (2018a). *G Class Blimp* (pp. 1–23). https://en.wikipedia.org/wiki/G-class_blimp/ (accessed on 30 July 2020).

Wikipedia, (2018b). *K-Class Blimps* (pp. 1–12). https://en.wikipedia.org/wiki/K-class_blimp (accessed on 30 July 2020).

Wikipedia, (2018c). *Zeppelin Types-Different types of Zeppelins and Dirigibles* (pp. 1–4). https://en.wikipedia.org/wiki/Zeppelin (accessed on 30 July 2020).

Winston, G., (2018). *Russian Military Developing Blimps, Balloons and Dirigibles* (pp. 1, 2). https://www.warhistoryonline.com/war-articles/russian-military-developing-blimps-balloons-dirigibles.html (accessed on 30 July 2020).

Wrags, D., (2008). *Historical Dictionary of Aviation* (pp. 27–29). History Press, London, United Kingdom.

Yang, G., Liu, J., Zhao, C., Li, Z., Huang, Y., Yu, H., Xu, B., et al. (2017). Unmanned aerial vehicle remote sensing for field based crop phenotyping: Current status and perspectives. *Frontiers in Plant Science*, 1–12. https://doi.org/10.3389/fpls.2017.01111 (accessed on 30 July 2020).

Yogita, S., Harish, I., Prasanna, K. R., Sundar, R. R., Vignesh, P., & Vishnu, V. M., (2017). Advanced pesticide sprayer using blimp balloons. *International Journal for Research in Applied science and Engineering Technology (IJRASET), 5*,1232–1237.

Zeppelin Luftschifftechnik Gmbh, (ZLT), (2010). *Zeppelin NT* (pp. 1–3). https://wikivisually. com/wiki/Zeppelin_NT (accessed on 30 July 2020).

Zeppelin Luftschiff-technik GmbH (ZLT), (2018). *Zeppelin NT* (pp. 1–7). http://www. blimpinfo.com/wp-content/uploads/2012/01/Technical-data-of-the-Zeppelin-NT-07.pdf (accessed on 30 July 2020).

Zeppelin Science, (2018). *Zeppelin NT Airship: The Versatile Airborne Platform* (pp. 1–8). Zeppelin NT Friederchshafen, Germany. zeppelin-nt.de/de/zeppelin-nt/einsatzfelder/ sondermissionen.html/ (accessed on 30 July 2020).

Zhang, Q., (2015). *Precision Agricultural Technology for Crop Farming* (p. 360). CRC Press Inc., Boca Raton, Florida, USA.

CHAPTER 4

Aerostats, Helikites, and Balloons in Agriculture

ABSTRACT

Aerostats, helikites, and balloons are among the important aerial vehicles that need greater attention with regard to their potential uses in farming sector. They are supposedly economically efficient methods of obtaining aerial imagery of ground features, particularly, farm infrastructure, agricultural terrain and crops. Research efforts to adopt these aerial vehicles are feeble. Therefore, reports that evaluate their performances in the agrarian regions too are very few.

Initially, historical facts about aerostats, the recently devised and developed helikites and the time-tested kites have been mentioned. This is to provide a background. Definitions, availability of different types of aerostats and their classification form a very useful portion of the chapter. A few examples of both free-floating and tethered aerostats and helikites have been dealt. Their role in conducting military and civilian tasks has been included.

Like other aerial vehicles, for example, the fixed-winged and copter drones, these aerial vehicles, i.e., aerostats, helikites, and balloons too were first utilized by the military establishments of different nations. Again, the emphasis here is to draw experiences from such efforts in military realm and adopt them for peaceful efforts in furthering the cause of agricultural crop production. Accordingly, a few sections in this chapter is entirely allocated to role of aerostats, helikites, and balloons in military activities, border security and patrolling and in rapid reporting of disasters.

There are indeed several civilian tasks wherein aerostats, helikites, and balloons find utility. Often, these aerial vehicles are less costly and easy to operate. Commensurately, a few sections in this chapter highlight such civilian uses of the above aerial vehicles. The role of tethered and free-floating aerostats, helikites, and balloons in the collection of data about atmosphere, weather parameters, greenhouse gas emissions, and farming belts in general has been discussed. During recent years, a spurt in interest in usage of aerostats actually relates to its possible role in space science. A few examples that suggest use of aerostats in space technology are available in the chapter. The other topic that is attracting greater attention is the utility of aerostats in high altitude wind generation. It seems farmers can reap high benefits from lofting such aerostats fitted with light weight turbine to high altitudes. Such aerostats with turbine can provide electric power for 25 homes or 7–8 farms per unit. Wind power is environmentally safe and renewable which add to their advantages. Several examples pertaining to wind power generation by aerostat + wind turbine have been discussed in the text. Aerostats and helikites play an immensely useful role in telecommunication. Particularly, in rural districts, mountainous zones, sparsely populated deserts and other areas. They are being rapidly installed in various countries such as USA, Australia, South Africa, India, etc. This is to provide extended mobile phone and internet services to difficult to reach remote areas. Aerial imagery is the essential aspect of aerostats, helikites, and balloons that are adopted to study the ecological aspects of natural and agrarian zones. Such aerial vehicles have also found use in monitoring and tracing archaeological features from above the historical sites.

The change in crop production tactics towards obtaining highly accurate aerial imagery and relevant digital data for use on robotic ground vehicles such as planters, fertilizer inoculators, sprayers combine harvesters, etc. seems imminent worldwide. The digital data obtained about land, soils, crop growth, nutrient status and water stress index status using aerostats is most easily adopted on semi-autonomous or totally robotic precision farming vehicles. This aspect may attract greater attention, in future, from agricultural engineers and other farm experts. Aerostats, helikites, and balloons are among most easily opted aerial vehicles to obtain data about pest/disease attack of other natural disasters affecting a farm. Reports dealing with their utility in bird scaring has been dealt in detail. The utility of aerostats and helikites in monitoring forests and pastures have been mentioned briefly. A few relevant forecasts and possible future course for aerostats, helikites and kites in agrarian regions are also mentioned.

4.1 INTRODUCTION

The intention of this chapter on aerostats, helikites, and balloons is to list and discuss the potentially wide range of utility attributable to them. Particularly, in various aspects related to military, civilian tasks and more importantly, in the agrarian regions of the world. The focus is to provide a review of the entire gamut of applications of aerostats/helikites in the agricultural realm. These lighter-than-air vehicles are yet to make an impact in farming zones. As aerial vehicles above a farm, they have the potential to replace the presently popular fixed-winged and helicopter drones, at least with regard to collection of aerial data and general survey of farmland. Right now, their introduction into the 'agricultural sky' is rudimentary. Actually, many of the aerostat/helikite manufacturing companies and research groups in universities are standardizing procedures and matching computer programs. They are essentially modifying aerostats/helikites to suit the farmer's requirements. The idea is to provide farmers with continuous stream of aerial survey and surveillance data about crops and farmland, in general. Helikites are supposedly cheaper and very effective aerial vehicles. They may act as sentinel plus offer wide range of information, to farmers. Recent research thrusts indicate that aerostats, particularly, tethered aerostats with high altitude turbines are excellent sources of electricity, to farmers. The farm world may soon relish having an aerial turbine above their farms. Tethered aerostats are also excellent masts or towers that help to relay broadband internet signals. Aerostats also provide excellent data that are necessary, to conduct agronomic operations adopting, precision farming principles. Tethered aerostats/helikites are also capable of offering aerial data about changes occurring in the ecosystems, in general. Even archaeologists have found a great many applications in assessing the historical sites. Yet, most conspicuous fact, right now, is that utilization of aerostat-aided techniques in farming belts is still in its initial stages. There are only few research efforts published. General literature, research journals, and manuals about aerostat models suitable for agriculture are needed to popularize them.

4.1.1   HISTORICAL FACTS ABOUT AEROSTATS, HELIKITES, AND BALLOONS AS RELATED TO MILITARY AND CIVILIAN ASPECTS

Let us begin by knowing a few details about the history of aerostats, helikites, and balloons. The assumption here is that historical facts offer us a better perspective about the topic of this chapter. In 1783, the properties of hot air and hydrogen were harnessed, to provide the first 'lighter-than-air' vehicle. They were hot-air balloon and the hydrogen balloon, respectively. The discovery of how to generate hydrogen had come from the advances in chemistry. Hydrogen gas was being manufactured in France and other parts of Europe. In 1783, the Montgolfier brothers made the first ascent in a hot air balloon. Later, Joseph and Etienne Montgolfier designed and tested balloons that lifted a payload, including humans. In the same year, Professor Charles made the first ascent in a hydrogen balloon (Mines, 2016). The creation of balloons was followed almost immediately by the descent of animals placed in the payload area of parachutes, from the balloons. The response of animals to this new experience-i.e., descent in the air has not been documented.

In 1790, Garnerin, a Frenchman, made his first ascent in a hot air balloon constructed by himself. He went on to become the leading balloonist of his period. He gave demonstrations of balloon ascents at public occasions for a fee. He initiated the use of balloons for long distance journey (Australian Parachute Federation Ltd. 2017). Garnerin's wife, Jeanne Genevieve Labrosse (1775–1847) was the first woman to pilot a balloon. She was also the first woman to make a parachute jump, on 10 November 1798. Garnerin's niece Elisa Garnerin (born, 1791) was introduced to parachuting by Garnerin. She became a professional parachutist and carried out about forty parachute descents, in countries around the world between 1815 and 1836.

Tethered balloons of different shapes were in vogue during mid-19th century (see Vasconcelos, 2019). They were in use during American civil war. Tethered balloons were demonstrated by Thaddeus Lowe in 1861. Holmes (2013) states that, it almost coincided with onset of civil war and the chaos in the administration that ensued during the period. To demonstrate, Thaddeus Lowe himself ascended a balloon and kept floating at 500 m above the ground surface. He carried a set of Morse code equipment. A telegraph wire was strapped to aerostat's tethers. This helped him to surveillance and report a little early about the happenings on the ground. Particularly, regarding the concentration and movement of troops, etc. So, the first use of tethered balloons by the Washington's military set up of Unionists, began in 1861–1862 (Holmes, 2013). In the meantime, Thaddeus Lowe developed a fleet of eight military aerostats that could float at 5000 ft. above ground surface. He collected sufficient length of tethers and telegraphic wires and keyboard, to relay the 'Morse code.' Records indicate that on September 24th, 1861, Lowe could effectively use his tethered aerostat, to collect information about Confederate troops stationed at Falls Church, in Virginia. He could do it by staying afloat at 5000 ft. altitude and 3 miles away from troops that he monitored. It seems Unionists calibrated their guns accordingly and fired more accurately. Holmes (2013) states that, in addition to locations in Virginia, Lowe's aerostats were utilized during the siege of Yorktown in

May, 1862; at the Battle of Fair Oaks in May–June, 1862; at the crucial Seven Days Battle outside Richmond in July, 1862; also, at Battle of Fredericksburg in December 1862. He also witnessed from his aerostats, the famous rebel victory by Robert E. Lee at Chancellorsville, in May 1863. It seems, Thaddeus Lowe moved his aerostats, using horse-drawn carts, railroad wagons and boats.

Dimitry Mendeleev who devised the 'Periodic Table of Elements' available on earth was a practical scientist. He was among the Russians to develop an aerostat and try to use it to study meteorology. It seems his early attempts to float above cloud level was not successful. Later, in 1887, he was awarded a medal by France, for his effort to float in the sky, using aerostats (RT-Russian TV, 2019).

4.1.2   RECENT DEVELOPMENTS RELATED TO AEROSTATS, HELIKITES, AND BALLOONS

It is generally believed that, like airships, many other aspects of the lighter-than-air technology experienced a low level of interest, particularly, after 1930 AD. The interest in aerostats too suffered a set-back since 1930s (Aglietti et al., 2010). Aerostats experienced a kind of neglect even by scientists involved in aerial photography, weather reporting and atmospheric analysis. It is said that, aerostat technology *per se* was neglected. Aerostats did not receive high priority as a topic of great relevance or that needing research inputs, within the departments of aerospace engineering in different institutions worldwide (Dick and Robinson, 1992; Kirschner, 1986). Lighter-than-air vehicles (aerostats) were progressively neglected by the mainstream research in Aerospace Engineering, during the second half of the past century. This happened after having made remarkable technological progress that culminated in the 1930's with the construction of over 200 m long airships (Dick and Robinson, 1992, Robinson, 1973; Department of Defense, Washington D.C., USA, 2016). Since then, there have been only few developments of historical interest, but, of little significance (Kirschner, 1986).

We may note that during the past decade, there has been a resurgence of interest in many types of aerial vehicles such as parachutes, parafoils, blimps even kites (Krishna, 2019; Colozza and Dolce, 2005; Badesha, 2002). These aerial vehicles are being examined, tested rigorously and adopted in wide range of activities. They include from simple aerial photography, surveillance, weather reporting and recreational aspects to harvesting wind energy at high altitudes. Aerostats are being actively considered in space science and planetary explorations. The 'Agrarian Sky' may absorb innumerable aerostats/helikites in due course. The general market for aerostats, airships and parafoils/microlights has already increased, during the past decade. The forecasts are encouraging too.

Now, regarding helikites, the original helikite aerostat was designed and patented by Sandy Allsopp in 1993, in United Kingdom. Since then, Allsopp Helikites Ltd has been producing helikites for different purposes. Helikites are used by the military, police and other government departments. Helikites have been adopted for commercial uses such as aerial photography (Helikite Hotspots Ltd., 2019). Helikites are also being introduced into agricultural regions to conduct aerial survey and imagery.

During past couple of decades (i.e., from 2000 AD till date), achievements in the realm of scientific ballooning and space science has been significant. The development and launch of 'Ultra High-Altitude Balloon (UHAB)' for NASA is noteworthy (Aglietti, 2009; Aglietti et al., 2010; Bely and Ashford, 1995). These aerostats and balloons are meant for space science. They are large and can reach an altitude of 49 km from earth's surface. Yet another example is the effort by Japanese space scientists, in 2002, who have developed aerostats/balloons that reach 53 km altitude and pick a pay load of 10 kg. During past decade, i.e., 2005–2015, space researchers at Johns Hopkins Applied Physics Laboratory have conducted successful trials with high altitude aerostats/tethered balloons, for space to ground communication. In the past few years, tethered aerostats have been extensively used by US Air Force and Homeland Security Department, to conduct surveillance of ground surface activities and relay the information rapidly (Badesha, 2002; US Department of Homeland Security-Science and Technology Directorate, 2018).

Elegant tethered aerostat designs capable of accommodating wind turbines at high altitudes are being manufactured, since past few decades. The 'tethered aerostat tower' placed at 2000 m to 2 km altitude has a light-weight turbine that generates electric power. The electric power harvested is transmitted to ground station via electric cable attached to tethers. These are among most useful aerostats when energy is required in rural areas not supplied by general transmission. They are also adoptable in hill tracts and far-off detached zones like Antarctic or Arctic or desert zones (see Glass, 2018; Altaeros Inc., 2018). This is historically an important event considering that wind power is converted into electric power in remote locations. It is a hitherto untapped renewable source of energy.

Since the past decade (i.e., 2008–2018), tethered aerostats are being utilized to transmit telecommunication signals. They are making rapid in roads into mobile phone communication in rural areas. The telecommunication companies are adopting aerostat towers to help in 4G broadband telecommunication (Cadogan, 2018; Glass, 2018; Livson, 2016). At present, tethered aerostats/balloons are predominantly manufactured in countries such as USA, France, Russia and other European nations. Several developing nations too have industries that supply material necessary for production of tethered aerostats, helikites, and balloons.

4.2 WHAT ARE AEROSTATS, HELIKITES, AND BALLOONS?

The aerostats, helikites, and balloons are lighter-than-air vehicles. They possess an envelope (hull) that is filled with lifting, lighter-than-air gas such as helium. They could be free-floating or tethered to a ground station or a mobile unit like a pick-up van. They are not easily controlled, if they are free-floating forms. The tethered vehicles could be controlled regarding their altitude, location, data capture and relay, etc. The size of envelop and tether depends on the purpose. The set of sensors, radars, and chemical probes for which they act as platforms, too depend on the purpose.

4.2.1 DEFINITION AND DESCRIPTION OF AEROSTATS, HELIKITES, AND BALLOONS

A survey of published literature suggests that there are a few ways to define an aerostat. Aerostat is a tethered unmanned airship. An aerostat is a lighter-than-air vehicle that takes advantage of buoyancy to create lift. The term "aerostat" is derived from the "aerostatic" lift force that is created by the buoyancy of the aerostat's body. The body typically contains a lighter-than-air gas, such as helium, which creates the craft's buoyancy.

Tethered aerostats are permanently attached to the ground through a cable or tether. The tether permits to station them and maintain at a localized position, without the expenditure of extra energy. This allows them to remain airborne for long period of time. Therefore, it makes them suitable for applications that require high-endurance aerial platforms (Howard, 2007).

Tethered aerostats have been in use by military/defense establishments and security agencies since several decades. They can operate the aerostats at different altitudes from 10 m–50 m (low altitude tethered aerostats) to 5 km (high altitude tethered aerostats). Their sizes vary depending on the purpose for which an aerostat is envisaged. Tethered aerostats meant for monitoring of public event or display of advertisements or even agricultural purposes can be small at 10–15 m in length. Tethered aerostats could be kept in flight for longer durations of up to 30 days or even more. For example, in farms, perhaps we can keep a tethered aerostat in the sky for the entire crop season, to derive aerial images continuously. Payloads range from couple of kgs, if just aerial images are required to as much as 50 kg, if wide ranging weather analysis equipment, radars, samplers, etc. are to be placed in the sky. Higher payload capacity is required if a wind turbine is to be lifted to a high altitude. Such turbines are used to convert wind power into electricity (Glass, 2018; Altaeros Inc., 2018; Altaeros Energies Inc., 2019).

Helikites are a kind of tethered aerial vehicles. They are hybrids between a helium balloon and a kite. They are well stabilized aerostats. They utilize both the upward lift obtained due to lighter-than-air helium gas filled into balloon (polyurethane) and the kite that utilizes the thrust developed by the blowing wind. We may note here that, originally, 'helikite or helikites' is a trademark covered name to vehicles produced by a British-based company-Allsopp Helikites Ltd., located at Fordingbridge, in United Kingdom. At present, the term 'helikites' denotes an aerial tethered vehicle with balloon and a kite fused in one. A helikite offers definite advantages during aerial survey and photography.

Primarily, helikites could be very small, say, 3–4 m in size (i.e., size of helium balloon is just 3–4 m³). Some of these balloons are being operated in rural areas, right now. Helikites could also be large at over 75 m³. They are being used to surveillance large areas around harbors, industrial installations, infrastructure projects, large areas of farms, etc. Medium sized helikites such as 'Desert Star' of say 35 m³ were originally designed by the British company, for use by military establishments. They carry wide range of payloads such as sensors, radars, lidar, electro-chemical analyzers, atmospheric gas samplers, bird scarers, etc. (Allsopp et al., 2013; Droneco, 2019).

Helikite are elevated platforms. They are being sought by military and civilian agencies, including the agricultural farming community. As stated earlier, helikites are a unique combination of a kite and an aerostat (or blimp). Helikite utilizes both helium and wind to keep itself aloft in the atmosphere. It is supposedly a stable aerial vehicle compared to a simple balloon or an aerostat, left to float freely. They can offer relatively stable platform for sensors, samplers, and probes, up to 2000 m above ground level. Helikites withstand gusts of wind at high speeds of up to 70 kmph. They are supposedly most cost effective, in terms of endurance in the sky and the aerial images or digital data that is offered through their sensors.

Helikites are stable across a wide range of weather and geographic conditions. For example, British Antarctic Survey group has used them with great efficiency to produce aerial photography, continuously. The helikites are stable at –20 °C to –55 °C that prevails in the Antarctic region. Helikites withstand windy situations better. They lift about 6 kg in non-windy conditions and about 20 kg in breezy conditions (Rogers, 2001).

Helikites are less costly compared to drone aircrafts. Many of the models of helikites are easy to transport from one location to another within a farm. It could be transported in a box or in tethered condition. Some of the major advantages attributable to Helikites used in general civilian and agricultural settings are as follows: Automatic flight in all weather conditions. Since the helikite is stable in sky, it offers well focused aerial photography, without haze. In future, farmers may utilize helikites often to surveillance their crops, in the outfield. The helikites used in agricultural farms accommodate a range of different cameras as pay load. They say, helikites/aerostats are generally easier to handle than drone aircrafts. Therefore, farmers may prefer them. Helikites that are small to medium in size withstand disturbance of 20 kmph wind. Tethered helikites can be installed in few minutes. They can be managed by a very small crew of 2 or 3 technicians. Tethered aerostats and helikites are easy to operate in urban conditions, agricultural farms and in open public places. Helikites are less complex, if compared with blimps or manned aircrafts or drone aircrafts (see Persistent Surveillance Services, 2019; TCOM, 2019).

From the above narration it is clear that, Helikites seem to have a great future in agriculture. They seem to be potentially highly useful in farming. However, they are yet to be tested thoroughly and proven in practical farming conditions. Drone aircraft surveys for precision agriculture and forestry tend to be a single flight survey, at a particular time of the day. Arial survey is done using visual, infra-red cameras and LiDAR. Drones aircrafts provide only a snapshot showing plants in distress. So, it allows planning for remedial action. A permanent survey vehicle, such as a helikite (or tethered aerostat), would allow crop monitoring over 24 hours or longer. It allows the agricultural researchers to pinpoint many more areas for crop yield improvement (Valour Consultancy, 2019; Altave Inc., 2019a,b). It also provides an idea about temporal changes in crops, like progress in canopy growth, panicle initiation, spread of disease/pest infestation, accentuation of soil erosion, etc.

Let us compare tethered aerostats and helikites with airships (blimps). Incidentally, blimps and their role in agriculture has been discussed in greater detail in chapter 3 of this book. Tethered aerostats are relatively large. Their payload capacity is commensurately higher. Tethered aerostats could be floated at low altitudes, when high resolution imagery is required. They could be lofted at very high altitudes of 200 m to 2 km with sufficiently long tethers. Then, they provide coverage of wider area under surveillance. Tethered aerostats are large in size. They require a few technicians to manage them on a mooring station, when not in use and while operating them in the sky. Tethered aerostats need special trucks or vehicles that transport them to the place where they are to be floated into the sky. Tethered aerostats need fairly good weather conditions. Wind turbulence may affect them. A tethered aerostat can be floated only at certain locations keeping in view the exclusion zones such as airports. Tethered aerostats could be costly, particularly, if they are sophisticated and are destined to conduct several jobs, simultaneously (See Airstar Aerospace SAS, 2019a,b; Altave Inc., 2019a,b; Alteros Inc., 2018; Lambert, 2018).

Now let us consider the salient features of helikites that are getting popular acceptance in farmland and elsewhere, in other civilian aspects. Helikites vary in size. They could be small and used to cover very small zones, on the ground. They could also be large, when they are expected to be hoisted at high altitudes, to relay imagery and radar information to ground station. Helikites are relatively more stable than a simple blimp or tethered aerostat. Helikites can withstand stronger wind turbulence compared to aerostats or airships. Helikites may carry smaller payloads of sensors, radar and few chemical probes. Again, we have to travel to the point where helikites have to be hoisted into sky. Helikites too have exclusion zones (Allsopp et al., 2013; Allsopp Helikites, 2019c,d,e,f).

Airships or larger sized blimps are able to carry larger payload of instruments and personnel. They fly to the point of use, since, they have propulsion. They are fitted with petrol or lithium battery powered engines. They need good weather conditions to take-off and land. Airships are highly expensive in relation to helikites or tethered aerostats or balloons. Airships need trained pilots. Airships could be adopted for aerial survey and long-distance travel (Lockheed Martin, 2018; Airlander 10, 2018, AeroDrum Inc., 2018; Goodyear Inc., 2018; The Christian Science Monitor, 2018).

4.2.2   COMPONENTS OF A TETHERED AEROSTAT, A HELIKITE AND BALLOON

Like any other engineering contraption or a vehicle that is adopted to conduct certain specialized function of utility to humans, a tethered aerostat/balloon has certain basic components (see Figure 4.1). In the general course, these parts are required for it to be fully operational. If we try to conduct a specific task other than general aspects, then, accurate modifications are necessary. For example, in addition to floating in air, we can adopt aerostats, to conduct aerial survey and obtain photograph of ground features. A computer unit can store aerial data in digital format. In this case, we need appropriate sensors and computer chips to store the imagery. On a broad scale, the components of the aerostat (i.e., specifications), depend on the mission requirements (i.e., purpose) and the general geographic conditions.

In order to conduct the normal floating functions, an aerostat should be equipped with a balloon/envelope. Envelope (i.e., hull) could be of different sizes and shapes. Normally, the lifting gas used is lighter-than-air helium gas. In addition, a tethered aerostat system includes a truck/ or a trailer. This is to transport the envelop (see Plates 4.1 and 4.2). The aerostat system also needs a mooring station when not in use or when we wish to keep it in readiness, for use in the sky. A platform to conduct the lift-off is necessary (Figure 4.1, item 4). A tethered aerostat, as name denotes, has to be equipped with several tethers (strings). The tethers are required to keep the aerostat in place in the sky or to move it across different locations. Tethers are needed to tie it down to a mooring station (see Plate 4.3). Also, most importantly, tethers are used to transmit power and aerial data captured by the sensors placed in the payload area. A tethered aerostat needs winches for letting out, pulling or to apply correct tension of the tethers. We also need devices to deflate the aerostats envelop, whenever it is not in use (Homeland Security, 2017; Altave Inc., 2019a,b; Airstar Aerospace SAS, 2019a,b; Allsopp Helikites Ltd., 2019a; SkyDoc Systems INC., 2019; Aeroscraft LLC. 2019).

As stated above, aerostats are deployed by industrial, meteorological, agricultural and transportation departments. The accessories required depends on the functions envisaged for aerostats. For military and general surveillance, a set of high-resolution cameras, such as visual (red, green and blue bandwidth), infrared, red-edge and lidar is required. The payload area should also be equipped with radars, to detect other flying objects, such as, aircrafts, etc. To conduct weather-related operations and detection of atmospheric pollution, including greenhouse gas emissions, we need appropriate electro-optical and electro-chemical probes. They are: gas detectors for CH_4, SO_2, NO_2, NH_3, particulate matter, etc. Most importantly, the data collected by payload instruments has to be accurately and rapidly transmitted, to the ground station or stored in computer chips. For this to happen, we need excellent computers and internet-based data relay system. The ground station equipment usually includes instrumentation for radio control of aerostat movement. Also, to analyze data using different computer programs (Homeland Security; 2013; Altaeros Energies Inc., 2019; Allsopp Helikites Ltd., 2019a; Airstar Aerospace SAS, 2019a,b).

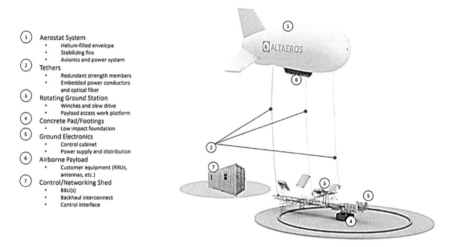

FIGURE 4.1   A generalized depiction of an aerostat and related installation.

Note: The aerostat can accommodate a range of optical sensors for aerial photography, infrared sensors to detect crops' water status, and lidar for sharp imagery. Several variations of electro-chemical sensors such as probes for NH_3, SO_2, CH_4, N_2O, NO_2, CO_2 and particulate fraction in atmosphere are also placed, in the payload area.

Source: Dr. Ben Glass, Altaeros Inc., Somerville, Massachusetts, USA.

Now, let us consider details about each major part of the tethered aerostats:

- *The Aerostat's Envelope (hull):* The aerostats' envelope holds the lifting gas (lighter-than-air gas), i.e., helium. It carries the lifting gas that generates buoyancy. Aerostat envelops could vary in sizes, shapes and design. Usually, size and shape of the envelope is decided based on the purpose. Wind speed is an important criterion that decides the shape of the envelope. The envelopes are designed to encounter least drag from the atmospheric turbulence (Homeland Security, 2017; Raina et al., 2017; Bajoria et al., 2017; Rajani et al., 2010). The cigar shaped oblong aerostats and spherical shaped ones are common. The size of the envelope is mentioned, using its length and breadth, or diameter and height, or length, breadth and height. The volume of envelope again depends on the function it has to perform and endurance. The volume determines the amount of helium it holds. A common small aerostat with 35 m³ volume is 5 ft. in length and 4 ft. in breadth. A very large envelop of 42000 cu.ft. volume will be 208 ft. in length and 70 ft. in breadth. A large aerostat can carry a pay load of 2000 lbs, up to an altitude of 15,000 ft. above ground surface (Homeland Security, 2013; Airstar Aerospace SAS, 2019a,b). The fabric used to make the envelope is very important. Most common fabric material used are the polyesters, polyurethane and polyvinyl sheets (see Dasaradhan et al., 2018). Sometimes the fabric is laminated to protect the envelope, from ill-effects of ultra-violet radiation. Some envelops are used in advertisement and amusement in public events. They come in different fancy shapes. They are flown only for a certain duration. They reach only low altitudes. Milkert (2014) states that, design of the aerostat's envelop should be commensurate with size of the advertisement screen and total payload weight. The lightweight LED screen with bright illumination and at least 12–18 m in height is required, if the advertisement has to be visible even from 4 km distance.

Over all, the key aspects of an aerostat's envelope are: the dimensions during transport from storage facility to launching point; knowledge about dimensions of aerostat when it is fully inflated is important; volume of the envelope when ready for lift off; the maximum wind speed allowed at the time of launching an aerostat; and operational wind speed allowed. We may note that, maximum operational altitude that an aerostat has to reach also determines the envelope's parameters. We should also consider the mean temperature in the atmosphere, while selecting envelope for an aerostat. The volume (capacity) in the pay load area is also important.

• *Lighter-Than-Air Lifting gas*: At present, most common lifting gas used in the aerostats is the helium. It is filled from a pressured canister. Usually, it takes only few minutes to inflate an aerostat with helium. Hydrogen, which is an inflammable gas is totally refrained from aerostats. Sometimes, coal-gas mixture is tried as lifting gas in aerostats (Homeland Security, 2013).

• *Winches:* A tethered aerostat's winches are spool-like devices used for letting out and pulling in the tethers, during launch or recovery of the aerostat. Winches are often used to adjust the tension of the tethers. The design of the winches may differ. In some cases, a wooden hand reel is operated manually. Larger aerostats are provided with powered winches. Powered winches could be computer controlled or pre-programmed. Winches used in an aerostat system often depends on envelope. Also, the type and number of tethers utilized in an aerostat. Winches should match the tethers used on the aerostat.

- *Tethers:* A tether stabilizes and holds the aerostat in place in the sky. The number, weight and material of tethers are important considerations. They say, adding an extra power line or a light cable to transmit electronic data to ground station, or adding another fiber may increase the weight of tether. Consequently, it may lead to lowering of pay load weight possible. The weight of the tether and number can affect the flight dynamics and the ease with which aerostats can be remotely guided. We may note that, to direct the aerostat, an operator may use tethers or control it thorough remote controller or broadband mobile technology. Basically, tethers are useful in mooring the aerostat to ground station, in transmitting electrical power to payload instruments, in transmitting electronic data and aerial images to ground computers.

- *Payload:* The payload instruments and their total weight depends on the mission requirements. The payload area itself could be small or large, depending on the model. The payload area and instrumentation may also depend on altitude and the weather/aerial imagery needed. The payload system is often integrated with other electronic circuits of the aerostat.

- *Ground Control Station*: The instrumentation within the ground station depends on the purpose. A ground station can be elaborate with radio control equipment, series of computers that process the data about weather and collect digital data about other ground features. In some cases, the ground station may be represented just by an 'iPad.' An iPad with appropriate computer programs to determine flight altitude, pathway, and to collect data may be more common, in agricultural settings. The cost of a sophisticated aerostat with ability for electronic surveillance, aerial photography, infra-red photography, LIDAR and a set of electro-chemical process may be high. For agricultural purposes, an aerostat with ability for aerial photography and collection of crop's data suffices. Of course, specialized computer programs are required for processing imagery and calculating fertilizer and water requirements of crops.

- *Mooring Station*: Mooring station is an important component of an aerostat system. Mooring station is required to store the aerostat when not in use or to keep it in readiness to launch. The mooring station, its type and model depend on the type of aerostat (Plate 4.3. The cigar type aerostats that are larger are often tied down, to a mast and held in readiness. The round type aerostats and those called as helikites are moored into specially designed stations. Aerostats may often require transportation from one location to another prior to launch. The transportation most commonly employed involves a truck, a pick-up van or any other movable vehicle (Plates 4.1 and 4.2).

PLATE 4.1   A small-sized mobile Skydoc aerostat system.

Note: Skydoc aerostats can be launched once the vehicle stops.

Source: Dr. Charlie Steffan, SkyDoc System LLC., Kansas, USA; Skydocballoon.com.

Now, let us consider the components of a helikite. The quality and durability of helikites, no doubt, is dependent on the use of correct material. The material required to make a helikite depends on the end use and durability envisaged. To design and develop a helikite, we need a helium holding balloon, a kite sail, a spar and a method of attaching these parts together (Rogers, 2001; Allsopp et al., 2013). A helikite requires good materials. For example, good helium retention capacity of the balloon is a necessity. Balloon should be flexible and light in weight. Preferably, transparent and easy to work on. Several types of materials have been examined for use as the balloon, to hold the helium gas. A few examples are light transparent polyester, metallized light polyester, scrimmed-polyester, soft polyethylene, metallized nylon, polyurethane, etc. It seems a combination of nylon/polyurethane makes a strong but light balloon material.

PLATE 4.2   A mobile aerostat that travels on a trailer.

Note: It is an alternative anchoring unit for this model of aerostat.

Source: Dr. Charlie Steffan, SkyDoc System Inc., Kansas, USA, www.skydocballoon.com.

In a helikite, the kites are normally made of rip-top nylon. It is a strong fabric. It is slippery to work. Sometimes, transparent polythene fabric is also adopted to make the kite. Tarpaulin is cheaper material but medium in weight. Cotton is easy to work, cheaper and biodegradable.

Spar is usually made of carbon fiber. Carbon fiber is strong, light in weight and flexible. Aluminum spars are also used. However, they are brittle and may brake without notice.

Fixing material used are nylon threads to sew the helikite. Sewing is usually done by skilled workers. They use heavy tailoring machines. Strong adhesive sticks are used to attach kite to the balloon. There are specific brands of glue used, to connect the kite to balloon. We have to be careful not to puncture the balloon while attaching them to kites. Velcro material is also used to attach the kite to the envelop.

They say, each 1.0 m³ of helium gas in balloon usually offers a lift for 1 kg weight. However, about 40% of lift is required to keep afloat the material of the helikite itself. About 5–10% lift is utilized to negotiate vagaries of the atmosphere. Therefore, under specifications most manufacturers state that helikite could lift about 0.5 kg by weight per 1.0 m³ helium gas in the balloon. Therefore, to lift a 25 kg payload, we need a helikite balloon that holds 50 m³ of helium. Rogers (2001) states that, compared to any other aerial vehicles such as blimps, free- floating aerostats or microlights, the helikites possess very small balloon per weight of payload lifted. They are efficient in terms of payload that could be easily lifted to 2000 ft. above ground level.

Most brands of helikites come with a carrier unit (an aluminum box). The carrier unit holds the balloon fabric, helium tanks, electric generator, tether, winch, sensors, etc. The carrier can be mounted on a truck or trailer.

4.3 CLASSIFICATION AND TYPES OF AEROSTATS, HELIKITES, AND BALLOONS

The tethered aerostats could be classified based on several different criteria. The criteria could be applied either singly or sometimes in combinations. Table 4.1 depicts one such simple classification based on size, purpose and endurance.

TABLE 4.1 Examples of Tethered Aerostats, Their Purpose, and Flight Endurance

Name	Purpose	Flight Endurance
Large sized:		
Condor	Military usage such as surveillance	Long term, 7–15 days
Eagle Owl	Tactical, military and strategic reconnaissance	Long term, 7–15 days
Medium sized:		
Jackdaw	Operative, civilian commercial	Short to medium term, 5–10 days
Small sized:		
White hawk	Tactical, commercial and general civilian tasks	Short term ≤ 5 days
Sky Star	Civilian tasks like advertisement, surveillance, etc.	Short term ≤ 5 days

Source: Airstar Aerospace 2019a,b; Dr. Kosberg, Sky Star Inc, Yavne, Israel.

4.3.1   A FEW EXAMPLES OF AEROSTATS USED IN MILITARY AND CIVILIAN TASKS

Let us consider a few examples of tethered aerostats produced and offered to public during recent years. Their specifications and utility should also to be noted. Most of the aerostats listed and described are versatile. They could be used in different situations. For example, they could be adapted to suit military, civilian, and agricultural activities. A few of them may also suit space science related projects, such as sampling and analysis outer atmosphere, etc.

4.3.1.1   CONDOR TETHERED AEROSTAT

Salient features of this model of aerostat adopted by military establishments are as follows: (a) Completely autonomous for a long-term use, (b) Embedding of any kind of payload, (c) High altitude for a broader spectrum and a quicker detection, (d) Resistance to harsh weather conditions, (e) Fail-safe: resistant to basic breakdown, (f) Persistence: up to 15 days.

PLATE 4.3   Condor: A Tethered Aerostat and Mooring Station

Source: Irene Guerrero, Aeroscraft Corporation, Montebello, California, USA; Also see Figure 4.1. by Ben Glass.

The specifications of Condor Aerostat are as follows:

Volume: 1600 m³; Flying height (AGL): Up to 1000 m; Dimensions: 11 m x 30 m; Payload: Up to 250 kg; Deployment: 4 hours, 6 to 8 operators; Logistics and Transport: Semi-trailer (aerostat + mooring station); Operational Wind Speed: Up to 130 km/h.

4.3.1.2 EAGLE OWL TETHERED AEROSTAT

Salient features of this model of tethered aerostat adoptable in different situations are as follows:

(a) It is capable of multi-roles. It is adaptable to a wide range of needs; (b) Broad embedding ability of the payload; (c) It is easy to deploy and to operate; (d) Resistant to strong weather conditions; (e) Fail-safe: resistant to basic breakdown; (f) Persistence: up to 7 days.

Specifications: Volume: 450 m³; Flying Height (above ground level): Up to 650 m; Dimensions :7 m x 19 m: Payload: Up to 50 kg; Deployment: 4 hours, 3 operators; Logistics and Transport: 20 ft. container (aerostat + mooring station); Operational Wind Speed: Up to 110 km/h.

4.3.1.3 JACKDAW TETHERED AEROSTAT

The Jackdaw tethered aerostat is manufactured by a French company named Airstar Aerospace SAS. The salient features of this tethered aerostat are as follows:

(a) It could be an all-in-one solution; (b) It is simple and reliable; (c) It is resistant to windy weather conditions; (d) It has innovative pressure control system; (e) Fail-safe: resistant to simple breakdown; (e) Persistence: up to 5 days.

Specifications: Volume: 90 m³; Flying height (above ground level): Up to 300 m; Dimensions: 4 m x 11 m; Pay load: Up to 15 kg; Deployment:1 hour, 2 operators; Logistics and Transport Platform truck (aerostat + mooring station); Operational Wind Speed: Up to 80 km/h.

4.3.1.4 WHITE HAWK TETHERED AEROSTAT

White hawk tethered aerostat is manufactured by the company Airstar Aerospace SAS located at Aygusvives, in France. The salient features of this model of tethered aerostats are as follows:

(a) Acceptable price to performance ratio; (b) Low logistical footprint and easy to deploy; (c) Automated take-off and landing system; (d) Compliant with European Aerospace regulation; (e) Fail-safe: resistant to simple break-down; (f) Persistence: up to 5 days.

Specifications of White Hawk Aerostat: Volume: 40 m³; Flying Height (above ground level): Up to 200 m; Dimensions: 3.4 m x 4.4 m; Payload: Up to 5 kg; Deployment: 45 minutes, 2 operators; Logistics and Transport: 5 m³ van; Operational Wind Speed: Up to 40 km/h.

Now, let us consider a different aerostat model. Skystar brand of aerostats are designed and manufactured by the subsidiary of Skystar Inc., called RT situated in Israel. There are at least 2 models of Skystar aerostats that are suited for military and civilian tasks, namely 'Skystar 180' and Skystar 300.' The basic Skystar system consists of a tethered helium-filled balloon platform. The mission payload comprises a stabilized day/night electro-optical (EO) sensor set. It has a tether that links the platform and payload to the ground control unit. The tether provides power and controls on uplink, and high-capacity video on downlink. The aerostats could be held on a transportable cart. They are easy to deploy. It seems Skystar aerostats are relatively cost effective. They reach an altitude of 1000 – 1500 ft. above ground surface. The aerostat model withstands wind turbulence of 40 knots while on surveillance above military barracks. The aerostat material resists temperature changes between –35°C to +45°C (see Aeronautics, 2019a,b; Shephard News Team, 2018; Plate 4.4).

Now, let us consider helikites that are actually a hybrid made of a kite's sail and balloon's envelope. Helikties are generally more stable in the air. The tethered helikites are often versatile and can be floated at both low and high altitudes, depending on the requirements. There are a few companies that are manufacturing these helikites, like Allsopp Helikites in Great Britain. There are also several models of helikites to suit to the purpose. For example, there are small helikites flown only to 5–10 m altitude and those commonly floated at 200 to 300 m. They are also few that can reach over 1000 m altitude above ground level.

Skyshot Helikites are a very popular series of aerial vehicles with various custom modifications added. Such modifications allow the lifting of compact digital cameras. There are presently three basic types: namely a)Standard 'Skyshot' Helikite Aerostat; (b) New 'ActionCam' Skyshot Helikite aerostat and (c) New radio-controlled Skyshot Helikite with pan/tilt rigs.

Let us consider the main features of a typical medium-sized helikite such as 'Desert Star Helikite.' 'Desert Star' helikites have numerous special features that supposedly allow them to lift more payload in a stable manner, fly in windy conditions and to a greater altitude.

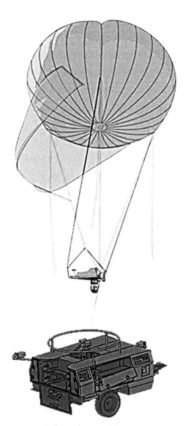

PLATE 4.4   A Skystar Aerostat: a tethered aerostat.

Note: The Skystar 180 and Skystar 300 both models of aerostats are primarily meant for military and civilian surveillance.

Source: Dr. Taly Kosberg, Skystar Inc., RT., Yavne, Israel.

The salient parts of Desert Star Helikite (or any normal helikite) has been numbered and shown in Plate 4.5. Their names/description are listed as follows. It is based on the literature (user manual) provided by Allsopp Helikites Ltd. (2019 a, d):

1. The envelope is a spheroid balloon made of 'ultra' outer protective cover. The cover is protective. It allows envelope to withstand harsh weather and gusty winds. It is dynamically shaped.
2. It has kite sail that provides extra lift under windy conditions. The kite is made of nylon. It allows the helikite to fly at an angle.
3. It has large stiff keel. It provides stability under high winds.

4. The variable bridle is said to be a unique system that allows the volume of the balloon (envelope) to vary. This trait allows desert helikite, to fly at greater height where there is reduced air pressure. It also helps to withstand day and night temperatures that occur at high altitudes.

5. The main spar is made of carbon fiber. It forms the main structure of the helikite. It helps to make helikite semi-rigid. It offers rigidity to withstand high winds.

6. The central carbon-fiber keel spar also adds to rigidity. It makes helikite stay steady in the sky despite wind. So, it reduces the vibrations that otherwise affect quality of aerial photography.

7. Lower keel spar is also made of carbon fiber. It offers stiffness to keel. Increase steadiness in the wind. It also supports camera mounting and gimbal.

8. Front attachment point: it provides immensely strong attachment between the envelope (balloon) and kite.

9. Handling Line. This facilitates easy and safe launch and recovery in any weather, by just one person.

10. Central Tie-Down Point. The handling line is tied to this strong point. It shelters electronics, computers and radar equipment. It is generally protected from sun and rain.

11. Upper Keel Payload Bay. This bay supports lithium batteries. It has custom-made pouches for electronic equipment and iPads.

12. Kite Front Attachment: The kite is strongly attached to the outer protective balloon cover. The area of attachment is larger. So, the strain gets distributed uniformly to greater area.

13. Side Tie-Down Point. Helikite can be tied securely down onto its' launch pad in all conditions when it is not flying.

14. Rapid Inflation Helium Input Valves: The helium filling hose helps to fill the lighter -than-air gas quickly.

15. Rapid Deflation Plugs: These are reachable through Velcro pockets in the outer balloon cover.

16. Balloon Service Access: It allows the inner balloon to be taken in and out of the balloon cover if required.

17. Stern Antenna Mounting Position: This aspect makes helikite a good platform for radio equipment. The antennas are kept perfectly vehicle while in flight.

18. Large Velcro Spar Closure: It makes insertion and removal spar easy and quick.

19. OPTIONAL 'Universal Camera Mount' (UCM): This is required only for cameras and wind-meters. It is a solid flat plate onto which

cameras can be securely fitted. Other types of aerostats have cameras that dangle from loose lines. Cameras could be very unstable. The unique UCM keeps cameras exceptionally steady even in high winds (See Plate 4.6 top row).

20. OPTIONAL 'Auto Emergency-Cut Down Device': This device has ensured that no Desert Star Helikites has ever been lost during operations.

21. None.

22. OPTIONAL 'Steady Cam Rod': It offers stiffness below camera payloads. Therefore, increases stability of both the helikite and camera. It is not needed for any other payloads and can be removed easily when not in use.

23. OPTIONAL Night-Time Warning Lights: This is required only for operations at night. The battery powered LED lights can operate un-interrupted for many days or weeks, and are held securely in special pouches.

PLATE 4.5   A Desert Star Helikite and its parts.

Note: The salient parts of Desert Star Helikite are marked. They are also described in the following paragraph.

Source: Sandy Allsopp, Allsopp Helikites Ltd. Fordingbridge, Hampshire, United Kingdom.

PLATE 4.6   Aerial imagery of farmland obtained using Allsopp Helikites Ltd.

Note: Top: Three photos depict Helikites with photographic equipment (sensors) placed at different spots on the envelop; Middle: The photograph is an example of aerial imagery of farms obtained, using cameras on helikite aerostats; Bottom: An infra-red imagery of wheat field derived using camera on Helikite. Infrared image helps farmers in detecting variations in water status of crops and drought effects, if any. Sometimes, cameras are located at different angle in order to obtain wider area within the photograph. Generally, helikites/aerostats can accommodate payload that includes different visual, infrared and lidar cameras.

Source: Sandy Allsopp, Allsopp Helikite Ltd, Fordingbridge, Hampshire, England, United Kingdom.

Some of the features stated by the manufacturer for helikite aerostats, in general, are as follows(Allsopp Heliktes, 2019a): Helikite can be utilized in all weather conditions. They are safe to use. They are quick to hoist and adopt for aerial photography of farm/crops. They offer steady flight. Therefore, the aerial imagery is of high quality. Cameras on helikites are simple to use (Plate 4.6). They could be pre-programmed helikites can be flown anywhere. The aerial imagery is generally accurate. There are no legal restrictions to use helikites. Of course, there could be many more advantages attributable to small helikite aerostats depending on the location, purpose and timing.

4.4 AEROSTATS, HELIKITES, AND BALLOONS: THEIR UTILITY

Following is a list of applications possible using different models of aerostats. A single aerostat model may not be able to conduct all functions and be used at all junctures. A few models may fit best, to conduct a single or a few, or a set of applications more accurately, easily and with better efficiency. We have to pick the best model of aerostat for the said purpose. However, in general, most companies that produce aerostats usually state following functions/application as possible, using aerostats. The applications of aerostats and balloons itself could be grouped, firstly, into few sets (see Airstar Aerospace SAS, 2019a,b).

There are indeed several types of unmanned aerial vehicles that were originally adopted for use by the military establishments. For example, drone aircrafts were predominantly utilized to conduct reconnaissance and bombing of enemy positions. Similarly, parafoils and blimps too found greater acceptance in military campaigns. In the present context, we are concerned with tethered aerostats. Aerostats too were employed initially to conduct surveillance of military camps and happenings on the ground near battle fields.

No doubt, we can classify the applications of aerostats firstly as those relevant to military and border security services. The other major group of applications deals with spectral observations made using sensors placed on the aerostats. Third group of applications include surveillance of disaster-prone zones, rapid first response and relay of useful information, to the ground station. Following are the various *groups* and individual applications of aerostats (see Airstar Aerospace SAS, 2019b):

Military and Border Security: (a) Military control (intelligence); (b) Anti-drone reconnaissance; (c) Surveillance of sensitive sites (nuclear facilities, offshore platforms, etc.); (d) Aerial observation of industrials sites, airports, harbors; (e) Surveillance of borders, roads and vehicular traffic; (f) Aerial photography and surveillance of public events; (g) Urban crowd control; (h) Surveillance of border posts and adopting anti-smuggling procedures.

Aerial observation and Data collection: (a) Applications include scientific missions, collection of weather data, study of climatic patterns; (b) Mapping of ground surface and topographic features, using sensors; (c) Aerial observation of agricultural land holdings and crops. (d) Collection of data about crops that is necessary to judge fertilizer, irrigation and pesticide needs.

Applications of Aerostats in situations of Emergency: (a) Reporting natural disasters; (b) Detection and assessment damage due to natural disasters; (c) Co-ordinating rehabilitation programs in a disaster affected location; (d) Spotting crime zones and tracking vehicles; (e) Detection of smuggling; (f) Adoption of aerostats in electronic warfare, etc.

Aerostats and helikites are manufactured by several companies across different continents. Each of the manufacturers may emphasize a set of applications for aerostats/helikites. Following is a list of applications proposed by an aerostat surveillance services provider (Persistent Surveillance Services, 2019). Each aerostat/helikite company lists several uses but they do excel in a few. Perhaps, they bestow less attention to a few others shown as routine. The uses of aerostats/helikites are also classified based on convenience, purpose and customer base in view. Let us consider an example. A private aerostat company classifies major uses of helikites as follows (see Persistent Surveillance Services, 2019):

(a) Inspections and Monitoring: Refinery inspections, oil and gas pipeline inspections, power line inspections, wind turbine inspections, solar energy farm inspections, railway and roads inspections, bridge safety inspections, river and canal inspections, assessment of structural safety of embankments, flooding monitoring, and erosion monitoring.

(b) Surveying: Buildings, bridges, railways, road, ports, building sites; land resources, and water bodies.

(c) Geo-Mapping: Asset management, tailings assessment, ore assessment, rockfall prediction, volume, calculation, environmental monitoring, roads, rivers, bridges, and canals.

(d) Safety Assessments: Flood mapping, and waste management.

(e) Agriculture: Detection of drought, irrigation leak detection, detection of irrigation blockage detection, and weed mapping.

(f) Research: Geography, meteorology, archaeology, ecology, zoology, and botany.

(g) There are also passenger carrying helium balloons (Lindstrand Technologies, 2019).

Now, let us consider applications of helikites. Helikites or helium-filled aerostats with kite are getting common in aerial photography and surveillance tasks. The helikites are useful in long term surveillance of military installations, buildings, industries, public places, roads, etc. There are indeed several companies now engaged in the production of helikites-aerostats. The applications of helikites are wide ranging. Helikites, along with other types of aerial vehicles such as blimps, aerostats and tethered balloons are slowly gaining acceptance. Otherwise, helikites were neglected. Again, major uses of helikites could be grouped. A generalized list of possible applications of helium kites (small aerostats) as stated by manufacturers are as follows:

(a) Aerial photography of real estate, industrial sites, dwelling localities, land resources, dams, irrigation projects, natural vegetation, etc.

(b) Aerial photography of land holdings and accurate demarcation of properties.

(c) Aerial imagery of geographical sites, land and topographic features, imagery of rivers, water bodies, mountain ranges, etc. Detection of atmospheric pollution and greenhouse gas emissions. Study of land, animal and bird migratory trends. Study of surface features of oceanic regions. Aerial imagery and study of archaeological sites. Collection of data pertaining to weather patterns and atmospheric disturbances. Detailed analysis of crops and their growth pattern.

(d) In agrarian regions, helikites are useful in a series of farm related activities. Firstly, helikites offer aerial images of land, soil and crops' condition, continuously or intermittently. The aerial imagery drawn using a set of sensors, such as visual (R, G and B), infra-red, red-edge, and lidar helps farmers. Aerial imagery helps in judging crop's growth (NDVI), leaf chlorophyll content (i.e., crop's N status), water stress index, pest/disease attacks if any, soil erosion, flood and drought related loss of crops, etc.

(e) Helikites are utilized in environmental analysis. They have been utilized to get aerial data about oil spills, their size and deleterious effects. Helikites have provided images of the effects of over grazing, soil erosion, floods and drought effects on natural vegetation.

(f) Helikites are useful during planning of large infrastructure projects. Aerial imagery from helikites are useful in town planning and during construction of large projects relevant to industries, irrigation, town infrastructure management, etc.

(g) Helikites have a role in policing and urban traffic control. They have been used in crowd control during public events. Monitoring sports events, buildings, monuments, thermal imaging of large building blocks, detecting wildfires in forests are few other uses of helikites.

(h) Helikites have been utilized by military establishments to monitor camps, vehicular and troop movements, and surveillance of installations.

The usage of aerostats/helikites naturally depends on the geographical region and the problems faced. Aerial surveillance, photography or any other similar task could be the most immediate requirement. For example, the immediate goal of the joint Soviet-Bulgarian project envisages use of aerostats in the fields of astronomy, meteorology, aeronomy, and remote sensing. It envisages a number of experiments concerning applied ecology, navigation, aero- and space equipment, etc. (Bonev, 1994). Recently, aerostats fitted with high altitude wind turbines are producing electricity from wind power. Here, the aim is to generate power for farmers and rural community. Aerostats are most sought-after aerial vehicles when one wants to survey disaster affected zones and report the digital information rapidly, to ground station. Aerostats have been in use by the meteorological departments. They are used to carry instruments that record atmospheric parameters such as temperature at different altitudes, relative humidity, gaseous emissions, pollution levels, etc. Agricultural farmers are among the most recent section of entrepreneurs who have adopted aerostats and helikites. Let us discuss each the many uses listed above and many more that are still to be standardized.

4.4.1   AEROSTATS, HELIKITES, AND BALLOONS IN MILITARY

Let us now consider potential uses and applications of aerostats, helikites, and balloons in greater detail. Aerostats/helikites have been utilized since decades. Aeronautic engineers have revised, modified and upgraded the design and materials used on aerostats. They have improved the performance of aerostats and helikites. Accordingly, utility of these aerial vehicles too has spread out into variety of professions. There are innumerable applications for aerostats in military establishments. Also, in border security and general aerial surveillance.

4.4.1.1   AEROSTATS, HELIKITES, AND BALLOONS IN MILITARY ESTABLISHMENTS

Like other aerial vehicles, aerostats, helikites, and balloons were in greater demand. They were sought and used by military establishments of different nations. Much of the initial developments in aerostat technology was performed by military engineers. Their aim was to standardize aerostats for rapid deployment and accurate use, by the army personnel. Again, many of the improvements appeared to have taken place in developed nations. At present, most of military establishments possess a range of aerostats and helikites. They are being used in establishing communication networks and offering crucial aerial photographs taken both during day and night. Aerial images and video recordings depict the happenings on the ground. Most often, aerostats appear to have been utilized as sentinels and surveillance gadgets floating above the military barracks and installations.

Aerostat manufacturers often design them for specific tasks such as military reconnaissance, aerial photography, monitoring military convoy movement, surveillance of military camps, etc. Offering aerial images helpful in arranging supply and logistics is important. Let us consider an example of an aerostat model built to suit the military functions. 'Skycrow' is an aerostat for military establishments (see Aeros, 2019a; Plate 4.7). It is an aerostat model meant for tactical purposes. It is ideal for military camps and establishments to float it in sky above their camps. It can also be used when commercial surveillance missions are required over, say, industries, dam projects, public places, etc. The main features of this tactical aerostat-'Sky crow' are: (a) It can be operated by singe person, (b) It is an effective soldier centric aerostat, (c) It has plug-in-play payload options, (d) The entire aerostat is stored in a pelican storage for rapid shipping and (e) Soldiers could be trained to float and use aerostats in short time. Regarding specifications: The 'SkyCrow' military aerostat is 10.5 m, i.e., 30 ft. in length. It is relatively a smaller aerostat. Its' operational altitude is 305 m above ground surface. It carries a payload of 9 kg. The payload includes surveillance cameras and early warning radar systems. The aerostat is fairly stable and withstands 40 knots wind speed disturbance. Aeros 1170 and 21 M are other models that could be used in military (Plate 4.7).

Lighter-than-air technology-based aerostats are currently in use by the military establishments of many nations. Although not manufactured indigenously by each nation, they are being procured for use in military reconnaissance. For example, recent reports state that even small nations in

Central America are adopting aerostats. The aerostats are very useful during surveillance of military camps, movement of trucks and supplies. Also, in checking drug smuggling at borders in Central America. Recently, a few nations in Central America procured the small sized helium aerostats known as 'SkyStar 110.' It is a micro-tactical aerostat. It is transported on mobile trucks before the actual launch. It needs only two people to launch and use this aerostat. It takes only 15 minutes to assemble and loft the aerostat. The aerostat has a set of sensors such as visual, high-resolution multi-spectral, an infra-red sensor for photography during night and lidar. The small aerostat could be utilized regularly for aerial reconnaissance by the military. It takes a payload of 20 kg. The aerostat system operates continuously even when wind speeds are 40 knots. It provides surveillance up to a height of 1000 ft. above ground level. Overall, it is clear that small aerostats are of great utility during military reconnaissance.

PLATE 4.7   Tactical Aerostats for military surveillance and reconnaissance; Left: Sky Crow; Right: Sky Cobra.

Source: Irene Guerrero, Aeroscraft Corporation, Montebello, California, USA.

Military establishments of many European nations have adopted the usual 'cigar shaped' large sized aerostats. They are large, say, more than 30 ft to be stable in tactical operations by military. In the sky, they are unwieldy at times, particularly, when weather and wind conditions are uncongenial. Also, they need a group of 10–15 soldiers or police, to maintain, float and control them. They are also costly. In contrast, small helikite aerostats are also ideal for military usage. Therefore, at present, many nations in Europe are trying to adopt smaller helikites in larger number. Helikites are smaller in size, easier to manage in the sky for longer period, and less costly to maintain. Helikites stay well stabilized in the sky. These helikites offer excellent imagery of ground features, military convoy movement, etc. The radars help in detecting other flying objects. They can be flown at relatively lower altitudes of 1500 ft. above ground level. Let us consider an example. Allsopp Helikites produced in Great Britain are smaller, stable and easy to operate. These tactical helikites are now being used in several nations. The helikites float at altitudes beyond the reach of the ground-based small arms. They are excellent for aerial photography of military set up, army barracks, supply lines, etc. As stated earlier, Allsopp Helikites are managed by just two soldiers/police. They can be moved on mobile vans on the ground, prior to launch. The Skyhook helikite aerostats can be assembled and launched in 30 minutes by 2 soldiers (Allsopp Helikites, 2019b).

The 'Skyhook Helikites' are smaller. They are 14 ft in length and 11 m^3 in volume. They can be transported even in inflated condition, since they are small. They can withstand harsh weather and wind speeds of 40–60 kmph. They can stay for longer duration in the sky. They possess gimbals that keep sensors in well stabilized condition. Therefore, aerial imagery is distortion-free. The skyhook helikites are fitted with a set of visual sensors, radars and electronic instruments to relay the images, instantly. It has radars to fore-warn about other flying objects. Hence, they are adopted for tactical operations. These skyhook aerostats are easily launched and retrieved, using remote controllers or they could even be pre-programmed. Such small helikites could be adopted for surveillance even in the mountainous region. They can be lofted to 5000 ft. from ground level to surveillance army move-ment in the mountainous region (Allsopp Helikites Ltd., 2019c, d; SkyDoc Systems INC., 2019).

The United States Army employed several small sized helikites to surveil-lance the terrain in the Afghan campaign. The use of versatile small helikites was necessitated, because, the traditional large-sized blimps/aerostats were not easy to inflate and use it above the several small outposts. Insurgents used very small military outposts. So, US Army needed small helikites that were easy to install at several places. The intention was to monitor the movement of insur-gents, their vehicles and the large number of enemy outposts. The program to surveillance using small helikites was named 'Small Tactical Multi-Purpose Aerostat System (STMPAS).' Under this program, US Army utilized small aerostats such as 'Desert Star Helikites' (The Editor, 2012; Plate 4.5. Desert star). They were equipped with necessary sensors placed in gyro-stabilized gimbals and rapid data relay equipment. The small helikites were generally floated at 1500 ft. altitude. This is to keep helikites away from enemy ground snippers. From that altitude they dispatched information to ground control about movement of insurgents and their attempts to place IEDs (bombs). They say, the small tactical helikites were versatile. Helikites are highly effective in Afghan war (see Allsopp Helikites Ltd., 2019e; Drummond, 2005).

Small helikites have certain clear advantages in terms of ease of operation, obtaining aerial imagery and conducting radar-aided reconnaissance. The Australian Army, it seems, first tried to float a large aerostat to gather continuous imagery of terrain. The idea was to surveillance large areas in one-go. However, that seemed not so easy in terms of logistics and cost effectiveness. Large aerostats had to be transported using special trucks and needed loads of helium gas. Plus, it could not be repeatedly lofted at different places. So, the Australian army resorted to small helikites (e.g., Desert Star) that were floated, at just 1000 to 1500 ft. above ground level. These are small helikite aerostats that could be repeatedly deployed. Aerial imagery from sensors gave a clear view of the terrain, up to 49–50 km in radius from the helikite. The helium requirement was less. The images were all GPS tagged and easy to read (see Allsopp Helikites Ltd., 2019f).

There are few other models of aerostats that are currently in use. For instance, the TCOM 28M Operational Class Aerostat is a versatile, battle-proven, rapid deployment, persistent aerostat. It is good for surveillance of areas on ground, air, and sea. It is equipped to detect and track targets in the most challenging terrain. TCOM 28M aerostat is actually mid-sized, wide-area surveillance vehicle that can be customized for a variety of mission types (Missy, 2018).

Next, the Joint Land Attack Cruise Missile Defense Elevated Netted Sensor System (JLENS), is a system of two aerostats. They float at 10,000 feet above in the air. The aerostats are filled with lighter-than-air helium gas. Each one is nearly as long as a football field. They carry powerful radars that can surveillance and protect a territory roughly the size of Texas from airborne threats (Raytheon, 2019).

4.4.1.2   AEROSTATS, HELIKITES, AND BALLOONS IN BORDER SECURITY AND POLICING

Aerostats could be classified based on purpose, its size and a range of characteristics related to their manufacture and use. Here, under this section we are concerned with tethered aerostats that are predominantly utilized in border patrol, security, law enforcement, disaster control, etc. The United States Department of Homeland Security's Science and Technology Directorate classifies tethered aerostats based on following purposes (see Homeland Security, 2013). They are;

(a) Emergency Management: i.e., Fire control, man-made disaster relief, natural disaster relief and search and rescue;

(b) Homeland Security: Land border security, harbor and coastal security, infrastructure security, sensitive surveillance;

(c) Law Enforcement: Crowd management, drug interception, vehicular traffic monitoring.

Tethered balloons have been utilized to conduct border surveillance and monitor activities at check posts. There is a program called 'Eye in Sky' that envisages use of tethered aerostats to monitor ground activities (US Department of Homeland Security-Science and Technology Directorate, 2018; US Customs and Border Protection, 2013). Tethered Aerostat Radar System (or TARS) is a system of surveillance aerostats. It oversees ground locations and traces flying aircrafts in the border zone of southern USA. It includes a series of tethered aerostats, placed at 10,000 ft. above ground surface, from Yuma in Arizona to Lajas in Puerto Rico. Each sweep of the radar on the aerostat covers an area of 200 km^2 under the aerostats. These aerostats help the border patrol and US Customs Department with forward information. They say, among various surveillance methods, tethered aerostats are cost effective. Tethered aerostats provide constant watch over border posts. Further, digital data could be relayed to other aerostat towers through internet and facilitate rapid exchange of information. The first TARS was set up by U.S. Air Force at Cudjoe Key, Florida. Next, a TARS was commissioned in 1983 at Fort Huachuca, Arizona. From 1988 to 1991, the U.S. Customs Service established more TARS sites at Yuma, Arizona, and three sites in Texas, including Marfa, Eagle Pass and Rio Grande City. By the end of 1994, additional TARS balloons were floating in Florida, Texas, Puerto Rico and even the Bahamas. The U.S. Air Force managed the TARS program until July 2013. Later, it was utilized entirely by Customs and Border Patrol Department (US Customs and Border Protection, 2013,2014; US Department of Homeland Security-Science and Technology Directorate, 2018; US Embassy in Philippines, 2014). Relocatable and tactical aerostat towers (tethered aerostats) have also been deployed by other agencies such as US Food and US Drug Control Departments. They use aerostats to monitor cross border trafficking, if any.

There are indeed innumerable methods to surveillance international borders. Most of them aim at answering the defense requirements, by using various aerial vehicles such as drones, blimps, aerostats, etc. Aerostat networks have also been established just to monitor borders and control human trafficking. Such aerostat networks help in rapid exchange of crucial information, by border check-posts. Aerostats located on mobile pick-up vans and fitted with surveillance cameras have also been utilized at the US-Mexican border. In addition, aerostat towers and other types of observation posts (towers) have been integrated into border security system. The radar system placed on these aerostat network helps police to surveillance vehicles/humans for up to 200 km range from aerostat locations. The information derived via aerostats actually fore warns of illegal border crossing of human migrants and suspicious vehicular movement, if any (Trumble et al., 2018; Lambert, 2018).

Border security seems to be a pre-occupation of certain military establishments. It happens to be so when borders between nations are porous or is troubled with intrusions. Let us consider an example. In case of border between India and Pakistan, the situation demands regular surveillance and high-resolution aerial photography, all through the day and night. It seems this need is being answered by Indian military by deploying Russian made aerostats, all through the Northwest border. It allows the military a high vantage point to observe the border activity of the other nations. Here, it helps in early warning and rapid reaction, to thwart any untoward activity in the border. They are using Israeli SPYDER missiles to stop intrusions (Sputnik News, 2019). The defense ministry of India has recently adopted aerostats in different locations. The aerostat called 'Nakshatra Aerostat' monitors the activity across the northern border (Cholan, 2017).

Russia's RosAeroSystems is among the leading aerostat manufacturers worldwide. Many of their models are primarily meant for military usage, particularly, in border security and surveillance. Reports suggest that Russian RosAerosystems has developed a very large aerostat. It is to be utilized by the Chinese for their border security operations. The large aerostat is said to act as early warning system. It has a series of radars that detect any movement across the border. The aerostat reaches a height of 5000 ft. above ground. It has long endurance of 30–35 day in the sky. RosAeroSystems of Russia is also a supplier of aerostats for defense to other nations such as Czech Republic, Italy, France, Hungary, etc. where border security problems need such large aerostats (RIA NOVOSTI, 2019).

The 28M Class TARS (Tethered Aerostat RadarSystem) is a self-sustained, rapidly deployable, unmanned lighter-than-air platform. It can rise to an altitude of 5,000 ft. It is tethered by a single cable. It enhances its capability in maritime intelligence surveillance and reconnaissance, by effectively detecting maritime and air traffic within the country's coastal waters. Moreover, it could also be utilized in Humanitarian Assistance and Disaster Response (HADR) operations. The TARS include a weather station that provides telemetry data. It helps in monitoring of ambient temperature, pressure, wind speed and other pertinent parameters, to successfully operate the system. It is clear that such aerostat networks are essential to many nations. Particularly, nations that have coastal areas and borders to be made secure.

4.4.1.3   AEROSTATS IN DETECTING AND RAPID REPORTING OF DISASTERS

The deployment of aerostats across various disaster-prone locations on earth could be a good idea. Aerostats are cheaper and efficient systems that offer immediate relay of information. They can be used to report about several different kinds of disasters. Some of these disasters occur in the general realm, while others are confined to agricultural cropping belts. A few examples are complex emergencies, conflicts, drought, earthquakes, floods, landslides, nuclear disasters, storms, tsunami, plant pests and diseases, transboundary animal diseases, etc. (FAO of the United Nations, 2019)

Let us consider an example, wherein, a tethered aerostat and a drone aircraft has been utilized in combination, to derive first pictures and digital information about a site. A location in South Florida, near Miami suffered a fire disaster. Aerostats kept lofted at different heights are usually excellent in detecting ground surface features and fire mishap, its extent and intensity. An aerostat provides constant imagery of the disaster zone, its expanse and speed with which the fire is progressing. Reports from a rescue group 'Training for Airborne Responses' located in South Florida states that, Tethered Aerospace System (TAS)' that was placed lofted provided multiple sets of data/imagery of fire disaster (Staff Writers, 2019). Such data could be utilized while deploying UAVS (drone aircrafts), to obtain greater details and to co-ordinate fire extinguishing procedures. The TAS and Drone combination was also able to provide real-time video coverage of fire damage, to the command center. It is said that tethered aerostat relays and supplies an un-interrupted view of the fire affected disaster zones. Drones that fly past pre-selected spots then offer greater details of extent of fire. However, fire extinguishing personnel opine that after receiving imagery of disaster zones, it is the logistics and swift actions that help us in controlling such disasters (Staff Writers, 2019). Now, extrapolate this situation to farm setting. Fire related disasters are not uncommon in farms, on cropped fields or to field vehicles. Vehicle breakdown too is not uncommon. Perhaps, farms with a lofted tethered aerostat to surveillance and report the happenings on the ground will be able to pick and send images of disaster quickly, to ground control computers. TAS should be kept floating for entire crop season, if fields are to be safeguarded from disasters. The TAS can actually offer first reports of farm related disasters such as fire, floods, extensive soil erosion, crop loss due to storms, drought, pest/disease attack and wild animals, or loss of border fencing even breakdown of farm vehicles, etc. Therefore, a tethered aerostat could be of immense utility to farms with large acreage and rural districts, in general.

4.4.2   AEROSTATS, HELIKITES, AND BALLOONS, AND THEIR UTILITY IN CIVILIAN ASPECTS

At this juncture, we have to note that, several types of aerial vehicles, say, blimps, parafoils, parachutes, aerostats, helikites, and balloons, even more recently the unmanned small aircrafts (drones) were all adopted firstly, for use in military. Later, most of them were utilized to conduct a series of civilian tasks. *For reasons unknown, these UAVs were not immediately adopted by the farming community.* The research thrust and necessary modifications to suit their use in agriculture and their production in large numbers got delayed. At present, there is greater interest in adoption of aerostats and helikites. They are being evaluated in wide range of civilian tasks, including those related to global crop production. Several types and models of freely floating aerostats, tethered aerostats and tethered helikites are being designed. They are getting evaluated stringently in urban and rural locations. They are being evaluated for their accuracy in providing aerial imagery, allowing payloads capable of wide range of tasks such as photography, relay of messages, carrying telecommunication instruments, radar usage and detection, collecting atmospheric samples, even to accommodate bird scaring equipment. Recently, they also have been adopted to generate electric power by converting wind energy. A high-altitude tethered aerostat tower supports turbines, to generate electric power (Glass, 2018). Researchers have experimented and documented the major usages, advantages and disadvantages of aerostats and helikites. Then, compared them with other types of unmanned aerial vehicles (Shakatreh et al., 2018). Such information will be useful while judging and selecting the most suitable aerostat or helikite model, for use in civilian applications.

At the same time, we may note that there is multi-purpose tethered aerostat/ helikite. For example, in South Africa, they have developed a multi-purpose tethered aerostat. The helium-filled envelop supports several different types of sensors for aerial photography and surveillance. This aerostat developed by the Council of Scientific and Industrial Research, South Africa is utilized by agencies involved in military surveillance, urban policing, agricultural agencies, wildlife monitoring, border surveillance, mining and advertisement, etc. The tethered aerostat can be placed at different altitudes, based on the purpose. The tethered aerostat has also been adopted along with other microwave towers, to transmit radio signals and help in mobile phone services (DefenseWeb, 2017). There are several models of tethered and un-tethered balloons that are utilized for advertising various products. Some are multipurpose and they can be used for carrying passengers (untethered free-floating balloons (Lindstrand Technologies, 2019). Overall, multipurpose tethered balloons may get accepted better than other UAV aircrafts.

4.4.2.1 *TETHERED AEROSTATS, HELIKITES, AND BALLOONS IN AERIAL PHOTOGRAPHY AND SURVEILLANCE*

Aerostats are getting noticed as useful aerial vehicles during the conduct of wide range of civilian tasks. They are being sought for aerial surveillance, mapping urban assets, relay of information by acting as microwave radio relay centers, and for the conduct of scientific research related to space science, natural vegetation, etc. Aerostats could be deployed in agrarian regions in good numbers. The aerostat services could be hired by farmers who need accurate aerial photographs and digital data about crop's performance. In due course, we may come across aerostat companies vying for farmers. No doubt, farmers have to select appropriate aerial vehicles. A few farmers may use drone aircrafts. A few others may receive data on a short notice from companies that use aerostats or helikites. Private companies with tethered aerostats placed at higher altitude and covering large areas of cropping zones may serve the farmers in that area. (e.g., Vigilance Group, B. V., Netherlands). (Vigilance, 2019)

Tethered aerostats/balloons were used during Olympic games held in 2016 at Rio de Janeiro, in Brazil. The aerostat company, 'Altave Inc.' offered balloons with internet and other telecommunication equipment and sensors, including infra-red cameras. The events were monitored at all locations and relayed to public. It costed 24.5 million US$ to monitor, capture images and relay the entire event via broadband, to Brazilian public and those in other regions of the world (Altave Inc., 2019a,b; Vasconcelos, 2019; Cortland Inc., 2019).

Aerostats play a vital role in obtaining details of ground features and mapping the terrain. They offer some of the best geo-spatial 'intelligence information,' to different agencies (Shukla, 2018). At present, there are indeed many brands and models of aerostats and helikites. They could easily serve in continuous aerial surveillance, and in procuring high-resolution photography of proceedings of public events. They can be utilized to conduct intelligence gathering during the conduct of public events. Let us consider an example. RT's SkyStar aerostats are apt to conduct aerial surveillance of public events. It seems such aerostats were regularly used to secure major public events such as FIFA's Foot Ball world cups, Olympics games, Tennis tournaments, etc. Aerostats are cost effective and safe. So, it seems, the Tokyo 2020 Olympics too is going to be monitored and aerial imagery (still and video) will be offered, using aerostats (Shukla, 2018). To a large extent, *the utility of aerostats depends on the sensors and radar system adopted. Plus, the computer programs that swiftly process the imagery.*

4.4.2.2   AEROSTATS IN WEATHER REPORTING, ENVIRONMENTAL ANALYSIS AND QUANTIFYING GREENHOUSE GAS EMISSIONS FROM AGRARIAN REGIONS

Atmospheric conditions and gaseous emissions that occur in agrarian regions have their impact on the crop growth and productivity. There are several procedures involved during crop production that may induce release of particulate matter and greenhouse gas emissions. For example, just ploughing, earthing-up or spraying chemicals may transitorily increase dust, particulate matter and greenhouse gases (GHG). Such effects could be observed to a certain level in the atmosphere immediately above cropped fields. It is common in many regions of the world, to prescribe burning of cereal stubbles, sugarcane stubbles, woody residues, etc. These procedures can cause a perceptible increase in particulate matter (PM), elemental carbon, organic carbon, total carbon, volatile organic compounds, poly-aromatic hydrocarbons (PAH), CO, CO_2, CH_4, NH_3, and poly-chlorinated dibenzo-dioxins (PCDD) and dibenzofurans (PCDF), in the atmosphere above fields (Holder et al., 2016, 2017; Liu et al., 2016; Aurell and Gullett, 2013; Aurell et al., 2015). Clearly, burning stubbles affects atmospheric quality within the agrarian region.

Holder et al. (2017) have recently conducted a detailed study in Pacific Northwest (Oregon), where in, stubbles of crops such as wheat and pasture grass (seed blocks) are burnt. Such a prescription affects greenhouse gas emission and particulate matter in the atmosphere. Sampling the atmosphere at different levels from ground surface to 300 m above ground level is important, to understand the natural distribution of emission factors and particulate matter. Therefore, Holder et al. (2017) have adopted 3 different sampling methods. They are the usual ground sampling (established laboratory sampling method), aerostat-sampling of atmosphere above the crop, at 50 and 300 m altitude and aeroplane-aided sampling of atmosphere. To obtain aerostat samples of aerosol, they fixed the full complement of sampling equipment and certain analyzers to the aerostat tether, at prescribed height. In many cases, the values for aerostat collected samples were higher than that found in ground or airplane drawn samples. The above studies demonstrate that aerostats could be useful, in monitoring the atmospheric quality in cropping belts. This is in addition to their use in aerial photography, bird scaring and even wind power generation above farms.

Knowledge about carbon di-oxide levels and its exchange between biosphere and atmosphere is important. It is an important process in the carbon cycle on earth. Photosynthesis by natural vegetation and agrarian regions affects the carbon cycle. Burning of fossil fuels and removal of biomass for various human activities does affect the carbon-di oxide cycle in nature. It is said that measurement of CO_2 emissions from small pockets of cropland as well as larger expanses of agroecosystems is crucial. It helps us to obtain better understanding of the carbon cycle in nature. 'High Altitude Balloons (HABs)' serve a very useful purpose in collecting atmospheric samples and in detecting CO_2 levels in the atmosphere, at various altitudes above agrarian regions (Bouche et al., 2016; Pocs, 2014; Poeplau and Don, 2015). Further, they say that HAB method measures carbon di-oxide mixing ratio at different altitudes in the atmosphere above agroecosystems. It provides details of vertical profile of CO_2 distribution. It seems the tethered high-altitude balloon aided analysis of CO_2 distribution in the atmosphere is relatively cheaper. It has no assumptions regarding homogeneity of CO_2 distribution. The high-altitude balloon technology has been adopted to obtain data about CO_2 distribution above cropland in the Great Plains of USA. Bouche et al. (2016) further state that, results from tethered aerostats/balloons have been verified using other methods, such as flux towers or satellite-aided measurements (Guan et al., 2016). Reports state that, globally, about 7.8 PgC was emitted into atmosphere via fossil fuel burning (Ciais et al., 2013). In the global carbon cycle, photosynthesis through agriculture and other land management practices is important (Krausmann et al., 2016).

Regular monitoring of carbon di-oxide efflux and fixation into biosphere, particularly, in agrarian regions could be done, using HAB methodology. It gives us an idea about environmental perturbations, particularly, those related to carbon cycle. Further, Bouche et al. (2016) point out that, knowledge about seasonal variations in carbon fixation and biomass production in agroecosystems could be gained via CO_2 measurements done, using high altitude balloon technology.

Tethered balloons or aerostats have the ability to hover at 2 km altitude above the ground surface. They provide us with valuable information about weather parameters, trace gas exchange and general vegetational changes. Vierling et al. (2006) has described a system that is termed 'Shortwave Aerostat-mounted Imagers (SWAMI). The tethered aerostat carries equipment such as dual channel video camera, visual still cameras (red, green and blue sensors), spectro-radiometer, thermal infra-red camera, and probes to detect trace gases such as CH_4, N_2O, NO_2, SO_2, CO_2, etc. In addition, this system helps researchers with sharp hyperspectral imagery of ground features (like, natural vegetation, industry, large agricultural expanse, etc.). For such high-altitude balloons/aerostat, ground control is essential. Its functions relate to data acquisition, storage, gas exchange calculations, and pointing sensors to obtain high- resolution images of terrain. These are done using radio control instruments at the ground station. It is said that this system (SWAMI) is useful in studying greenhouse gas emissions at different altitudes in the atmosphere. When extrapolated, data procured can be analyzed and effects of greenhouse gas emission can be deciphered, at agro-ecosystem levels. We can forecast and warn of any detrimental effects of excessive emissions from larger cropping zones (Klemas, 2015; Shaw et al., 2012). Further, it has been reported that SWAMI can be adopted to obtain digital surface models (DSMs) of cropping expanses, using the stereo photographic equipment placed on the tethered balloons. The DS help in relating gas emission data with changes in natural landscapes, vegetation and cropping expanses. For example, Kushida et al. (2009), state that they could relate the gas emission and exchange data with changes in the forests, using DSM of forest stands. Overall, tethered balloons floated at different altitudes over agricultural and coastal regions are useful. They stay in air for a long duration. They could be effectively utilized to monitor changes in coastal farming zones and even assess productivity of crop land along the coasts.

Tethered aerostats of different sizes are amenable to hold a tethersonde. For example, helikite such as 'Skyhook' (11 m³ volume) can be attached with a tethersonde. Smaller aerostats of just 2 m³ volume too could be used to tie the tethersonde to its tethers. The tethersonde is used to record weather data such as wind speed, air temperature, humidity, visibility, gaseous components and polluting compounds, etc. The main advantage of attaching a tethersonde to aerostat is that, it helps in recording atmospheric data at different altitudes, by sliding the tethersonde to different levels (see Allsopp Helikites Ltd., 2019i).

Agricultural crop production and dairying enterprises are among the important sources of methane emission. In some countries about 50% of CH_4 emissions attributable to agriculture are accounted by dairying. For example, a study of emission profiles for methane in the dairy belt of New Zealand shows that, it could be affecting the atmospheric parameters and environment above farming zones (Gimson and Uliasz, 2003). Often, it is said that emissions of CH_4 from grazed pastures or dairying zones are affected by the grazing intensity and breeze that blows over the pastures. Measurement of CH_4 at different levels above the pasture surface indicates that CH_4 profiles differ at levels of 10 m height. The tethered aerostats/helikites could be used to collect samples of air above pastures, of course, with due consideration to the wind breeze across pastures. They say, it is preferable to obtain gas samples using a sampler attached to tethersonde at least 10 m difference in altitude. Tethered aerostats/helikites allow us to get air samples at farm scale. Samples could also be drawn from areas where cattle sheds are situated, and dairy cows are sheltered. Here, the samples are drawn at even lower heights of 1.5 m, 3 m and 6 m above the ground surface. Gimson et al. (2015) have further analyzed the samples using gas analyzers (see Gimson and Uliasz, 2003; Lassey et al., 1997; Lowe et al., 1994). Results showed that emission of CH_4 is affected by weather condition and wind velocity. Also, the density of cattle in the area. The CH_4 profiles clearly depict difference in emission across locations. Therefore, tethered aerostats or helikites mounted with air samplers could be a good idea, to get a clearer knowledge about greenhouse gas emissions, particularly, CH_4 above pastures and cattle dairying zones.

Strand et al. (2016) have studied the effect of prescribed burning of forest stand and understorey vegetation on smoke plumes, their size, amount of particulate contamination of atmosphere and greenhouse gas emissions. They employed instruments that were placed on tethered aerostats at different altitudes. Their aim was to measure atmospheric parameters and gas emissions at different altitudes. They measured emissions such as carbon dioxide, carbon monoxide, methane CH_4, and particulate matter. They noticed that emission of greenhouse gas and particulate matter were high in forests. They were usually twice that recorded for understorey grass. Tethered aerostats could monitor the emissions on a continuous basis until the fire and emissions subsided. There are reports that show us that, tethered balloons with necessary accessories to estimate volatile organic emissions could be adopted above the forests. for example, Spirig et al. (2004) have recorded emission of volatile organic compounds from a boreal forest (see Plate 4.15). They were able to detect a series of volatile compounds emanating from forest vegetation.

4.4.2.3 AEROSTATS, HELIKITES, AND BALLOONS IN FARMING: A GENERALIZED VIEW

There is every indication that, in future, the *'agricultural skyline'* above a location/farm will harbor, either transitorily or for a long stretch of time, UAVs such as tethered aerostats, balloons, blimps, etc. Most immediate reason would be to obtain high quality, high resolution aerial imagery of fields, soils and crops. Also, to surveillance farms. They may also oversee farm activity. The aerial images derived using tethered aerostats can also reveal details about crop's progress, its nutrient and water status, plus diseases/pest attack, if any (Digital Design and Imaging Service, 2019; Abrahao, 2011). For example, Plate 4.13 depicts a tethered blimp stationed above crops that are irrigated, using pivot-irrigation. The aerial imagery relayed to ground station iPad could be highly useful, to a farmer. It helps him to decide on the sequence, timing and intensity of agronomic procedures/precautions.

Recent innovations indicate that, aerostats could also be used by farmers, to generate electricity via high altitude turbine and solar cells on them. Such small quantities of electrical energy could be channeled to run a few farm gadgets like irrigation pump sets/lights, threshers, driers, etc. (Gonzalez, 2015; Altaeros Energies Inc., 2019). Lithium batteries of farm vehicles and other gadgets could be recharged with electricity, using aerostat-turbine generated energy.

Aerostats are being considered for use as super towers to relay wireless communications in farming zones. A single super tower at 1500 m above ground level is said to replace 20–30 traditional microwave towers. The tethered aerostats are cheaper by many folds compared to traditional towers. Plus, they could be moved from one location to another much easily and quickly (Altaeros Inc., 2018; Cadogan, 2018, Livson, 2016; Steadler, 2018).

Researchers are presently contemplating on use of tethered aerostats that could be installed right at the beginning of the crop season. These aerostat towers mounted with full set of cameras that operate at visual, infra-red and red-edge band widths and lidar would be allowed to stay till the end of the crop season. The cameras record useful spatio-temporal changes in the fields.

The digital data pertaining to variations in crop growth, soil fertility, soil moisture content, disease/pest and weed infestation, etc. would be relayed to ground station dealing with precision agriculture. The general forecast is that aerostat could be important in procuring accurate data for farmers, to conduct precision agriculture (Aeros, 2019 b, c).

Aerostats could be adapted to scare the birds away from farms with crops. Aerostats may also help farmers in making aerial surveillance of bird infestation. Birds usually attack panicles/fruits during the ripening period. Tethered aerostats could be effective as bird/animal scarers in farmland. Reports suggest that, aerostats that possess longer endurance could be fitted with very light acoustic and laser emitting gadgets. They could deter and scare birds. So far, reports suggest that, bird control through aerostats could be effective on avian species that otherwise reduce grain yields of cereals, peas, other field crops, and fruit trees (Bajoria et al., 2017; Bird Control Systems Ltd. 2018). Control of birds (pests) via aerostats is environmentally safe. Also, it does not involve destruction of pest species. It only deters and drives birds (pests) out of cropped fields that are at ripening stage.

A series of aerostats placed at selected spots in an agroecosystem can be helpful in recording weather parameters, atmospheric pollution, if any, green-house gas emissions and carbon fixation in the agroecosystems. It includes both natural vegetation and specific crops/cropping systems (Bouche et al., 2016; Krausmann et al., 2016; Ciais et al., 2013). In fact, measurements of NDVI and carbon fixation in agrarian belts have improved our understanding of carbon cycle in cropping zones.

Aerostats could be utilized to study the crop disease/pest attack using aerial imagery. We can decipher the rate of spread of disease/pest in the farmland. Aerostats can also help us in detecting the propagules of disease-causing fungi, bacteria and actinomycetes. Microbial load and its diversity in the atmosphere above the crops could be studied using traps, i.e., petri-plates with appropriate selective nutrient media. We can fill petri-plates with selective medium for plant pathogens, if the intention is to detect occurrence of propagules (spores, hyphae, sporangia, etc.) in the sky above crop's canopy. In case the density of propagules of infective agent is high, then, we can forewarn farmers of the impending disease.

Here, we may note that, above list provides merely an idea regarding various uses of aerostats/helikites in farming. Table 4.2 provides a list of agri-cultural uses of aerostats/helikites. It is indicative of the wide range of utility that aerostats/helikites have in farming. However, detailed discussions about role of aerostats/helikites in conducting individual agronomic procedures are available in the following paragraphs under Sections 5.2–5.8. Overall, there is a strong forecast that, in future, tethered aerostats may flourish in the sky above farmland (Table 4.2).

4.4.2.4   *TETHERED AEROSTATS IN HIGH ALTITUDE WIND ENERGY GENERATION*

To a great extent, agricultural crop production depends on availability of power. Particularly electrical energy to conduct many of the field and post-harvest operations. In a farm, electrical energy is usually generated using portable generators. Otherwise, often, electric power is transmitted to different farms from distant locations. *In situ* power supply via batteries and generators is a common source worldwide. During recent years, there has been a tendency to introduce aerostat technology into many aspects. Wind power generation using tethered floating aerostats with turbines is one such new idea. It is being tried by developed nations. Farmers may benefit utmost by these high-altitude tethered aerostats that can harbor wind turbines. Farmers may derive *in situ* power for their farms. Often, the same aerostat may also carry equipment for broadband mobiles and internet connections. In addition, tethered aerostat could be utilized to obtain aerial photograph and spectral data about crop's growth (NDVI), crop's-N status (leaf chlorophyll, green index) and water crop water stress index. Right now, use of wind power in farming enterprises is at low level. This is true in many agrarian regions of the world. In future, we may see more agricultural farms resorting to wind power generation, say, using aerostats with wind power generator fitted to them. *The multi-purpose tethered aerostats may literally rule the 'agricultural sky,' as time lapses.*

Airborne wind power generation is supposedly a relatively recent trend. It is yet to make a mark on a worldwide basis (Bull and Phillips, 2018). It exploits high altitude wind energy to generate electricity. It is a renewable energy source. There are two designs possible. Kite based and aerostat-based systems. The aerostat systems use a balloon to achieve buoyancy to the turbine. The other one is the traditional aerostat wind power generator where balloon forms part of the turbine. For example, Altaeros traditional wind power generating aerostat has a balloon that surrounds the turbine (Glass, 2018).

TABLE 4.2 Aerostats and Helikites Utilized During Crop Production and Related Procedures

Aerostat/Helikite Model/ Company	Remarks relevant to Use of Aerostats, Helikites and Tethered Balloons in Farm procedures	References
Tethered Aerostats		
Compact Aerial Photography System MA–01-09E/ Aero Drum s.a.r.l Belgrade, Serbia	Aerial imagery and surveillance of crops. Monitoring disease droughts, floods, soil erosion, disease/pest damage in fields. Monitoring farm vehicles and agronomic activity in agricultural experimental stations	Mijatovic, 2017
'White Hawk' Airstar Aerospace, Ayguesvives, France.	Aerial photography and survey of land resources, waterways, crops, etc. Monitoring natural vegetation	Airstar Aerospace SAS, 2019a,b
Skystar 180 and 300 / Aeronautics, Yvone, Israel	Primarily meant for military surveillance and reconnaissance. However, could be modified to obtain aerial photography of landscape and crops, water ways. It can be used to monitor crop growth status. It can be used to monitor farm vehicles etc.	Aeronautics, 2019 a,b
Altaeros BAT Wind Turbine/ Altaeros Inc. Somerville, Massachusetts, USA	This high-altitude wind turbine generates wind power. A tethered aerostat is the platform for wind turbine. Tethered aerostat holds wind turbine at 2000 m altitude. It generates electricity for farms and other customers	Altaeros Inc., 2018; Glass, 2018
Tethered Aerostat Altaeros Inc. Somerville, Massachusetts, USA	It is used for aerial imagery and spectral analysis of crops using R, G, B, Red-edge and Lidar sensors. It may also carry probes to detect greenhouse gas or collect sample from above the crops Samples are utilized for analysis of CH_4, NH_3, NO_2, N_2O and particulate matter.	Altaeros Inc. 2018
Helikites		
Desert Star Allsopp Helikites, Fordingbridge, Hampshire,	Aerial photography of farms, land resources, soil, topography of fields, crops and cropping systems. Monitoring crops for biomass accumulation (NDVI), Crop's water stress (CWSI), and leaf chlorophyll (crop's-N status).	Allsopp Helikites Ltd. 2019 a, d;

TABLE 4.2 (Continued)

Aerostat/Helikite Model/ Company	Remarks relevant to Use of Aerostats, Helikites and Tethered Balloons in Farm procedures	References
United Kingdom	Detection of disease/pests on crops. Monitoring farm vehicle activity and progress of agronomic procedures	Allsopp et al., 2013
Skyshot Helikite (Standard Hybrid) Allsopp Helikites Ltd, Fordingbridge, Hampshire, United Kingdom	It is deployed above farms. It offers aerial imagery of land resources, water bodies, Cropping systems, etc. It is an all-weather helikite. It could be utilized to monitor and collect digital data about crops tested/evaluated in agricultural experimental farms.	Allsopp Helikites 2019 b, c.
Light Helikite/ Carolina Unmanned Vehicles, Raleigh, NC, USA	Monitoring crop belts and forest stands for growth, biomass accumulation rates, leaf chlorophyll status, diseases/pests, soil erosion, gully formation, floods and droughts, if any	Carolina Unmanned Aerial Vehicles Inc. 2019
Perigrine Helikites/ Bird control Systems Ltd. Fordingbridge, Hampshire United Kingdom	They are predominantly used to scare birds. They have bird scarers attached to tethers. They easily ward-off bird species such as rooks, crows, wood pigeons, starlings, sparrows, pheasants, geese and herons. A single helikites protects an area of 15-25 acres of field crops such as wheat, pea, maize, brassica etc. or 5-7 acres of fruit crop orchards. These helikites are utilized to scare-off birds that destroy grain crops during seed set. For example, during pearl millet panicle/grain maturation. Also, during sunflower head/seed maturation	Bird Control Systems Ltd. 2018; Perigrine Ltd., 2018; Bajoria, 2017
Tethered Balloons		
Tethered Balloon Altave Inc. San Jose dos Campos, S. P. Brazil	Tethered balloon floats at 200 m above fields/crop canopy. It offers aerial imagery and digital data, using a series of sensors. The aerostat (or balloon) tower helps in broadband telecommunication in rural regions of Brazil	Altave Inc. 2019b

At present, the share of energy generated using wind power is low in many regions of the world. For example, in USA, even by 2030, it seems only 20% of total energy generated would be derived through wind energy. Of course, there are other regions such as in Netherlands, wherein electricity generated via wind power is much greater. However, there are reports that globally, energy that could be generated using wind power is much greater. At low altitudes below 200 m above ground surface, the wind power that could be harnessed is about 400 TW. However, if we reach higher altitudes, say, between 200 m to 20 km above ground surface, the theoretical limits and potential for harnessing wind power increases to 1800 TW. They say, it is a huge reserve of energy that needs to be harnessed carefully and efficiently. It suffices to supply electrical energy to 18 times the present needs of human habitation on earth. Wind power is quite a large source of electricity for it to be ignored. Yet, we are slow to harness it (Cassimally, 2012; Inman, 2012; Marvel et al., 2012). Agricultural farms may benefit from *in situ* wind power generation through a turbine lofted into sky above the farm.

PLATE 4.8   A helium aerostat with wind turbine and ground station.

Source: Dr. Ben Glass, Altaeros Inc., Somerville, Massachusetts, USA; Also see, Gonzalez, 2015, National Science Foundation, Washington D.C., USA.

There is no doubt that we need appropriate methods and vehicles to reach the high altitudes, to harness the wind energy. Then, it has to be relayed to ground stations and then to other locations. Aerial vehicles such as blimps, kites, kytoons (kite plus balloon hybrid) and aerostats have been tried as possible vehicles. They could be adopted to locate a turbine that will exploit wind power and generate the electricity. At this juncture, we may note that wind energy is a renewable source. Therefore, it is an attractive proposition to explore. There are reports that, aerostats with modified shapes that allow them to carry an electricity generation set-up has been tested, by several companies (Hassan, 2011; Glass, 2018; Plates 4.8 and 4.9).

PLATE 4.9   Altaeros BAT wind turbine: A helium aerostat with wind turbine to generate electricity.

Note: Altaeros BAT can lift Telecommunication, Internet and Camera equipment alongside the turbine, to provide additional services for customers. Probes to estimate greenhouse gas emissions and bird scarers too could be attached to tethers, at suitable levels. The addition of payload equipment does not affect its performance.

Source: Dr. Ben Glass, Altaeros Energies Inc., Massachusetts, USA; https://phys.org/news/2015–02-turbines-electricity.html.

The 'high altitude wind power' is defined as harnessing power from wind that blows at higher altitudes in the sky. The power generated is relayed using tethers or cable, to ground stations. We may note that wind velocity recorded at higher altitudes is usually much higher than at altitudes close to ground. The wind velocity is a crucial aspect of aerostat-mediated high-altitude wind power generation. A doubling of wind velocity increases the power generation potential by 8 times the original. Tripling of wind velocity means 27 times more power could be generated. There are several methods through which we can harness wind's kinetic energy and convert it to electrical energy and transmit (Wikipedia, 2019). Here, we are concerned more with aerostats with lighter-than-air system. It adopts a tethered aerostat to act as a platform for a turbine that harnesses wind's kinetic energy and produces electrical energy. Incidentally, heavier-than-air methods that involve tethered parafoils have also been utilized to harbor a wind turbine.

Reports suggest that, at present, many of the companies producing wind energy turbines are trying to build aerostats. These aerostats operate efficiently, at an altitude above 500 m from ground surface. The idea is to harness wind power efficiently compared to that possible at low altitudes. For example, Altaeros Energies Inc., (Massachusetts, USA) is a company that is concentrating on manufacturing aerostats with ability for wind power generation, at altitudes higher than 500 m (Altaeros Energies Inc., 2019; Plate 4.8). Indeed, there are many aerostat/turbine manufacturing companies that are competing with models of aerostats plus wind power turbines. Usually, they operate at altitudes 150 m and above (Gonzalez, 2015; Plate 4.9). There are also buoyant wind turbines that keep afloat through an aerostat, at altitudes above 600 m. As stated earlier, wind turbines that operate at high altitudes above 300 m are usually more efficient, in generating electricity and transmitting it, to ground station (Gonzalez, 2015). Further, Gonzalez (2015) states that, most high-altitude wind turbine manufacturers are competing to build taller turbines, to harness more powerful winds that occur above 500 feet, or 150 meters. However, Altaeros Energies Inc, a company dealing with high altitude wind turbines is going much higher with their novel 'Buoyant Airborne Turbine (BAT). It is a company that manufactures wind turbines mounted on aerostats. The aerostat is a helium-filled balloon. Usually, it floats usually above 2000 ft. from ground surface. It converts wind power into electrical energy. Similar models of airborne wind turbine residing on a tethered aerostat has been tested in other locations. For example, at Limestone, in Maine state of USA, they tested a 60 ft. diameter aerostat-borne wind power turbine. The tethered aerostat wind power generators initially look like balloons floating in the sky. However, they carry wind turbine plus other usual accessories such as, sensors for crop analysis and obtaining aerial imagery of terrain and its features (McGonegal, 2013). Therefore, it becomes further clear that, such aerostat-borne wind power generators could be most useful to individual farms that need energy, for various agricultural activities. For example, to energize irrigation pump sets for lift irrigation, processing seeds, cold storage facilities, recharging batteries of electric vehicles, etc. We may note that a single aerostat borne wind turbine can generate 30 to 100 Mw of energy. The energy is usually transmitted to ground station using electric lines attached to tethers. Incidentally, cost of wind power generation seems to have fluctuated depending on location. In Britain, for example, wind power was not remunerative for some years. However, recent reports suggest that the wind power units are economically viable. We have to obtain clearer and authentic evaluations about feasibility and economic gains possible, using aerostat mediated high altitude wind power generation (Harris, 2019). The opinions expressed for other aerial gadgets (e.g., small drone aircrafts) may also apply equally to high altitude tethered aerostats with wind turbines. That is, as we mass produce these 'aerostats plus wind turbines' the initial cost of purchase and installations may reduce, perceptibly.

Tethered aerostats have also been utilized to generate electrical energy through solar cells. The envelope has the solar cells incorporated on the surface. Solar cells are shown to sun at proper angles, for a long stretch of time. The solar energy generated is transmitted via tethers and cable to the ground station (Aglietti, 2009; Aglietti et al., 2008a; 2008b; Aglietti et al., 2010; Redi et al., 2010). There is also a report about a hybrid aerostat with facility for wind power generation plus solar cells. It generates electricity, of course, depending on sunshine hours (Hosakoti, 2019).

We should note that, Agrarian farms within plains region that get lashed with wind breeze lavishly at different altitudes, those in the coastal belts, or those in windy high mountainous region could benefit much from the aerostat technology with wind power generation facility.

Overall, specific problems that aerostats have to negotiate has to be answered by researchers. For example, as we increase altitude, the length of tether increases. Risks of wind turbulence and damage to aerostat or wind turbine could increase. The material utilized to build aerostat may have to withstand certain vagaries related to weather, changes in temperature and atmospheric pressure. The high-altitude wind power generation will also need methods that allow very good control of the aerostat and its movements in the high-altitude region. The context of this book relates more to aerostats and their role in farming enterprises. Therefore, to utilize this technology, farm engineers may have to rapidly devise the needed changes in aerostats. As such, currently, aerial robots (drone aircrafts) that offer aerial photography and digital data about crops are being adopted worldwide. So, an aerostat that harbors a wind turbine of suitable size, a set of electro-optical and electro-chemical sensors may actually be a composite that is most useful, to a farm enterprise. Ultimately, we can think of farms with two versatile and most useful vehicles, one that does innumerable tasks on the ground (a tractor) and one that stays afloat (a tethered aerostat) and offers farmers the required electrical energy and most crucial data about their crops, incessantly. Simultaneously, aerostats could be good sentinels operating from the sky. The same aerostat could also host bird scarers and probes to detect atmospheric quality. As stated earlier, the tethered aerostats of different sizes and capabilities are set to modify the 'agrarian skyline.' Then like agricultural crops, agricultural soils, agricultural economics, and agricultural extension, we may have to think of a faculty of studies termed 'Agricultural Skys.' It is imminent.

4.4.2.5   ROLE OF AEROSTATS AND HELIKITES IN ESTABLISHING TELECOMMUNICATION NETWORKS

Reports emanating from UNESCO and 'International Telecommunication Union' states that, at present, there are about 3.9 billion people comprising largely the remote farming communities that need internet access. Among various options, tethered balloons and aerostats acting as taller towers of 200 m to 20 km altitude are preferable. They carry telecommunication equipment (Vasconcelos, 2019; Cadogan, 2018; Airstar Aerospace SAS, 2019a,b; National Instruments, 2019). Cadogan (2018) states that, aerostats are being deployed in USA, to improve wireless communication. Aerostats help to reach countryside and remote locations through mobile. The aerostat-based 'Super Tower' series helps in high speed internet and mobile phone connections (Glass, 2018). We may note that, for many of the precision farming related operations, incessant internet connectivity and mobile phone operation are essential. Almost all farm vehicles used in precision farming or even otherwise, need continuous GPS connectivity. Therefore, in future, almost all farms may need aerostat based wireless connectivity. Aerostats (super towers) located at 1500 m above ground level may be most useful to rural regions, anywhere in the world. Similarly, reports suggest that, in Ireland, there are initiatives to develop 'Super Tower'- i.e., aerostat-based internet connectivity in the entire rural district. A recent report from Himalayan region in India suggests that, they have installed aerostats to develop internet network, to reach far flung mountainous districts and hill country farming zones (Cadogan, 2018; Hindustan Times, 2018; Grace, 2013; Lambert, 2018). Again, in France, companies such as Airstar Aerospace SAS has been installing tethered aerostats, to build mobile connectivity in the rural farming belts (Airstar Aerospace SAS, 2019a,b; see Plate 4.10).

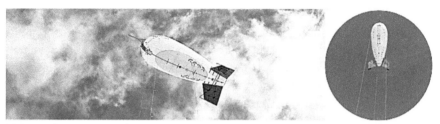

PLATE 4.10   A Tethered Aerostat acts as platform for telecommunication instruments

Note: A tethered aerostat is cost effective. It is autonomous in flight. It helps to develop ubiquitous mobile phone connectivity in rural areas. It requires only a fraction of cost compared to traditional towers.

See also, https://www.forbes.com/sites/roberthof/2015/06/02/googles-project-loon-internet-balloons-coming-to-u-s-eventually/#1f96875825d6/

Source: Dr. Ben Glass, Altaeros Inc., Somerville, Massachusetts, USA

In the remote regions of Australia, Aerostats have been investigated and proposed as possibly very good candidates, to establish internet/ mobile networks. About 70% of land area is mainly desert and feebly populated farming zones. They are still to be covered via broadband mobile network. They say, aerostats could be installed to act as mobile towers. Such a method could bring most, if not all farm belt in Australia within 4G mobile coverage. In terms of population, about 2.3 % of remotely placed farming community and those in desert fringes may need telecommunication networks established on tall aerostats (Livson, 2016). They have tried to adopt aerostat placed at 900 m above ground level. It is said that a tethered aerostat placed at 50 m height covers a radius of 35 km and an area of 4,000 km²; a tethered aerostat placed at 150 m height covers 56 km radius and 10,000 km² in area; a tethered aerostat placed at 300 m height above ground covers 77 km radius and 19,000 km² area. A tethered aerostat placed at 900 m above ground covers 149 km radius and 70,000 km² area, in terms of mobile signals and connectivity (see Livson, 2016). Reports by 'Aerostat All Australia' suggest that, among the three types of aerial vehicles, namely; (a) platforms with propulsion; (b) free floating balloons and (c) tethered aerostats, the last one, i.e., aerostat seems most suited. It acts as 'super towers' to relay signals for internet operation. We may note that, at present most farming vehicles and agronomic operations are controlled using mobile networks, particularly, precision farming techniques. A GPS guided autonomous tractor, seed planter or a fertilizer inoculator may all need internet connectivity, at all times. The type of aerostat adopted needs to be apt to the requirements of farming community. Three types of aerostats encountered commonly are the cigar type, jelly fish type and tethered helikites- a hybrid of kite and balloon. Each type of aerostat has its advantages. Special types and shapes of aerostats have also been evaluated by 'Aerostat All Australia.' Overall, it said that Australia plans to cover the unserved regions and population with mobile networks using aerostats. Aerostats based network seems to cost several times less than a traditional steel or concrete tower-based network. Further, aerostat-based telecommunication towers can be established at a rapid pace compared to traditional networks (Livson, 2016).

Now, let us consider efforts to use tethered balloon and aerostats to provide internet access, to farmers and general public, in remote locations of Latin American nations such as Brazil, Peru and Argentina. Vasconcelos (2019) has reported that a project named 'Loon' has been successful in providing broadband access to remote areas in Brazil, Peru and New Zealand (South Pacific) (Hof, 2015; Vasconcelos, 2019). It adopts a tethered jelly-fish model balloon, to pick signals from ground and relay it to other locations. The balloon tower is taller than usual telephone ones (20 m). Aerostat tower reaches a height of 200 m from ground level. For example, in Brazil, Altave Inc., a company dealing with tethered aerostats for broadband transmission provides models that float at 200 m above ground. It transmits internet signals to farming community without much disturbance or noise in the signals (Altave Inc., 2019a,b; Plate 4.11). Some of balloons used in the project 'Loon' fly at 7 km to 20 m above in the stratosphere. The tethered balloons harbor electronic devices that serve as cell tower. It aids in high speed internet connection in remotely placed regions. Vasconcelos (2019) further reminds us that, such tethered balloons and aerostats are in use in other nations, such as USA. In USA, NASA is actively providing improved models of aerostats and telecommunication accessories.

Helikite aerostats could be very small in size. They may hold just 30 g of helium gas and fly at low altitudes. For example, Allsopp's Lightweight Helikite is among the smallest of helikite aerostats. They can fly at 1000 ft. altitude after inflation with helium gas. They are usually compact. When not in use they can be folded into small box. These small helikites are stable in the sky and withstand up to 25 kmph wind disturbance. They are used for amateur radio, falconry and position marking (Allsopp Helikites Ltd.,

2019d). Sweney (2019) states that, a parallel network of helikites with necessary telecommunication instruments and electronics can be of great help to
rural and far-off locations, in case there is an emergency or there is a breakdown of general telecommunication network. Small helikites could be useful
in such situations.

4.4.2.6   AEROSTATS AND BALLOONS IN SPACE SCIENCE STUDIES

Aerostats and balloons have been put to good use even in the harsh weather
conditions of Antarctica. Several different balloon campaigns have taken
place in this ice capped region. The balloon usually carries a payload of
visual, high resolution multi-spectral, infra-red and lidar sensors, magnetometers, gas analyzers, etc. The array of payload instruments, of course, depends
on the aerial measurements that is intended. Balloons have been flown to
different altitude above the ice-capped region. For example, a mission by
NASA in 2014, derived useful data by flying the balloon up to 23,000 ft.
above ice surface (NASA Balloon Program Office, 2017; Eggers, 2014).
Further, it is said that 'Super Pressure Balloons' are the latest in balloon
technology. They allow long endurance, i.e., ultra-long duration missions of
100 days or more, at constant float altitudes due to the pressurization of the
balloon. Forecasts state that 'Super Pressure Balloons' are going to be useful,
for conducting scientific investigations in the sky above Antarctica and in
the near-space environment. The balloon missions in Antarctica also carried
different payloads. For example, SPIDER is a balloon-borne polarimeter.
It was utilized to obtain high fidelity images of the sky above the ice cap.
Scientists have utilized different types and sizes of balloons over the Antarctica. For example, a report suggests that, certain very large balloons are so
large that they can hold about 200 large blimps (by volume) (Eggers, 2014).

The European Space Agency has contemplated to send an 'Orbiter' and
expects to lower down an aerostat, balloon or blimp, to survey the 'Titan'
(Saturn's moon) surface and atmosphere (Wall, 2010; Heun et al., 1998).
There is literature concerning aerostats as possible vehicles, to explore and
gather scientific information about other planets, their surface and atmosphere. There are well educated guesses, theoretical evaluations, lists of pros
and cons and modifications required to aerostats, in order to use them on
planets. Feasibility about use of aerostats in space science, particularly, on
other planets has been discussed (see Bryce, 2018).

PLATE 4.11   Altave Tethered Balloon for rural communication.

Top left: A tethered balloon; Top right: A tethered balloon plus telecommunication equipment and ground station; and Bottom: A ground control computer station to process imagery and relay internet signals to surroundings.

Source: Altave Inc., San Jose dos Campos, SP, Brazil.

As stated earlier, Aerostats and helium-filled balloons have been investigated as possible vehicles in the atmospheres of planetary bodies. There are suggestions that aerostats could be kept floating in atmospheres of planets. So that, they could be used to report about planets' ground surface. Aerostats on planets and moons of Jupiter and Saturn have been touted to serve as outposts, when interplanetary vehicles travel in the solar system. We may note that, aerostats encounter different levels of gravity, atmospheric composition, wind currents, drags, etc. Each planet's atmosphere may require a different but accurate design of aerostat/balloon. The lighter-than-air gas that is filled into envelope of aerostat has to be chosen correctly. We may note that, lighter-than-air gases may already be in good proportion in the atmosphere of certain planets. Therefore, making it difficult to float aerostat/balloons. For example, a combination of methane, hydrogen, helium, ammonia, etc. may make it difficult for helium-filled aerostats to float. Pure hydrogen that is really lighter than all other gases is required (Bryce, 2018). There are reports that space scientists are active in developing suitably large and high-altitude blimps, aerostats and autonomous parafoils. They intend to place them floating over Martian atmosphere. Aerostats possess longer endurance in the planet's atmosphere. Therefore, they may relay information about planet's ground surface, atmosphere and outer space regularly, to earth's stations (Elfes et al., 1998, 2000, 2003) and Ramos et al. (2000).

4.4.2.7   AEROSTATS IN ECOLOGICAL STUDIES: A FEW EXAMPLES

No doubt, there has been a spurt in the use of tethered aerostats and helikites to conduct ecological studies. In the following paragraphs, only few representative examples have been discussed. We should note that, in addition to tethered aerostats, other aerial vehicles such as drone aircrafts, parafoils, blimps and other types of aerial vehicles have been used as platforms. Particularly, to place optical sensors and chemical probes meant for ecological studies (Klemas, 2015; Shaw et al., 2012; Krishna, 2019). The basic idea is to observe visual changes on ground and relate it to environmental parameters such as gas emissions, burning, erosion, oil spills, pollutant discharges, etc. For example, Klemas (2010, 2013) has shown that, data from remote sensors could be utilized to study the dynamics of coastal land, its vegetation, species diversity, greenhouse gas emissions, impacts of precipitation and sea waves on erosion, etc. The stereo photography obtained from tethered balloons help in devising accurate remedial measures to the coastal border, in case deterioration is detected. (Klemas, 2010, 2013, 2015). Further, Klemas (2010) states, that oil spills in the coastal belts or even in high seas could be monitored using tethered aerostats floated at low or high altitudes (2 km above ground/sea surface). Such aerial imagery helps us in studying the progression of oil spill fronts and loss of seawater quality. The impact of oil spill on sea fauna could be evaluated, using data from aerostats.

In USA, White and Madsen (2016). have obtained extensive aerial photography of South Carolina coastal zones using helikites. Their aim was to study the effect of hydrology and demarcate the coastal area into different ecological zones. We may note here that, aerostat methods could be applied to practical farming. 'Management blocks' could be formed, using aerial photography that depicts the hydrological aspects of a field/farm. Inherent soil moisture levels could be a good criterion to demarcate 'management block.' Accurate zonation is an essential step during precision farming.

Now, let us consider examples from marine ecosystem. Small aerostats are compact, easy to inflate and float repeatedly from aboard the ships. They have a role to play in studying the maritime ecology, both near the shores and in the high seas. For example, there is a program called 'Ocean Eye' conducted by the 'Norwegian Clean Seas Association.' It directs its investigation on oil spills and their effects on fauna and flora on ocean surface. They adopt small helikite(aerostats) to visualize, measure the size and record dynamics of small oil spills that occur, on the surface of sea. Usually, such small oil spills go undetected, if we search for them from the sea surface. Therefore, a small helikite that stays lofted in the sky even as the boat or a ship transits the high seas is necessary (See Allsopp Helikites Ltd., 2019g). Surveillance of maritime ecosystem is essential. They say, pristine undisturbed marine ecosystems occur both near the coasts and in high seas. There are several methods by which we can keep a vigil. We can avoid intrusion of such ecosystems by fishing boats or any other disturbing sea farers. Among the gadgets, they say, tethered aerostats or helikites mounted on ships could be effective. Particularly, in relaying information about disturbances to marine ecosystems (Brooke et al., 2010, FAO 2001).

As stated earlier, tethered aerostats could be of immense utility while obtaining detailed imagery of sea surface. The sensors on tethered aerostats aboard the ships provide excellent data about the rate at which the oil slick is getting dispersed. Yet, we may note that, measuring the rapid dispersion tendency of oil slick at the interface of sea and air is not easy. The velocity of wind and oil slick dispersal could vary at short distances and intervals. Recently, after the oil spill at the northern Gulf of Mexico, there have been many experiments conducted. The aim was to assess the extent of oil slick, and the rate at which it might disperse. The detriment to fauna and flora of sea surface, no doubt depends on rapid spread of oil slick (Carlson et al., 2018; Richardson and Stommel, 1948; Matsuzaki and Fujita, 2017; Miyazo and Isobe, 2016). Carslon et al. (2018) has made a series of measurements of oil slick using aerial photography. A 'Ship Tethered Aerostats System' was adopted to get accurate data about oil spill, its size at various spots in gulf. The aerostat was equipped with Canon EOS 5DSR 50.6-megapixel camera. Such data from sensors is useful while assessing and forecasting the ecological disturbance. Particularly, effect of oil spills on the biotic portion of the sea.

Tethered aerostats and helikites are getting greater attention from scientists involved in variety of topics in ecology. It is not just the aerial photography that helikites produce using the sensors. Ecologists involved with the study of migratory trends of birds and insects are banking on detailed information that radars and sensors placed on the helikites could offer. The rapid movement of insect groups are being tracked by the radars on a tethered aerostat. The infra-red images produced by NIR sensor can be of utility to judge the nocturnal movement of insects and birds. Aerostats/helikites allow us to study the flight pattern and speed of insect/bird migration at high altitudes. Hitherto, this was not possible easily (CANR News, 2019). We should note that, when we consider, insect pests affecting agricultural crops, an accurate knowledge about their migratory trends and rate of spread is of utmost utility, to farmers. They then can adopt spray schedules appropriately.

Teams at the Smithsonian Conservation Biology Institute and National Geographic Association Washington D.C. have attempted to study the migratory habits of prairie dogs. They have also aimed at enumeration of the population of prairie dogs plus 100 other species of animals that loiter in the prairies of Central USA. Firstly, they adopted a helikite of 2 m³ size and payload capacity of 0.5 kg. It is a small helikite. As stated earlier, helikite is a combination of helium balloon and kite. It has better stability in the air. The net result is that helikites offered haze- and aberration-free digital imagery of prairie dogs and other species. Further, they have aimed at studying the impact of ambient weather and diseases on the migratory habits of prairie dogs (National Geographic, 2019).

As stated earlier, tethered aerostat towers serve as platforms for a series of sensors. They have the ability, to continuously observe the happenings on the ground. Therefore, they could be adopted to study a wide range of natural phenomena. We can observe the entire ecosystem, land resources, natural vegetation, crops, animals, water bodies and their interactions with weather, etc. In England, researchers have tried to assess the effects of cattle grazing pattern on other aspects of the ecosystem. To observe the cattle trails and grazing locations, Bond (2018) used a lighter-than-air vehicle, namely, a small helikite. The sensors on the helikite provided detailed imagery and maps of topography, soil quality, vegetation and the Chalk stream. Further, Bond (2018) reports that, cattle grazing has both short term and long-term effects on the Chalk river and its nutrient levels. Cattle grazing near the stream, firstly increased leaching of phosphate and other nutrients into the stream. Phosphorus loading of stream water was higher in areas that were frequented by the cattle. Nutrient loading due to cattle defecation also had direct impact on algal bloom, microbial load and growth of weeds in the stream.

Helikites could be playing a major role in understanding the desert, its flora and fauna. Helikites placed at different locations may provide us most valuable data pertaining to desert ecology and sustenance. Let us consider an example. Helikites are good candidates to obtain aerial imagery of Nebkhas (i.e., phytogenic mounds) that occur in the deserts of Saudi Arabia and other West Asian nations. We can keep a thorough vigil and record details of nebkhas that spread across vast stretches of desert. Our knowledge hitherto was restricted to formation of nebkhas by individual phytogenic group and its consequences on trapping sand and dust, on wind erosion or on dune formation. Nebkhas are ecologically important set of phytogenic mounds. They withstand sand burial and trap sand mass in and surrounding their canopies. Quets (2016) states that, we have some knowledge about mechanisms involved with individual nebkhas, their formation, distribution and sizes. However, knowledge about its consequences to desertification on a wider horizon is lacking. High altitude aerostats/helikites with ability for continuous imagery of nebkhas, both during day and night is indeed a good idea. It helps in knowing natural mechanisms that control the distribution pattern of nebkhas. Nebkhas could be playing an important role in maintaining landscapes, the flora and fauna considering that parched deserts are scarce in vegetation and moisture. Quets (2016) further states that, ecologically, these nebkhas supporting regions are indicative of the state of desertification. They could be placed between dry grass lands with scarce vegetation to regions affected by severe desertification. Nebkhas, it seems, are a novel phenomenon that has impact on various geomorphological changes that occur transitorily or permanently in the deserts.

The 'agricultural sky' may be defined as the area just above the ground (field soil) or crop's canopy if the field is supporting a crop. The atmosphere above the crop canopy is a big repository of dust particles, gaseous material, microbial flora, insects that fly, birds and other flying animals, etc. The 'Agricultural sky' is also a source of moisture (dew) and dissolved nutrients, although feebly. Incidentally, during recent years, the same agricultural sky is also the region that is utilized by a range of unmanned aerial vehicles, piloted vehicles, aerostats, helikites, and balloons. Here, in this section, we are concerned with the spatio-temporal fluctuations in microbial flora of the atmosphere above the crop's canopy. We also try to understand the microbial diversity and their populations in the atmosphere above crops. At this juncture, we should note that tethered aerostats and helikites could be floated at different heights and air samples drawn from different altitudes. The tethers could also be fitted with petri-plates or similar traps with 'selective nutrient media' to trap the microbes distributed in the atmosphere above the crop. The microbial load and diversity could be analyzed and estimated, using the rRNA sequences. Microbes found in air can also be identified using simple morphogenic charts. To quote an example, microbial DNA extracted from air samples have also provided useful insights into microbial diversity and population levels at different levels in the atmosphere. Tethered aerostats and helikites are most useful when we aim at analyzing the population levels of microbial pathogens of crops. We can also trap the propagules of pathogens (e.g., spores, hyphal bits, etc.) floating in the air, using selective spore germination media. We can place the traps at different altitudes on tethers of aerostats.

To quote an example, Spring et al. (2018) studied the microbial flora above Kalamazoo in Michigan, using samplers placed at different altitudes on the tethers of an aerostats. Airborne, bacterial and fungal populations were estimated. The results showed that *Proteobacter spp.* were most predominant flora in the atmosphere. They have concluded that aerostats with air samplers attached could be used to study the microbial ecology of atmosphere, say, 2–3 m above ground or several meters above a crop's canopy. They have pointed out that we know very little about the microbial interactions that occur in the atmosphere just above the urban locations or agricultural crop fields. Knowledge about microbial load, particularly of the plant pathogens has great applications, during adoption of prophylactic sprays of fungicides, bactericides, etc. An aerostat or a helikite placed permanently or for long duration in the sky above the crop fields may serve excellent purpose, in tracking spores of fungi or bacteria that cause disease. Agricultural drone aircrafts, parafoils, microlights have also been adopted by researchers, to study the microbial flora and gaseous components of atmosphere above crop's canopy (see Krishna, 2018; Krishna, 2019).

4.4.2.8 AEROSTATS, HELIKITES, AND BALLOONS IN ARCHAEOLOGICAL STUDIES

Historically, tethered aerostats, helikites, balloons, kites and a few other aerial vehicles have been adopted, to study archaeology. Low altitude aerial imagery provided by the aerostats are of great use to archaeologists who analyze topography and other physical features. There are several computer programs (e.g., Agisoft's Photoscan, Pix4D Mapper) to process and develop the ortho-images. The imagery could be simple visual photographs. Infrared images that could be obtained, both, during day and night or high-resolution LIDAR images could also be processed. There are also reports, wherein 3D stereo-images of archaeological sites have been obtained with great accuracy, using the helikites (Kostyrko et al., 2015).

Cambridge University Archaeologists have employed small helikite aerostats to study the Egyptian historical sites, in a place known as Armarna. This location houses archaeological ruins and is supposedly the birthplace of Tutankhamen-the famous pharaoh. Researchers used a small aerostat (73 m^3 in size) fitted with stereoscopic visual still and video cameras. Reports indicate that the helikite aerostat performed perfectly, despite hot weather conditions. It offered sharp stereo-images of the entire historical site at Armarna, in Egypt. Images could be transmitted easily to laboratories for further processing (Allsopp Helikites Ltd., 2019h). We should note that, small helium-filled aerostats are cheaper. They are amenable for repeated use. Since, they could be flown at low altitude, they offer high resolution stereo-images of good quality. Parcak (2016) states that satellite imagery helps in studying ancient sites in Egypt. However, helikites that float at low altitude over the archaeological sites may provide well focused and accurate details of the site. Aerial photography obtained from sensors on helikites may supplement the data obtained through other methods.

Verhoeven et al. (2009) explain aerial archaeology as an aspect of science that includes entire gamut of acquisition and inventory to ultimate mapping of details of archaeological site *per se*, from a vantage aerial location. Aerial picturization of archaeological sites with an aim to obtain overview of the entire site and note the features has been in practice for several decades. They have mostly used high perched platforms, mountain peaks, aircrafts, drones, balloons, aerostats, kites, etc. In the present context we are focused on use of tethered aerostats and helikites that are adopted, to obtain aerial imagery of historical sites (Verhoeven et al, 2009; Riley, 1946).

Verhoeven et al. (1994, 2009) state that, for a considerably long stretch of period we were obtaining images using visual range bandwidth cameras (red, green and blue). However, during recent years aerial imagery from visual, infrared, red edge and lidar sensors are possible. Infar-red imagery provides some special insight about the ground features, vegetation and temperature fluctuations around the archaeological sites (Verhoeven, 2007; Verhoeven et al., 2009). Further, they state that using tethered aerostats we can analyze the vegetation that occupies the archaeological sites. We can study their spectral signature. It includes the infrared spectral signatures. We can get detailed data about the 'crop marks,' the spectral signatures of crop species that inhabit the sites, crevices,' the pits, hillocks and deterio-rated portion of archaeological locations. For example, they say, vegetation colonizing the various strata of archaeological sites containing Neolithic and Roman era has been possible, using tethered aerostats mounted with visual, multispectral, infrared and lidar sensors (Verhoeven and Loenders, 2006; Verhoeven, 2012).

Several types of aerial vehicles that are autonomous, semi-autonomous or remote- controlled, using radio signals have been employed to make aerial survey. Such aerial imagery helps us to obtain still and video imagery of archaeological sites. Veerhoeven et al. (2009) say that, helikites with helium balloons are among the recent aerial vehicles to be adopted, for archaeological studies. The helikites fly at low altitude and carry sensors that operate at visual bandwidth, infra-red and high-resolution multispectral wavelength. Sensors can provide us with valuable information about the historical site. Such small helikites overcome the difficulties we otherwise encounter, if large blimps or aerostats are adopted. Small helikites can be repeatedly lofted into sky above the archaeological sites and aerial photography could be made for longer durations.

Rebecca (2016) states that, tethered aerostats, tethered balloons, kite-balloons or helikites have found applications during study of archaeological sites. Usually, these aerial vehicles are fitted with cameras that obtain high resolution imagery. The digital data could be relayed instantaneously to the ground station (say an iPad), to process the ortho-images. Stereo-images and 3D data are of immense utility to historians too could be obtained. Let us consider an example. Whittelesey (1970) utilized an aerostat tethered to ground near archaeological sites in Greece, Italy, Turkey and Cyprus. Several cameras were mounted on to tethered aerostats and held at an altitude of 50–70 ft. above ground. This is to obtain aerial images. Myers (1978) has utilized tethered balloons/helikites to map the archaeological sites in Italy and Greece. Stereo-graphic cameras have offered excellent images. Such images provide an overview of the entire historical site. Otherwise, only the required portion could be mapped with clear GPS tags. Aerostats have also been utilized to conduct aerial imagery of iron-age sites in South Africa (Noli, 1985).

4.5   ROLE OF AEROSTATS, HELIKITES, AND BALLOONS IN AGRICULTURAL CROP PRODUCTION PRACTICES

4.5.1   AERIAL PHOTOGRAPHY OF FARM INSTALLATIONS AND CROPS

The first step during development of a new farm is to survey and make a detailed topographic map of the entire region. Farmers often divide their land into management blocks based on a few criteria. This procedure makes it easy for them to manage farm resources. For example, management blocks are made using topography, soil types, soil fertility trends, water resource and irrigation lines, etc. Aerial photography using helikites/ aerostats could be most useful, to obtain an idea about topography. Often, these maps are GPS tagged and highly accurate. Helikites mounted on to ground vehicles have been adopted, to conduct aerial survey and scan the topography of land surface. Water resources could be mapped in detail using lidar photography (Sara, 2013; Environment Agency, 2007). Such maps are essential prior to land development for agriculture. The maps showing natural vegetation, wild growth and weeds could be helpful in judging the cleaning up procedures. Such cleaning is needed prior to land development. Topographic maps can be used to prepare management blocks. Helikites with appropriate sensors and ortho-image processing programs can provide better topographic maps to farmers (LIDAR USA, 2019). Topographic maps of even the streams that supply irrigation water to fields could be developed. For example. 3D maps of streams have been developed using helikites (Fonstad et al., 2013; Ozbay et al., 2017). They adopt 'structure from motion (SfM)' methods and photogrammetry, to develop 3D maps of farms, streams, water resources, vegetation, valleys, gullies, hillocks, boulders, etc. Helikite derived topographic maps with GPS co-ordinates are of immense utility to farmers. Particularly, to those initiating land preparations and setting up irrigation lines. The helikite-derived high-resolution maps are getting common due to availability of sensors technology and ortho- rectification computer programs (Nandi et al., 2017).

Helikites with facility for aerial imagery (sensors: visual, infrared, red edge, lidar) have been adopted by the British Environment Agency. The aim of aerial photography is to provide a visual representation of the land surface over time, using aerial techniques. The topographic maps showing land deterioration, if any, are useful to farm agencies. Aerial photography via helikites is also needed while conducting large or small scale, land restoration programs. Usually, helikite can fly at up to 1000 ft. in altitude. A digital camera can be fixed on the helikite, to provide aerial photography of the agrarian zone. Of course, the topographic map could also be utilized, to devise land reclamation programs (Environment Agency, 2007). Fonstad et al. (2013) have reported that, tethered helikites were useful in assessing topography of riverine zones. They used lidar and other methods to obtain aerial images of topographical features and native vegetation in the riverine region.

Aerostats, both tethered and untethered ones could be of immense utility, while obtaining aerial photography of crops. They offer continuous video data too. The tethered aerostats could be left to stay in air with sensors, for longer durations of several hours in a day or even weeks together. No doubt, helium-filled balloons find practical use during agricultural crop production. They say, on a windless day, a helium balloon performs better. Weather and wind speed are crucial aspects for balloon-aided aerial photography. Helium balloons may drift and become unstable. Kites usually descend if wind and weather are uncongenial. Helikites are endowed with greater stability in the air, despite wind disturbance. Helikites with helium-filled envelope and tethered to ground or crop field are again useful. Particularly, while conducting aerial surveillance of crops. The sensors present in the helikites can relay large amounts of data about the standing crop below. The data sets obtained using sensors on helikites are usually the NDVI, GNVI, leaf chlorophyll index (i.e., plants'-N status), crop's water stress index, disease/pest incidence or disasters such as soil erosion, floods, drought, etc.

Let us consider structural details about a typical cigar shaped aerostat that is adopted for aerial photography of general terrain and its features. The same aerostat is also used to obtain imagery of farms and cropped fields. The ortho-images derived from sensors are processed and enhanced, using computer programs such as Agisoft's Photoscan or Pix4D Mapper, etc. The imagery is further analyzed to understand the variations in soil fertility, nutritional requirements of the crops, water status and irrigation needs. Aerial imagery also helps to detect insect attack/fungal diseases, if any, etc. Aerial imagery from such aerostats are useful in rapid detection of farm related disasters such as floods, droughts, soil erosion loss of fencing, etc. The 'AEROS 3200' manufactured by Aeroscraft Corporation LLC., Montebello, California, USA is one such aerostat. It is 39 m (120 ft.) in length. It has an endurance of 16 days in air. Helium is the lifting gas. It lasts for longer duration. It tolerates wind speed of 80 knots without getting disturbed during aerial photography. The operational altitude is 1500 m (4500 ft.) from ground level. It has several options for payload area sensors (Pantuso, 2019). The major sensors adopted during flights over farms are the electro-optical sensors such as visual cameras, infra-red cameras, red-edge cameras and LIDAR sensors. When used for reconnaissance, the aerostat is also equipped with GMTI radar that detects flying objects from 140 km distance and ground targets from 90 km distance. Maritime targets could also be detected using the aerostat. A few other noteworthy features are remote controlled winches, payload mounting, full-fledged ground station with image processing iPad, and relocatable mooring tower.

Overall, it is said that an aerostat (e.g., Aeros-3200) is cost effective in terms of regular surveillance of crops and obtaining crucial crop growth data. The aerostat has longer endurance. Hence, aerial imagery becomes efficient. The sensors allow us to obtain images of ground features for longer period (Pantuso, 2019: Plate 4.12).

PLATE 4.12   AEROS 3200.
Left: An aerostat floating low over farmland. *Right:* An aerial photograph of farm installations and fields derived using the aerostat.

Note: The Tetracam sensor is hanging below the envelope of the aerostat. It helps farmers with visual, infrared and LIDAR photographs of farms and crops. AEROS 3200 could be useful to farmers while conducting agronomic operations adopting precision farming principles.

Source: Irene Guerrero, Aeroscraft Inc., Montebello, California, USA.

4.5.2   ROLE OF AEROSTATS AND HELIKITES IN ASSESSING GROWTH AND NUTRIENT STATUS OF CROPS

We should note here that nutrient stress, pertaining to major nutrients (N, P, (k) is among most important constraints to crop production. The crop growth and yield formation are highly dependent on maintaining optimum levels of major nutrients in soil. At any stage, the crop growth measured as NDVI and crop's N status measured as leaf chlorophyll using spectral analysis is of utmost utility to farmers. The NDVI tells us about the rate of biomass accumulation (i.e., photosynthetic carbon fixation). Leaf chlorophyll content indirectly suggests about the crop's N status. There are indeed innumerable methods that allow us to measure crop's N status. Some of them involve destructive sampling of crops in the field and chemically estimating N, in different parts of the plant. A popular method is to use hand-held leaf color meters or leaf chlorophyll meters. They help us in gaining insights into variation in crop's N status across the field. However, it is tedious. It needs skilled technicians, and accurate mapping of distribution of N in crops, in different parts of the field. Precision farming methods often need detailed data about crop's N status, in the entire field. This could be tedious and difficult if manual methods are adopted. Aerostats or Helikite towers fitted with sensors could be a good method, to obtain continuous data about crop's N status. It is common in most of the agricultural experimental stations to evaluate the response of various crops and their genotypes to supply of nutrients, particularly N. Nitrogen is required in relatively higher quantities compared to other major elements. Let us consider an example from maize/bean growing regions of Brazil. The reports by Abrahao (2011) states that, bean crop responds to supply of fertilizer N, done at differential rates. The response is reflected as different levels of leaf chlorophyll content. The leaf chlorophyll content is often easily measured by using sensor (visual, red, green and blue). Response of beans to fertilizer could be studied, using spectral data from sensors on aerostat/helikite (Abrahao, 2011). This study clearly reveals that we could depend on aerial imagery of legume crops such as bean or lentil. Particularly, to assess response of crop to supply fertilizer N, accordingly. There are computer programs that allow the farmers to supply exact quantities of fertilizer-N after assessing N status in leaf. The aerostat mediated methods are also useful when farmers adopted precision farming methods. The sensors could provide detailed digital data about the variation in crop/soil-N status. Such data could then be used in the variable-rate applicators.

4.5.3 AERIAL PHOTOGRAPHY AND COLLECTION OF CROP'S PHENOMICS DATA

The collection of crops' phenomics data using aerial vehicles is relatively a new aspect of crop physiological research (CIMMYT, 2014). It has great relevance to rapid and accurate selection of high-performance genotypes of say, wheat, maize, rice, legumes, oil seeds, etc. Phenomics data obtained using aerostats could be useful to both large farming companies and individual farmers. Particularly, while scheduling various agronomic procedures. Phenomics data and digital surface models (DSMs), together, may be of help to farmers during crop production. We may note that evaluation of crop's phenotype is an important step during crop improvement. Plant breeders do collect data about a series of plant characters. Several types of ground and aerial vehicles have been adopted to secure such phenomics data. Adoption of ground-based methods may involve tedious field work. Often, it involves collection of data by skilled farm technicians. UAVs such as fixed-winged drone aircrafts or helicopters or autonomous parafoils may also help in rapid procurement of phenomics data of crops (see Krishna, 2018, 2019). Phenomics may involve repeated collection of data for the entire field and in short intervals. In such cases, aerostats/helikites that stay permanently, all thorough the season for long duration could be of immense utility to plant breeders. Tethered aerostats and helikites of different sizes are available. The payload too varies depending on the extent of data collection envisaged. The sensors placed on the tethered aerostats could provide them crucial phenomics data all through the season. Aerostats have been adopted to obtain NDVI data of cotton crop (Haire, 2004). Infrared sensors on aerostats have offered excellent data about crop's water stress index. Such data is highly pertinent while scheduling irrigation accurately (Jensen, 2017; Ritchie, 2008). Aerostats/helikites are significantly low-cost methods compared to ground methods or UAV aircrafts. The main drawbacks with aerostats/helikites are that they could get unstable, if weather is too windy. It takes time to lower down and re-install aerostats (White et al., 2012). Overall, tethered aerostats or helikites installed at different altitudes above the fields may become common while collecting phenomics data of crops. Such data is useful while deciding quantum and timing of inputs. For example, while channeling fertilizer irrigation and other inputs.

Hernandez (2015) states that, in California, farmers are trying to adopt advanced methods. They are experimenting with robots to conduct seeding. At the same time, helikites that offer visual and infra-red imagery of crops at different stages is being adopted. It has been done to monitor agronomic procedures and to forecast crop yield. Specific computer programs are adopted to develop forecasts about crop yield. An array of mobile applications is used to control the agronomic operations in the field. At this juncture, we may note that, phenomics data derived using tethered aerostats/helikites also helps in fixing yield goals and forecasting harvest levels.

4.5.4   AEROSTATS AND TETHERED AEROSTAT TOWERS IN PRECISION AGRICULTURE

Precision agriculture involves analysis of soils, crops, moisture and detrimental factors (insects/disease/weeds) if any, keeping in view the spatial and temporal variations in grain productivity. It also considers the intensity factor. We usually judge variations in growth, leaf chlorophyll content, disease/pest infestation, water status, etc. During practical farming, adoption of variable-rate techniques does require a series of data about the crop (see Krishna, 2012, Stafford, 2005; Sharma, 2017; Zhang, 2015). Primarily, it depends on accurate knowledge of spatio-temporal variations in soil fertility and crop growth pattern. The inputs are essentially channeled at variable-rates. This is to counter and remove the variations and reach the yield goals. Measurement of crop's parameters at various stages using field scouts is not an easy task. It requires skilled labor and monetary investment. In large farms, it involves use of sophisticated and costly equipment such as tractors/planters with variable-rate mechanisms for seeding, fertilizer application, herbicide and pesticide application, etc. It is essential for vehicles used in precision farming to possess GPS connectivity. Recent trend is to obtain data using small drone aircrafts with sensors that measure NDVI (growth/biomass), leaf chlorophyll content (crop's N status), CWSI (crop's water stress index), spectral data about disease/pest situation, weeds, etc. Aerostats of different sizes and capabilities are being tried briskly by various companies and agricultural research institutions. Aerostats used are generally small. Tethered lighter-than-air (LTA) stationary platforms that can support a variety of payloads, including multi-spectral imaging systems are useful. Payloads up to 20 lbs for precision agriculture applications are easily accommodated on a tethered aerostat (Aeros, 2019b).

The aerostats/helikites can help the farming enterprises during surveillance and data capture about crops. Otherwise, manual operations to get the same amount of aerial data could be enormously costly. Reports suggest that aerostats (tethered or free) and helikites are excellent aerial vehicles that aid in rapid capture of data pertaining to crops. They offer digital data about crops from vantage levels and points in the sky. Such data could be processed using appropriate computer programs. Then, the digital data can be directly used in the variable-rate applicators, during precision farming (Aeroscraft LLC. 2019).

A recent trend with regard to use of aerial vehicles is to set up 'Aerostat Towers.' An 'aerostat tower' can be installed at the beginning of crop season. This is to help in performing procedures relevant to precision farming. The aerostat essentially consists of a full set of visual, infra-red, lidar, and gas emission sensors depending on the need for accurate data. The set-up of aerostat tower in the field allows constant surveillance. Plus, it offers continuous collection and relay of data about crops, to iPad (ground control station). Many of the agronomic prescriptions could indeed be done rapidly taking note of crop's growth status and yield goals. Appropriate computer programs to correct the nutrient and soil moisture deficiencies are needed. According to Aeros (2019b), an aerostat tower meant for precision agriculture, usually, consists of definite accessories/facilities.

Let us consider an example. An aerostat tower is a lightweight equipment that could be lifted into sky with a tether. It can be handled by a single operator within minutes (Aeros, 2019c). Its salient features are that it offers continuous surveillance. It is easily transportable. Its linear weight is less than 1.3 lb per foot. Its operating height is 70 ft. above crop's canopy. Its maximum payload capacity is 75 lbs. Its capabilities are that it offers persistent still and video image coverage of crops. It provides data tagged with precise GPS coordinates. It can operate at wind speed of 45 kmph without suffering distortion of images. It offers a series of data useful in the analysis of crops. The digital data about crops could be utilized directly in the variable-rate applicators. It is a useful equipment during crop management. An aerostat tower can be kept floating for several days at a stretch, in the cropped fields (see Plates 4.13 and 4.14).

There are private aerostat owning companies that offer specific services to customers (Persistent Surveillance Services, 2019). They serve by providing useful aerial imagery and digital data that could be used after processing. Usually, image processing too is done by the same agencies. Such aerostat agencies may become common in the major crop production zones worldwide, in due course. They could be serving large farming companies and farmers with necessary data, to run the variable-rate seeders, fertilizer inoculators, sprayers of herbicide, pesticide or fungicides, etc. Precision farming needs support from aerostat companies to provide accurate digital data about spatial variations occurring in the cropped fields. Agronomic prescription depends on availability of computer programs that utilize aerial digital data and offer accurate agronomic prescriptions. This aspect of developing appropriate computer programs that utilize digital data from aerostat and convert them to realistic and accurate agronomic prescriptions needs greater attention from agricultural researchers.

PLATE 4.13   A tethered balloon capturing details of infrastructure and crops below using different sensors- a simulation
Source: Mijatovic, A. AerDrum Ltd, Belgrade.

4.5.5   ROLE OF AEROSTATS IN MAINTAINING AGRICULTURAL EXPERIMENTAL STATIONS

Aerostats, by definition are tethered blimps. In the Chapter 3 on blimps under Section 5.8 a few remarks and examples have been made available. It relates to role of tethered blimps (aerostats) lofted for longer durations above agricultural experimental stations. As such, aerostats can play a vital role in managing agricultural experimental station. The tasks may relate to general surveillance necessary for watch and ward of the experimental fields and infrastructure. Aerostats can offer 24 hr surveillance and imagery of the entire crop material. Intrusion, if any, by animals, disasters like soil erosion, stagnating water, loss of plant density, etc. can be detected. In fact, there are a few agricultural experimental stations such as CIMMYT at El baton, in

Mexico, and few locations in East Africa (e.g., Tanzania, Kenya) that have already tested tethered blimps. They have monitored the performance of wheat genotypes in terms of growth parameters (NDVI, leaf chlorophyll, i.e., plant-N status) and drought tolerance (i.e., crop's water stress index) (CIMMYT 2012,2014; Tattaris, 2014, 2015). Aerostats with a series of sensors provide excellent spectral data of wheat genotypes that are attacked by rust fungus (*Puccinia striformis f. tritii*) and those that are tolerant. So, crop breeders may evaluate a very large set of germplasm lines for disease resistance of say any field crop. The evaluation of crops could be done much easily without resorting to tedious scoring, using field workers. Aerostat networks too have been adopted by wheat agronomists. They have tried to relay the spectral data and details about growth performance of genotypes, to all the locations (experimental stations) that are interconnected.

PLATE 4.14 Tethered Aerostats above farmland.

Note: Aerostat towers could become a common feature among farms adopting Precision Farming practices. Aerial photography derived from such low floating tethered aerostats can be of use to farmers during formation of 'management block.' Incidentally, formation of management blocks is essential first step during adoption of Precision Farming.

Source: Mijatovic, A. AerDrum Ltd, Belgrade.

Aerostats have been used to monitor insect pest attack. Aerostat networks could also be useful in collecting data about the spread of major disease caused by plant pathogens. Such aerostat networks could help in early warning of spread of diseases and insects. For example, spread of insect vectors such Russian aphid that spread wheat viruses could be detected and information could be relayed to all aerostat stations. In the general course, aerostats lofted and kept for longer duration say 15–30 days or even a few months serve the farmers and agricultural researchers usefully. The imagery related to land and soil type, water resources, irrigation lines, cropping systems, stage of each crop, progress of agronomic procedures, movement of farm vehicles, harvest and its transport could all be carefully monitored using iPad (ground station computer).

4.5.6 DETECTION OF FARM RELATED DISASTERS USING AEROSTATS AND HELIKITES

Agricultural cropping zones encounter several different pests/diseases that lead to decreased productivity. Many a times pests/disease may reach disastrous proportions leading to massive loss of forage and grains. We may note that, pest build-up and its rate of spread in a cropping zone could be monitored. There are pests like locusts that spread rapidly and cause crop disasters. Therefore, Food and Agricultural Organization of United Nations (Rome, Italy) has installed several monitoring stations over the entire region, particularly, where locust is severe. For example, desert locusts are monitored regularly. Farmers are fore warned of impending danger. They are asked to conduct remedial measure forthright without delay. At present, such monitoring of locusts is done using skilled farm scouts, specialists, airplane campaigns, sometimes small drone aircrafts, etc. Here, we should note that aerostat towers with sensors and radars to detect locust clouds could be a good idea. The aerostat towers could relay accurate data about the stage of the crop, density of locusts, their feeding rate and habit, expected loss of vegetation and grains (FAO of the United Nations, 2019). Aerostat derived data could add enormously to the Desert Locust Information Services (DLIS). Aerostats have not been deployed so far in the locust endemic zones. However, it is mandatory to know the spectral properties of locust clouds, plants that are healthy and those destroyed by the locust infestation. Knowledge about spectral signatures is essential. To a certain extent, we can assess the locust infestation using aerial photography, i.e., by adopting visual band width sensors placed on aerostats/helikites.

Food and Agricultural Organization (FAO) of the United Nations, Rome, Italy lists a series of pests and diseases that could reach disastrous proportions. They could affect individual farmers and the entire agroecosystem, if they go unattended. Timely remedy depends on close and continuous monitoring of pests/disease. Aerial photography, that shows spectral signatures of healthy and disease/pest attacked plants/crops could be compared with data available in the data bank. Some of the pests listed by the FAO need emergency remedial measures. Therefore, we need a network of aerostats/helikites with sensors that capture accurate data. We need computer programs that process the imagery and develop maps, depicting the extent of damage to crops. Aerostats placed at appropriate 'hot spots' in a particular agroecosystem or larger stretches of agricultural zones could then share information, rapidly. Later, they can arrive at accurate remedial measures. At present, FAO, Italy lists following disease/pests as ones that need continuous monitoring and emergency precaution/remedial measures. They are: Wheat rust disease across different continents, Cassava viruses: Cassava streak and Cassava mosaic in Africa, Fall army worm in entire African continent, Desert Locust in Africa, etc. In future, aerostat derived data and network supported by them could play a vital role, in thwarting losses due to major pests/disease. Many of the maladies spread rapidly. They are actually transboundary pests/diseases affecting several different nations at a time, depending on the cropping area. Hence, it is believed that aerostat networks could be useful in transmitting information regarding the rate of spread of the disease. They also offer accurate maps to farm agencies in several different nations at a time.

Aerostats are useful UAVs that can provide us with accurate aerial images of natural disaster affected zones. Arial imagery may pertain to flood damage in cropped fields, large soil erosion, gully formation and loss of topsoil, droughts, large patches of disease/pest attacked zones, loss of riverine embankments, wildfires and fire affected zones in natural vegetation zones or even cropped fields. Tethered aerostats placed at high altitudes or even helikites can detect wildfires in their initial stages and relay warming signals. The sensors help us in gaining aerial photography that tells us about the total volume of fire affected forest stands or natural vegetation Gabbert (2011). There are aerial surveillance and measurement techniques such as SfM (Structure from Motion) and LIDAR that offer excellent data about wildfires. Helikites are useful in offering data that helps to decide on the fire mitigation procedures and in scheduling them. The remedial measures could be targeted much better, using aerial photography from helikites (Gabbert, 2011; Fonstad, 2013).

4.5.7   AEROSTATS TO CONTROL BIRD AND ANIMAL PESTS AFFECTING AGRARIAN REGIONS

Tethered aerostats have been adopted to scare away birds and other animal pests that affect the agricultural crops. Reports from South India suggests that the aerostats that float at around 10 m-100 m above crop canopy are fitted with electro-optical, acoustic and laser-based sensors. The acoustic machines make sounds similar to those made by birds or animals during situations of scare. They create noise that literally scares the bird/animal pests to move away. Ultrasonic sound and lasers are also used to make birds fly away from the fields. The aerostats used in the field are relatively small at 3–4 m in diameter (Hemanth, 2016). The aerostats could be held tethered to farm vehicles and moved to right spots in the field. An aerostat once flown can be held in the field for 3–4 weeks at a stretch. Therefore, an aerostat placed at the beginning of seed set in cereal/legume fields is very useful in scaring birds that destroy ripe panicles. Such aerostats can also supply images of crops round the clock from boot-leaf stage till panicle/grain maturity.

Damage to standing crops in fields arise due to variety of biotic factors such as insect pests, disease causing microbes, small animals, birds, etc. Methods adopted to reduce such menaces too vary depending on the destructive agent. Here, we are concerned with how tethered balloons and aerostats fitted with acoustic and laser guns are utilized to scare away birds, rodents and small animals such as macaques, boars, foxes, etc. Field crops such as cereals, legumes, oil seed crops and horticultural fruit crops suffer bird damage, to a great extent. The damage to crops often occurs during panicle ripening and seed set or fruit ripening. Bird damage at harvest is a worldwide menace on cereals. A few estimates done in USA, in particular, suggest that bird damage to crops can be severe in some states, while it could be feeble to moderate in other locations. Overall, crops valued at about 737 million US$ was lost due to bird damage in 2011 (Anderson et al., 2013; Bajoria et al., 2017). In India, standing cereal crops at 'seed set stage' do experience moderate to severe damage. It leads to reduction of grain yield. Bajoria et al. (2017) state that, crops such as sorghum and pearl millet may lose 25–65% grains, if unattended during panicle ripening. Similarly, sunflower heads at seed set are vulnerable to bird damage. Seed loss due to bird damage ranges from 20–65% in case of sunflower. It is interesting to note that tethered aerostats scare away birds and reduce damage to a great extent. Further, often the birds get scared to return to the same fields even after removing the tethered aerostat with bird scaring acoustics. This happens at least for a while, say, a couple of weeks. There are instances, wherein, bird menace could be controlled using acoustic noises created by machines placed on the tethered aerostat (Haque and Broom, 1984; Shirota et al., 1983; Gilbert et al., 2003). A few evaluations indicate, generally, bird scaring conducted using helikites are superior to the traditional methods that adopt propane canon, sonic wailers or flashing scarers placed on ground in the cropped fields (Gaskite Helikites Ltd., 2019).

There are a few companies that specialize in the production of 'Hawk Kites' and 'Helikite-aerostat' hybrids that could be utilized for aerial bird control. The bird scarers employ acoustics and lasers, to deter birds. These Helikite-aerostats have been employed to drive away bird species such as pigeons, seagulls, geese, parrots, rooks, sparrows, finches, blackbirds, etc. (Perigrine Ltd. 2018). Reports state that, helikite-aerostats have been adopted to control bird menace in locations such as agricultural farms, airports, land-fills, dump yards, etc. Such bird scaring aerostats have been employed in agricultural farms within many countries such as United Kingdom, United States of America, Canada, European nations, Asia and Fareast (Perigrine Ltd. 2018; Bird Control Systems Ltd., 2018; Gilbert et al., 2003). Here, we may note that aerostat mediated bird control is environmentally safe and does not involve use of harmful chemicals. They say, the method is less costier than other known bird control measures. The birds are not sacrificed. They are only fended away from crops during grain formation period.

The helikite aerostats used in some European nations are light weight. They have provided good control of birds that infest cereal crops and peas. Aerostats with bird scarers have been adopted in different climatic conditions, say, in hot and humid conditions and in moderately temperate climates. Helikite aerostats can withstands wind turbulence of up to 25 mph (i.e., force 6). They have been adopted to control bird damages on vegetable, arable crops and fruit trees (Bird Control Systems Ltd. 2018; Table 4.3). These aerostats floating at 200 ft. above crop's canopy can scare away the bird pest in an area equivalent to 25 acres.

'Vigilante Helikite' is an example of aerostat utilized for bird scaring. They say, several hundreds of such bird scaring aerostats are in operation in European agrarian regions (Bird Control Systems Ltd. 2018; Perigrine Ltd. 2018). Vigilante model has been adopted to control birds infesting fields with vegetable, fruit trees and arable crops, particularly cereals. Vigilante is a lightweight aerostat. It is easily portable using any small farm vehicle. It stays afloat for 3–4 weeks to coincide with most vulnerable period during the crop season. Its polythene envelope is disposable after the use. Helikite is cheaper than most other methods. The gas usage is low. It costs only few cents per day. Reports suggest that helikite aerostats have been thoroughly tested since 1998 in the farms located within European community region (Bird Control Systems Ltd. 2018). For example, the Allsopp Helikites have provided good control of birds affecting field crops. A single aerostat is tethered and installed for 9–10 ha of field with cereals plus pea intercrop. The aerostats can be used for 25–30 days without birds getting habituated.

Let us consider another example. In Nigeria, use of helikites to deter birds that destroy grain crops is supposedly an advanced methodology in the rice producing regions (Ejiogu and Okoli, 2012; Gallagher et al., 2002). For example, scientifically designed and patented Vigilante Helikite is a bird control system that works well over a long period of time. Birds find it difficult to overcome the innate terror of predatory hawk's sounds that helikites create (Ejiogu and Okoli, 2012).

TABLE 4.3   Helikites in Bird Control Practices in Farmland

Bird Species (pests)	Crops Protected	Crop Area Protected (in Acres Helikite^{-1})
Rooks/Crows	Emerging cereal seedlings, maize, strawberries	10–15
Wood Pigeons	Emerging cereals, oilseed rape, peas, beans, maize	15–25
Starlings	Cherries, strawberries, fruit trees	3–7

Source: Peregrine Ltd., 2018; Bird Control Systems, 2018; Gaskite Helikites Ltd., 2019.

Note: Major bird species deterred using helikites are: Small birds, Cormorants, Local Geese; Migratory Geese, Pheasants, and British Herons.

Helikites could be a good idea to deter the birds from damaging high value crops, particularly, at the heading and grain maturity stages. In USA, sunflower is one of the crops that suffers bird damage when the heads show seed set and progress towards grain maturity. The sunflower heads loose filled seeds to different extents. It depends on the population of birds that flock into fields and the remedial measures adopted.

Seamans et al. (2002) have reported that birds such as gulls could be restricted from nesting, using helikites. Such trials dealing with effect of helikites on gulls were conducted in ponds of different sizes and in locations such as Albany, in New York, near Lake Erie in Michigan, at Erie county in Ohio, etc. As an extension of this idea, they also evaluated the effect of helikites on the birds that affect sunflower productivity. In all cases, helikites reduced nesting and loitering of gulls in the vicinity. The experimental results showed that in plots with helikites installed, the bird damage was 8 %. However, if helikites were not used as a treatment, then, 26% grain loss was recorded from sunflower heads. It is said that we should conduct studies in order to standardize the best height at which the helikites could be floated. The noise deterrent, if any, that could be adopted has to be appropriate. The number of helikites required per unit area to deter different kinds of birds also needs experimentation. We also have to overcome problems such as birds getting used to the presence of helikites.

Bird damage to blue berries is increasing in many parts of North America and Europe. In response, there are several methods adopted by farmers, to thwart bird damage to fruits. In USA, they say, bird damage could be severe on blue berries. The damage level reaches almost 30% loss of fruit yield. On a yearly basis, about 10 million US$ is lost due to bird damage to blue berries. Noise makers and bird scarer-guns (sound-based) are common. However, birds do get accustomed to noise-based scarers, after a while. Most recent alternative that is gaining in acceptance is the helikite-based sound maker that mimics sounds of birds of prey. For example, in Washington State (Pacific Northwest USA), they are adopting helikites with bird scarer placed in the payload. Helikites fitted with noiseless laser scarers are also being searched (Whatcom Co. 2008; Engineering 360,2019). Demonstrations to farmers from different states in Central Great Plains indicates that, helikites with bird scarer devices can control the menace effectively. Helikites are effective within an area of 5–10 acres per helikite tower (Missouri Alternative's Center, 2019).

There is a renewed interest to grow soybeans in the weather and general agricultural conditions prevailing in the United Kingdom. Experimental trials are being conducted repeatedly, to gain better insights into the problems that soybean crop may face. Methods to overcome those constraints are being devised and tested. It seems, damage to soybean due to pigeons could be severe. The damage occurs mostly during early seedling stages. So, farmers need to keep a watch and ward-off birds for a small period until the plant gains in stature. A few farmers have reported that they are using helikites placed at low/medium altitudes, to divert the attention and ward-off pigeons. They are lofting helikites with search lights as payloads. The rotating search lights distract the pigeons and they are kept away from soybean fields (Case, 2019).

Santilli et al. (2007) states that, most of the bird scarers used are acoustic or visual tools. They are effective only for short periods and their efficiency is confined to small areas of influence. Hence, they examined the effect of a combination of a kite plus a Mylar helium filled balloon- "HelikiteO.' The idea was to scare away crows, pigeon and other bird species that feed on corn seeds and small sprouted seedlings. They also tried the HelikiteO in sunflower fields that were having seeds at maturity. Overall, they conducted 4 trials spread out in an area of 26 acres. Then they estimated the bird damage to corn and sunflower crops. Santilli et al. (2007) have reported that, helikiteO could reduce loss of sunflower seeds. It could keep pigeons and crows away from sunflower crop. The damage to sprouting corn was 19% in areas away from helikite, but, only 4% just closer to helikite.

4.5.8   ROLE OF AEROSTATS IN FOREST PLANTATIONS

Globally, large areas of forest stand are being monitored regularly, using satellite imagery. Satellite-derived images are used to map the extent of forest stands, its species diversity, biomass accumulation pattern, disease/ pest attacks, if any, logging and transport, etc. Forest management and scheduling various procedures may depend to a large extent on the aerial photographs supplied via satellite agencies (see Krishna, 2016, Rassinima-licki, 2019). The satellite images lack in sharpness and resolution. They are not available easily. We cannot repeatedly obtain satellite imagery, because of re-visit times of satellite that may not coincide with our needs (see Krishna, 2016). Therefore, to overcome these difficulties, it is preferable to operate a few tethered aerostats or helikites above the forest stands. Then, note the forest maps, forest species diversity, biomass accumulation pattern, carbon accumulation trends based on location and tree species, detect forest maladies if any, monitor soil erosion, floods and drought related loss of forest tree plantation, monitor movement of wood out of the plantation, etc. Tethered aerostats that are kept afloat at 2000 ft. above the forest stand may be helpful (Plate 4.15; Altave, 2019a,b; Boulanger, 2019; Bonev, 1993; Carolina Unmanned Aerial Vehicles Inc., 2019).

Helikites are useful in making aerial survey and collecting samples of insects that damage forest nursery and stands in the outfield. Let us consider an example. Regular spruce budworm (SBW) outbreaks are responsible for significant damage caused to Canadian forests. Tethered aerostats and helikites

PLATE 4.15   A tethered aerostat (helikite) floating above natural vegetation.

Note: Tethered aerostats are among most simple unmanned aerial vehicles capable of excellent aerial photography. They also help in regular surveillance of natural vegetation and forest stand. They can monitor activity in the forests. If chemical probes are attached, then, they can detect greenhouse gas emissions from forest stands.

Source: M. E. Rogers Carolina Unmanned Vehicles Inc., Raleigh North Carolina, USA and Sandy Allsopp, Allsopp Helikites Fordingbridge, Hampshire, United Kingdom.

with accessories (nets)for sampling of spruce bud worm clouds have been utilized. The Spruce budworm that migrate over the forest plantations are sampled at various altitudes (400 to 600 m), to understand the pattern of flight of the clouds and population intensity of Spruce Bud worms. Remedial measures could be adopted accordingly. Helikites could also be adopted to obtain aerial photographs and digital signatures of forest stands that are affected by disease/ pest and healthy ones. No doubt, spectral data from tethered aerostats helps in rapid diagnosis of maladies that may affect forest stands.

As stated earlier, tethered aerostats are utilized to collect atmospheric gas sample from different altitudes above forest stands. For example, in South-eastern USA, Strand et al. (2016) have clearly shown that tethered aerostats help in sampling atmosphere. The samples could be analyzed to decipher the levels of particulate matter and emissions of greenhouse gases such as CO_2, CO and CH_4.

4.5.9   AERIAL IMAGERY OF PASTURES

Pastures are among the major sources of carbohydrate and nutrients to several different species of domestic and wild animals. Pastures support diverse species of grasses, legumes and others. There are pastures developed specially to suit the domestic cattle, horses, sheep, etc. These pastures show enormous variability in growth, biomass productivity and nutritive value. Such spatial variability can be easily attributed to factors such as soil fertility variations, differences in nutrient and water availability to pasture grasses, genetic diversity with regard to nutrient accumulation pattern of mixed pastures, etc. Farmers often develop pastures after correcting such fertility variations. Aerial photography can help farmers to judge the variability in biomass accumulation pattern. Such data aerial data and maps could be utilized during management of pastures and domestic animals, horses, sheep, etc. These natural grass lands or those developed by farmers vary due to grazing, by domestic animals (cattle, horse, etc.). There is no doubt that farmers need a periodic report about growth and spatial variability of their pastures/grasslands. The nutrient deficiencies, droughts/dearth of soil moisture, disease/insect attack or even soil erosion needs to be monitored regularly. Aerial imagery using UAVs such as drone aircrafts, blimps, parafoils, aerostats or helikites, and kites are perhaps best methods available, currently. Detailed scouting of pastures using skilled personnel is costly. Whichever is the UAV, there is need for sensors like visual, infrared, lidar, etc., to study the pastures and map the variations in growth patterns.

Farmers with ability to adopt aerostats plus sensors can easily monitor the farm animals and judge their grazing pattern. They can modify the grass removal pattern, if needed. Dougherty et al. (2019) has utilized helikites to conduct an experiment. It aimed to study the grazing pattern of horses. The pasture was sub-divided into small plots to apply different treatments. Pastures were supplied with ammonium nitrate at 50 lb ac^{-1}. The helikite used to keep a vigil on pasture plots and provide aerial imagery. It was mounted with Sony DSC U30 camera. The helikite was stationed immediately above the pasture, at an altitude of 400 ft. The pasture actually supported horses. The imagery transmitted to iPad contained JPEG pictures and digital data showing the variation in growth (NDVI) of pasture grass and grazing pattern. Further, Dougherty et al. (2019) have concluded that helikites with SONY visual and infrared camera helped in regular observation of pastures/horses. It relayed digital data quickly to ground station. It offered NDVI data about pasture growth and biomass accumulation. At this juncture, we may note that, if the pasture is a composite mixture of grasses, legumes and other species, then, spectral reflectance properties and spectral signatures of the plant species that compose the pasture are required. We can then estimate the proportion of different grass species or legumes, using aerial spectral data.

Pasture production involves series of agronomic procedures beginning with seeding. The pasture monitoring and upkeep are important. Pasture quality depends on fertilizer and water supply at accurate levels. Pastures have to be monitored periodically for diseases and pests. They may also be affected by soil erosion. There are several methods of aerial survey of pastures, like using drone aircrafts to obtain imagery, aircrafts piloted over pastures, use of blimps to survey pastures and similar vegetation. Aerostats, helikites and kites could also be utilized. Pastures could be monitored using sensors such as visual (red, green and blue bandwidth) sensors mounted on helikites. Lidar too offers details about the pastures and surface features. In any case, farmers need data that clearly depicts biomass of pastures. It should tell them about the rate at which fresh biomass of pastures is being accumulated (Szpakowski, 2016). Usually, NDVI and leaf chlorophyll measurements suffice to get clearer idea about pastures. We should note that, farmers could periodically record data about their pastures using aerostats/ helikites with apt sensors. Remedial measure could then be accurate and timely.

Overall, we may infer that an aerostat tower at a height of 200–400 ft. above the pasture surface may be a good idea. It helps in pasture management by being a sentinel above the cattle/horse farms. It could inform the farmers about grazing pattern and movement of animals both during day and night. It offers most crucial data about spatial variation in biomass accumulation, leaf chlorophyll content (i.e., leaf-N status), water stress index. It also provides information about disasters such as excessive gazing caused soil erosion, flooding loss of hedges and borders, etc. The general forecast is that UAVs, particularly, small aerostats and helikites are efficient in terms of capital requirement and recurring costs. As stated earlier, skilled farm scouts could be costly. However, there is need for experimental comparison and regular experience of farmers to know more about economics of aerostats/helikites in pasture management and animal monitoring.

4.6 AEROSTATS, HELIKITES, AND BALLOONS IN AGRICULTURE: FUTURE COURSE

The overall market for aerostat systems was valued at US$3,604.7 million in 2014. It is anticipated to expand at a compound aggregate growth of 13.8% during the forecast period from 2015 to 2022 (Global Aerostat Systems Market Analysis, 2015). A different forecast states that global aerostat market is estimated to reach 10.95 billion US $, by the year 2021 (Markets and Markets, 2018; Gupta, 2018). Another report by Research and Markets (2018) states that, global market for aerostats was 6.94 billion US$ during 2017. It is expected to be 29.12 billion US$, by 2026. Visiongain's definitive new report assesses that the market for 'Military Aerostats' reached US$4,175m in 2016. The performance of the industry is forecasted to grow steadily, with an anticipated CAGR of 4.8% for the five-year period 2016–2021. This is expected to drive the industry to a value of US $5,289 million, by the end of 2021 (Visiongain, 2019). Therefore, aerostats/helikites are expected to become more common in future.

Vasconcelos (2019) states that, in 2019, the overall global market for aerostats/balloon is estimated at US$ 5 billion. It pertains to their use in rural communication systems. Further, it is estimated to increase to 11 billion US$, by 2021. During next few years, manufacture, installation and adoption of tethered aerostats in farming regions of the world may become a conspicuous activity. It involves active participation of both telecommunication and agricultural engineers. We may note that, as time lapses, most farm equipment in the field or even in farmhouses become more dependent on internet signals. Most farm vehicles and agronomic operation may depend immensely, on GPS-aided guidance in the outfield. Many of them are already controlled using mobile phones/iPad that need internet access. So, tethered balloons/aerostats and helikites may have a 'field day' all across the agrarian regions of the world.

A few of the leading companies involved with military and civilian aerostats with key financial investments and deriving exchequer out of its sales are as follows: Allsopp Helikite (Great Britain); ALTAVE (Brazil); ILC Dover Ltd.; Lindstrand Technologies Limited; Lockheed Martin Corporation (USA); Raven Industries Inc.; Raytheon Company (USA); RT; SkySentry; TCOM, L. P.; and Worldwide Aeros (see Visiongain, 2019). In future, agricultural aerostats/helikite industry may get further impetus. Accordingly, we can expect the list of aerostat-producing companies to expand. Major players within aerostat industries could hail from those who produce agricultural aerostats.

A report based on a regular market survey conducted since few years' states that, airships are sought in good numbers. They are preferred because they could be autonomous, pre-programmed in flight and offer excellent intelligence information. They withstand turbulence in the atmosphere that occurs at different altitudes. Among the airships, tethered airships are also expected to be popular during next few years. The demand is driven by need for surveillance, aerial photography and general intelligence information. In the realm of military, need for both low and high-altitude free-floating aerostats and tethered aerostats is based on the fact that, they are good deterrents to cross-border terrorism and military intrusions. The low fuel consumption and cost of operation are added advantages connected with aerostats. Aerostats are in use in greater number in North America, Europe and Fareast. North America shows the greatest potential for use of tethered aerostats (Research and Markets, 2018). During the most recent past, aerostat market in the Middle eastern region registered highest growth. At this juncture, we should note that, until now, military establishments and police have adopted aerostats. However, in near future, adoption of tethered aerostats could increase rapidly, in the agrarian belts. Farmers might adopt aerostats, to gather most of their crucial data about crops and productivity levels. Aerial photography obtained using aerostats becomes essential. Therefore, agricultural enterprises could be the major users of aerostat. Perhaps small aerostats, also helikites of small to medium sizes may get preferred in the agrarian belts. Tethered helikites may dominate the 'agricultural sky.' They would be helping farmers with aerial photography and digital data about crops. Based on such aerial imagery, the prescription of fertilizers, water, pesticides, fungicides, weedicides and monitoring the normal agronomic practices in the field are possible.

In future, there is a strong need to make use of tethered and free-floating aerostats more amenable. Aerostats should be easily accessible for farmers. They should be able to use them above their crop fields. Several mechanisms related to transport, launching, recovery and later storage until re-use needs to be streamlined. Tethered aerostats may need several computer programs to improve their use in agriculture. Particularly, those related to obtaining aerial photographs, digital data, use of digital data directly on a ground vehicle. Instantaneous transfer of digital data or storage on a chip might be crucial for farmers. Some of these should be made extremely easy for farmers. The current trend is to use broadband mobile controlled equipment. So, use of aerostats in farmland should be made easy. Farmers should be able to control them using a broadband mobile phone. Aerostats of different sizes, endurance and abilities are required for them to become common above crop fields.

The introduction of aerostats/helikites in agrarian belts is still rudimentary and sporadic even in developed nations. It has to be popularized. Farmers have to realize that aerostats are cost effective. They are relatively permanent fixtures in farms and easier to operate and reap benefits. They are cost effective in terms of aerial photography, survey, monitoring crops and in offering digital data for GPS-guided precision equipment. Once this happens, like drone aircrafts, the aerostats will be common and could be seen floating over most farms, all across the farming land. The 'agricultural skyline' is set for a change with aerostat towers.

Aerostats promise much more than drone aircrafts to farmers, once they are adopted by them. Aerostats are much simpler to handle than fixed-winged aircrafts or copter drones or parafoils or microlights. Perhaps, small aerostats are among the best bets for farmers since they are excellent sentinels. Phrases such as agro-aerostats, agricultural sky, aerostat data, aerostat imagery may become more frequent in the farming zones. Not just the skyline above farmland that mutates, but even the aerostat production industry will experience the ripples of popularization of small and large tethered aerostats/helikites in agrarian regions. We have to wait and watch the farm world and discern the changes forecasted regarding tethered aerostats. Tethered Aerostats have the potential to dominate the *'agricultural sky'* and *revolutionize* the way we conduct agronomic procedures and monitor our farms.

KEYWORDS

- **buoyant airborne turbine**
- **crop's water stress index**
- **digital surface models**
- **greenhouse gases**
- **high altitude balloons**
- **humanitarian assistance and disaster response**
- **poly-aromatic hydrocarbons**
- **poly-chlorinated dibenzodioxins**
- **poly-chlorinated dibenzofurans**
- **small tactical multi-purpose aerostat system**
- **shortwave aerostat-mounted imagers**
- **tethered aerostat radar system**
- **tethered aerospace system**
- **universal camera mount**

REFERENCES

Abrahao, A., (2011). *Leaf Nitrogen and Chlorophyll Contents and Yield of Common Bean Discrimination Using Low Cost Remote Sensing* (pp. 1–98). Universidade Federal de Vicosa, Minas Gerias, Brazil. PhD thesis.

AeroDrum Inc., (2018). *10 m Long Blimps* (p. 161). AeroDrum Inc., Belgrade, Serbia https://www.rc-zeppelin.com/outdoor-rc-blimps.html (accessed on 30 July 2020).

Aeronautics, (2019a). *Skystar Aerostats: Skystar 180* (pp. 1–3) https://aeronautics-sys.com/wp-content/themes/aeronautics/pdf/Skystar_300.pdf (accessed on 30 July 2020).

Aeronautics, (2019b). *Skystar Aerostats: Skystar 300* (pp. 1–3). https://aeronautics-sys.com/wp content/themes/aeronautics/pdf/Skystar_180.pdf (accessed on 30 July 2020).

Aeros, (2019a). *Innovation Never Stops* (pp. 1–8). Aeroscraft Corporation, Montebello, California, USA. http://aeroscraft.com/aerostat/4575666089/ (accessed on 30 July 2020).

Aeros, (2019b). *Precision Agriculture* (pp. 1–3). Aeroscraft. http://aeroscraft.com/precision-agriculture/4584020239/ (accessed on 30 July 2020).

Aeros, (2019c). *Aeros Tactical Tower in Agriculture* (pp. 1–3). http://aeroscraft.com/precision-agriculture/4584020239/ (accessed on 30 July 2020).

Aglietti, G. S., (2009). Dynamic response of a high-altitude tethered balloon system. *American Institute of Aeronautics and Astronautics Journal of Aircraft, 46,*2032–2040, doi: 10.2514/1.43332.

Aglietti, G. S., Markvart, T., Tatnall, A. R., & Walker, S. J., I., (2008b). Solar power generation using high altitude platforms feasibility and viability. *Progress in Photovoltaics: Research and Applications, 16,*349–359.

Aglietti, G. S., Markvart, T., Tatnall, A. R., & Walker, S. J. I., (2008a). Aerostat for electrical power generation concept feasibility. *Proceedings of the IMechE Part G: Journal of Aerospace Engineering, 222,* 29–39. doi: 10.1243/09544100JAERO258.

Aglietti, G., Redi, S., Tatnall, A. D., & Markyart, T., (2010). Aerostat for solar power generation. In: Rugescu, R., (ed), *Solar Energy.* INTECH Publishers, Croatia. doi: 10.5772/8076.

Airlander 10, (2018). *Airlander: Rethink the Skies* (pp. 1–12). Hybrid Air Vehicles, Bradford shire, United Kingdom. https://www.hybridairvehicles.com/ (accessed on 30 July 2020).

Airstar Aerospace SAS, (2019a). *Tactical Aerostats* (pp. 1–18). http://airstar.aero/en/tethered-aerostats-range/ (accessed on 30 July 2020).

Airstar Aerospace SAS, (2019b). *Aerostats in Telecommunication Services and Mobile Broadband Services in Rural Area* (pp. 1–12). http://airstar.aero/en/tethered-aerostats-range/#contact/ (accessed on 30 July 2020).

Allsopp Helikites Ltd., (2019a). *Aerial Photography* (pp. 1–7). Allsopp Helikite Ltd, Hampshire, England, United Kingdom, http://www.allsopphelikites.com/index.php?mod=page&id_pag=33 (accessed on 30 July 2020).

Allsopp Helikites Ltd., (2019b). *The Skyhook Tactical Helikite Aerostat* (pp. 1–5). Allsopp Helikites Ltd., Hampshire, United Kingdom, http://www.allsopp. co.uk/index.php?mod=page&id_pag=24/ (accessed on 30 July 2020).

Allsopp Helikites Ltd., (2019c). *Skyhook Helikites* (pp. 1–5). Allsopp Helikites Ltd, Hampshire, United Kingdom. http://www.allsopp. co.uk/index.php?mod=page&id_pag=10/ (accessed on 30 July 2020).

Allsopp Helikites Ltd., (2019d). *The Light Weight Helikites: World's Smallest Aerostat* (pp. 1, 2). Allsopp Helikites Ltd. Hampshire, United Kingdom, http://www.allsopp. co.uk/index.php?mod=page&id_pag=55/ (accessed on 30 July 2020).

Allsopp Helikites Ltd., (2019e). *Helikite Case Studies: US Army in Afghanistan* (p. 1). Base force protection. Airborne video surveillance 2012 ongoing. http://www.allsopp. co.uk/index.php?mod=page&id_pag=55/ (accessed on 30 July 2020).

Allsopp Helikites Ltd., (2019f). *Helikitecase Studies: Long-Range Digital Radio-Relay and GPS free Positioning* (p. 1). Australian Defense Force (ADF)/US Army. Australian Exercise Talisman-Sabre 2012. http://www.allsopp. co.uk/index.php?mod=page&id_pag=55/ (accessed on 30 July 2020).

Allsopp Helikites Ltd., (2019g). *Helikitecase Studies: Oil Spill Detection and Clean-Up* (p. 1). Arctic ocean, Norway 2012-Ongoing. http://www.allsopp. co.uk/index.php?mod=page&id_pag=55/ (accessed on 30 July 2020).

Allsopp Helikites Ltd., (2019h). *Helikite Case Studies: Aerial Photographic Survey of Ancient Armarna* (pp. 1, 2). Egypt. Archaeology Department, Cambridge University, United Kingdom. http://www.allsopp. co.uk/index.php?mod=page&id_pag=55/ (accessed on 30 July 2020).

Allsopp Helikites Ltd., (2019i). *Lifting Multiple Tethersondes for Meteorological Research* (p. 1). University of Millersville, USA. http://www.allsopp. co.uk/index.php?mod=page&id_pag=55/ (accessed on 30 July 2020).

Allsopp, S., Reynaud, L., & Mohorcic, M., (2013). *Integrated Project: ABSOLUTE-Aerial base Stations with Opportunistic links for Unexpected and Temporary Events* (pp. 21–25).

https://cordis.europa.eu/docs/projects/cnect/2/318632/080/deliverables/001-FP7ICT20118 318632ABSOLUTED23v10isa.pdf (accessed on 30 July 2020).

Altaeros Energies Inc., (2019). *Clean Energy* (pp. 1–3). Altaeros Energies, Somerville, Massachusetts, USA (accessed on 30 July 2020).

Altaeros Inc., (2018). *The Super Tower ST 20* (pp. 1–12). Altaeros Inc., Somerville, Massachusetts, USA. http://www.altaeros.com/technology.html (accessed on 30 July 2020).

Altave Inc., (2019a). *Altave Omni* (pp. 1–4). http://www.altave.com.br/en/produtos/altave-omni/ (accessed on 30 July 2020).

Altave Inc., (2019b). *Innovative Impact Technologia* (pp. 1–3). http://www.altave.com.br/en/tecnologia/ (accessed on 30 July 2020).

Anderson, A., Lindell, C. A., Moxcey, K. M., Siemer, W. F., Linz, G. M., Curtiss, P. D., Caroll, J. E., et al. (2013). Bird damage to select fruit crops: The cost of damage and the benefits of control in five states. *Crop Protection, 52,*103–109.

Aurell, J., & Gullett, B. K., (2013). Emission factors from aerial and ground measurements of field and laboratory forest burns in the south eastern USA. Black and brown in carbon, VOC, and PCDD/PCDF. *Environmental Science and Technology, 47,*8443–8452.

Aurell, J., Gullett, B. K., & Tabor, J., (2015). Emissions from south eastern US grasslands and pine savannas: Comparison of ground and aerial measurements with laboratory burns. *Atmosphere and Environment, 111,*170–178.

Australian Parachute Federation Ltd., (2017). *The Early History of Parachuting* (pp. 1–7). Australian Parachuting Federation Ltd, Sydney, Australia, https://www.apf.com.au/APF-Zone/APF-Information/History-of-the-APF/Early-History-of-Parachuting/default.aspx/ (accessed on 30 July 2020).

Badesha, S. S., (2002). SPARCL: A high-altitude tethered balloon-based optical space-to-ground communication system. *Proceedings of the SPIE-The International Society for Optical Engineering, 4821,*181–193.

Bajoria, A., Mahto, N. K., Boppana, C. K., & Pant, R. S., (2017). Design of a tethered aerostat system for animal and bird hazard management. *First International Conference on recent advances in Aerospace Engineering (ICRAAE),* 1–12.

Bely, P., & Ashford, R. L., (1995). High-altitude aerostats as astronomical platforms. *Proceedings of SPIE – The International Society for Optical Engineering, 247*8,101–116.

Bird Control Systems Ltd., (2018). *Helikites: Bird Control Research* (pp. 1–7). https://www.birdcontrol.net/helikites-for-farms/ (accessed on 30 July 2020).

Bond, T. A., (2018). *Understanding the Effects of Cattle Grazing in English Chalk Streams* (p. 448). Department of Geography and Environment. University of Southampton, England, https://eprints.soton.ac.uk/350656/1/__userfiles.soton.ac.uk_Users_gc10g12_mydesktop_TrevorBondThesis_TrevorBondThesis.pdf/ (accessed on 30 July 2020).

Bonev, B., (1993). A joint soviet-Bulgarian scientific program for free-flight and tethered aerostat observations. *Advances in Space Research, 13,*131–133 https://doi.org/10.1016/0273-1177(93)90287-L (accessed on 30 July 2020).

Bonev, B., (1994). A program for Scientific and Applied investigations using aerostat and complexes. *Advances in Space Research, 14,*177–179.

Bouche, A., Beck-Winch, B., & Potosnak, J., (2016). A high altitude balloon platform for determining exchange of carbon di oxide over agricultural landscapes. *Atmospheric Measurements Techniques, 9,*5707.

Boulanger, Y., (2019). *Research at Laurantian Forestry Center: Spruce Budworm* (pp. 1–8). Healthy Forest Partnership http://www.healthyforestpartnership.ca/en/news/ (accessed on 30 July 2020).

Brooke, S., Lim, T. Y., & Adron, J., (2010). *Surveillance and Enforcement of Remote Maritime areas (SERMA): Surveillance Technical Options* (pp. 1–48). Marine conservation Biology Institute. https://marine-conservation.org/media/filer_public/2012/03/26/serma_tech-options_v13.pdf (accessed on 30 July 2020).

Bryce, J., (2018). *Classic themes: Our Solar System* (pp. 1–123). Moon Miners manifesto, http://www.MMM-MoonMinersManifesto.com (accessed on 30 July 2020).

Bull, L., & Phillips, N., (2018). *Towards the Design of Aerostat Wind Turbine Arrays through AI* (pp. 1–3). Department of Computer Science and Creative Technologies University of the West of England, Bristol, United Kingdom. https://arxiv.org/ftp/arxiv/papers/1811/1811.05290.pdf/ (accessed on 30 July 2020).

Cadogan, S., (2018). *Aerostats Emerge in US for Rural Broadband Delivery* (pp. 1, 2). https://www.irishexaminer.com/breakingnews/farming/aerostats-emerge-in-us-for-rural-broadband-delivery-838133.html (accessed on 30 July 2020).

CANR News, (2019). University of Delaware researchers track fall migratory patterns of insects. College of Agriculture and Natural Resources, University of Delaware, Delaware, USA. https://canr.udel.edu/page/9/?menu=home/ pp. 2, 3 (accessed on 30 July 2020).

Carlson, D. F., Ozgokmen, T., Novelli, G., Guigand, C., Chang, H., Fox-Kemer, B., Mensa, J., et al. (2018). Surface ocean dispersion observations from the ship-tethered aerostat remote sensing system. *Marine Science* (pp. 1–8). https://doi.org/10.3389/fmars.2018.00479/ (accessed on 30 July 2020).

Carolina Unmanned Aerial Vehicles Inc., (2019). *Surveillance and Communication Solutions for Military, Law Enforcement and Border Control* (pp. 1–12). https://carolinaunmanned.com/products/ (accessed on 30 July 2020).

Case, P., (2019). *Video: Fresh Bid to Grow Soya in UK Conditions* (pp. 1, 2). Farmers Weekly. www.fwi.co.uk/arable/video-fresh-bid-to-grow-soya-in-uk-condition/ (accessed on 30 July 2020).

Cassimally, K. A., (2012). *Forget Aeolians, We Need Airborne Wind Farms to Harness Maximum Wind Energy* (pp. 1–4). https://www.nature.com/scitable/blog/labcoat-life/forget_aeolians_we_need_airborne (accessed on 30 July 2020).

Cholan, R. R., (2017). *Nakshatra Aerostat (an Unmanned Aerostat filled with helium gas)* (pp. 1–3). https://www.iaspreparationonline.com/nakshatra-aerostat// (accessed on 30 July 2020).

Ciais, P., Sabine, C., Bala, G., Bopp, L., Brovkin, V., Canadell, J., Chhabra, A., et al. (2013). *Carbon and other Biogeochemical Cycles* (pp. 465–570). Cambridge University Press, Cambridge, UK and New York, USA, doi: 10.1017/CBO9781107415324.015.

CIMMYT, (2012). *Obregon Blimp Airborne and Eyeing Plots* (pp. 1–4). http://www.cimmyt.org/obregon-blimp-irborne-and-eyeing-plots/ (accessed on 30 July 2020).

CIMMYT, (2014). *An Aerial Remote Sensing Platform for High Throughput Phenotyping of Genetic Resources* (pp. 1–24). International center for Maize and Wheat, Mexico. www.slideshare.net/CIMMYT/an-aerial-remote-sensing-platform-for-high-throughput-phenotyping-of-genetic-resources/ (accessed on 30 July 2020).

Colazza, A., & Dulce, J. L., (2005). *High Altitude, Long Endurance Airships for Coastal Surveillance* (p. 22). National Aeronautics and Space agency, Washington D.C. USA.

https://ntrs.nasa.gov/archive/nasa/casi.ntrs.nasa.gov/20050080709.pdf (accessed on 30 July 2020).

Cortland Inc., (2019). *Security in the Cloud* (pp. 1–4). Cortland Company, Sao Paulo Brazil. https://www.cortlandcompany.com/sites/default/files/project/files/project-altave-aerostat-case-study-us.pdf (accessed on 30 July 2020).

Dasaradhan, B., Biswa, R. D., Sinha, R. K., Kumar, K., Kishore, B., & Eswara, P. N., (2018). Review of technology and materials for aerostat application. *Asian Journal of Textile, 8*, 1–17. doi: 10.3923/ajt.2018.1.12.

DefenseWeb, (2017). *CSIR Develops Active Aerostat* (pp. 1–4). Council of Scientific and Industrial Research, South Africa. https://www.defenseweb.co.za/joint/science-a-defense-technology/csir-develops-active-aerostat/ (accessed on 30 July 2020).

Department of Defense, (2016). *Lighter than Air Vehicles* (p. 117). Washington D.C. USA. https://apps.dtic.mil/dtic/tr/fulltext/u2/a568211.pdf (accessed on 30 July 2020).

Dick, A., & Robinson, D., (1992). *The Golden Age of Great Passenger Airships* (p. 226). Penguin Random House, New York, USA.

Digital Design and Imaging Service, Inc., (2019). *Simulation: Tethered Balloon Scanning Crops with Visual Spectrum Cameras* (pp. 1–5). https://airphotoslive.com/portfolio/aerial-precision-agriculture-support-services/ (accessed on 30 July 2020).

Dougherty, C. T., Flynn, E. S., Coleman, R. J., Sama, M., & Stombaugh, T. S., (2019). *Remote Sensing of Equine Bermuda Grass Pastures from a Helikite* (pp. 1–5). Department of Plant and Soil Science, University of Kentucky, Lexington, Kentucky, USA. http://www.helikites.com/images/2015/gis/remote_sensing_equine_bermudagrass_c_e_dougherty.pdf (accessed on 30 July 2020).

Droneco, (2019). *Helikites Aerostat Capabilities* (pp. 1–3). http://thedroneco.com.au/helikite-aerostat/ (accessed on 30 July 2020).

Drummond, K., (2005). *Battle Kites Eyed for Afghan Spy Duty* (pp. 1–3). Wired.com. https://www.wired.com/2012/05/helikite/ (March 15th, 2019)

Eggers, J., (2014). *NASA's Scientific Balloon Program* (pp. 1–12). National Aeronautics and Space Agency, Washington D.C. USA. http://sites.wff.nasa.gov/code820/index.html/ (accessed on 30 July 2020).

Ejiogu, A., & Okoli, V. B. N., (2012). Bird scaring technologies in rice production. *Global Advanced Research Journal of Food Technology*, 1, 31–38.

Elfes, A., Bueno, S. S., Bergerman, M., & Ramos, J. J. G., (1998). A semi-autonomous robotic airship for environmental monitoring missions. In: *Proceedings of the 1998 IEEE International Conference on Robotics and Automation* (p. 1). Leuven, Belgium, IEEE. https://ieeexplore.ieee.org/abstract/document/680971 (accessed on 30 July 2020).

Elfes, A., Bueno, S. S., Bergerman, M., Paiva, E. C. D., Ramos, J. G., & Azinheira, J. R., (2003). Robotic airships for exploration of planetary bodies with an atmosphere: Autonomy challenges. *Autonomous Robots, 14*,147–164.

Elfes, A., Campos, M. F. M., Bergerman, M., Bueno, S. S., & Podnar, G. W., (2000). A robotic unmanned aerial vehicle for environmental research and monitoring. In: *Proceedings of the First Scientific Conference on the Large-Scale Biosphere-Atmosphere Experiment in Amazonia (LBA)* (pp. 220–229). Belém, Pará, Brazil, LBA Central Office, CPTEC/INPE, Rod. Presidente Dutra, km 40,12630–000 CachoeiraPaulista, SP, Brazil.

Engineering 360, (2019). *Agricultural and Farming Products and Equipment Information* (pp. 1–4). https://www.globalspec.com/learnmore/specialized_industrial_products/agricultural_farming_products_services_equipment/ (accessed on 30 July 2020).

Environment Agency, (2007). Geomorphological guidelines for river restoration schemes Final report: 2007 Rio House Waterside Drive, Aztec West Almondsbury, Bristol BS32 4UD Tel: 0870 8506506 Email: enquiries@environment-agency.gov.uk www.environment-agency.gov.uk © Environment Agency. pp. 1–74 (accessed on 30 July 2020).

FAO (2001). Report of the national workshop on fisheries monitoring, control and surveillance in support of fisheries management Goa, India. *Vessel Monitoring Systems* (pp. 1–77). GCP/INT/648/NOR Field Report C-7. ftp://ftp.fao.org/docrep/fao/field/006/ad495e/ad495e10.pdf/ (accessed on 30 July 2020).

FAO of the United Nations, (2019). *FAO Advanced Tools and Technologies for Locust Monitoring and Early Warning* (pp. 1–3). http://www.fao.org/3/a-i5344e.pdf (accessed on 30 July 2020).

Fonstad, M., Dietrich, J., Courville, B., Jensen, J., & Charbonneau, P., (2013). Topographic structure from motion: A new development in photogrammetric measurement. *Earth Surface Processes and Landforms, 38,*421–430.

Gabbert, D., (2011). *An Evaluation of Image-Based Techniques for Wildfire Detection and Fuel Mapping* (pp. 36–37). Department of Aerospace Oklahoma State University, Stillwater, USA.

Gallagher, K. D., Mew, T. W., Borromeo, E., & Kenmore, P. E., (2002). Integrated pest management in rice. In: *International Rice Commission Newsletter* (Vol. 51, pp. 23–29).

GaskiteHelikite Ltd., (2019). *Gaskite Aerostat Bird Control* (pp. 1–3). GaskiteHelikite Ltd. Summerland, Canada, http://www.gaskite.com (accessed on 30 July 2020).

Gilbert, L., Teasdale, J. R., Kauffman, C., Davis, M., & Jawson, L., (2003). *Characteristics of Sustainable Farmers: Success in Mid-Atlantic* (p. 24). USDA-ARS Sustainable Agricultural laboratory, Beltsville, Maryland, USA.

Gimson, N. R., & Uliasz, M., (2003). The determination of agricultural methane emissions in New Zealand using receptor-oriented modeling techniques. *Atmosphere and Environment, 37,*3903–3912.

Gimson, N. R., Brailford, G., Bromely, T., Lassey, K., Martin, A., Moss, R., & Uliasz, M., (2015). *Determination of Methane Emission from Livestock at the Farm Scale Sing Observed Meteorological and Concentration Profiles and Inverse Modeling Techniques* (pp. 1–5). http://www.allsopphelikites.com/images/2015/gis/measuring_methane_emissions_niwa.pdf (accessed on 30 July 2020).

Glass, B., (2018). In: Steadler, T., (ed.) *Soaring 'Super Towers' Aim to Bring Mobile Broadband to Rural Areas* (pp. 1–7). ITU News https://news.itu.int/soaring-supertowers-aim-to-bring-broadband-to-rural-areas/ (accessed on 30 July 2020).

Global Aerostat Systems Market Analysis, (2015). *Global Aerostat Systems Market Analysis 2015-Size, Share, Growth, Trends and Forecasts from 2016 to 2022* (pp. 1–8). https://www.prnewswire.com/news-releases/global-aerostat-systems-market-analysis-2015---size-share-growth-trends-and-forecasts-to-2022-300119708.html (accessed on 30 July 2020).

Gonzalez, C., (2015). *Floating Wind Turbines Bring Electricity to Where it is Needed* (pp. 1–3). National Science Foundation, Washington D.C. https://phys.org/news/2015–02-turbines-electricity.html (accessed on 30 July 2020).

Goodyear Inc., (2018). *Wingfoot One* (pp. 1–5). https://www.goodyearblimp.com/behind-the-scenes/current-blimps.html (accessed on 30 July 2020).

Grace, D., (2013). *ABSOLUTE Aerial Base stations with Opportunistic Links for Unexpected or Temporary Events* (pp. 1–18). https://cordis.europa.eu/docs/projects/cnect/2/318632/080/

deliverables/001-FP7ICT20118318632ABSOLUTED23v10isa.pdf/ (accessed on 30 July 2020).

Guan, K., Berry, J. A., Zhang, Y., Joiner, J., Guanter, L., Badgley, G., & Lobell, D. B., (2016). Improving the monitoring of crop productivity using space borne solar-induced fluorescence. *Global Change Biology, 22,*716–726. doi: 10.1111/gcb.13136.

Gupta, A., (2018). *Global Aerostats System Market Size Study, by Product Type (Hybrid, Balloon, Airship), by Propulsion System (powered, unpowered), by Class (compact-sized, mid-sized, large sized), by Payload (Communication intelligence, Cameras, Electronic intelligence) and Regional forecasts, 2017–2025* (pp. 1–28). Ken Research. https://www.kenresearch.com/defense-and-security/defense/global-aerostats-system-market/177430–16.html (accessed on 30 July 2020).

Haire, B., (2004). *Blimp Helps Fine-Tune Irrigation* (pp. 1–6). University of Georgia Extension Service, Georgia, USA. http://grains.caes.uga.edu/news/story.html?storyid=2045&story=Infrared-cotton/ (accessed on 30 July 2020).

Haque, A. K. M. F., & Broom, D. M., (1984). In: *Habituation* (pp. 1–7). Vigilante Helikite. Bird Control Systems, Hampshire, United Kingdom. https://www.birdcontrol.net/installation-instructions-vigilante (accessed on 30 July 2020).

Harris, M., (2019). *Higher, Cheaper, Sleeker Wind Turbine of the Future-in Pictures* (pp. 1–3). https://www.theguardian.com/sustainable-business/gallery/2017/may/08/renewables-wind-energy-turbines-tech-kites-drones-in-pictures/ (accessed on 30 July 2020).

Hassan, G. L., (2011). *Market Report on High Altitude Wind Energy* (pp. 1–87). http://www.gl-garradhassan.com/en/highaltitudewind.php/ (accessed on 30 July 2020).

Helikite Hotspots Ltd., (2019). *Airborne Wi-Fi* (pp. 1–3). https://www.helikite-hotspot.com/about/ (accessed on 30 July 2020).

Hemanth, C. S., (2016). *Bengaluru Scientists Gift Farmers an Aerostat to Scare Away Pests' Bird* (p. 1). Hindustan Times. https://www.hindustantimes.com/india/bengaluru-scientists-gift-farmers-an-aerostat-to-scare-away-pests-birds/story-ANBDtscfcZOpO47sVJ3gMM.html/ (accessed on 30 July 2020).

Hernandez, R., (2015). *Where Agriculture and Tech Intersect* (pp. 1–4). Santacruz Tech Beat. http://www.santacruztechbeat.com/2015/04/18/agriculture-tech-intersect-2/ (accessed on 30 July 2020).

Heun, M. K., Jones, J. A., & Neck, K., (1998). Solar/infrared aerobots for exploring several planets. JPL New Technology report: Npo-20264, Jet Propulsion Laboratory (JPL), Pasadena, CA, USA. *Autonomous Robots, 14,*147–164.

Hindustan Times, (2018). *Uttarakhand Launches Aerostat Balloon Facility to Boost Internet Connectivity* (p. 1). http://www.hindustantimes.com/dehradun/to-boost-internet-connectivity-uttarakhand-cm-launches-aerostat-balloon-facility/storyJnGnP1K5iaXd2TYInUBLOO.html/ (accessed on 30 July 2020).

Hof, R., (2015). *Google's Project Loon Internet Balloons Coming to US-Eventually* (pp. 1–5). https://www.forbes.com/sites/roberthof/2015/06/02/googles-project-loon-internet-balloons-coming-to-u-s-eventually/#1f96875825d6/ (accessed on 30 July 2020).

Holder, A. L., Gullet, B. K., Urbanski, S. P., Elleman, R., O'Neill, S., Tabor, D., Mitchell, W., & Baler, K. R., (2017). Emissions from prescribe burning of agricultural fields in the Pacific Northwest. *Atmospheric Environment, 166,* 22–33.

Holder, A. L., Hagler, G. S., Aurell, J., Hays, M. D., & Gullet, B. K., (2016). Particulate matter and black carbon optical properties and emission factors from prescribed fires in the southwestern United States. *Journal of Geophysical Research and Atmospheric Sciences, 121,*3465–3483.

Holmes, R., (2013). *The Drones of the Civil War: Meet the Hot Air Balloonist Who Convinced Abraham Lincoln to Use Aerial Reconnaissance* (pp. 1–7). https://slate.com/news-and-politics/2013/11/civil-war-hot-air-balloons-thaddeus-lowe-and-union-aerial-reconnaissance.html/ (accessed on 30 July 2020).

Homeland Security, (2013). *Tethered Aerostat System: Application Note* (pp. 1–18). SAVER. Us Department of Homeland Security's Science and Technology Directorate, Washington D.C. https://www.rkb.us/saver/ (accessed on 30 July 2020).

Hosakoti, S., (2019). *Aerostat Based Hybrid Power Generation* (pp. 1–12). https://contest.techbriefs.com/2018/entries/sustainable-technologies/9218–0702–184543-aerostat-based-hybrid-power-generation/ (accessed on 30 July 2020).

Howard, A. J. G., (2007). *Experimental Characterization and Simulation off a Tethered Aerostat with Controllable Tail Fins* (p. 124). Department of Mechanical Engineering, McGill University, Montreal, Canada, http://digitool.library.mcgill.ca/webclient/StreamGate?folder_id=0&dvs=1552406509939~264/ (accessed on 30 July 2020).

Inman, M., (2012). *Energy High in the Sky: Expert Perspectives on Airborne Wind Energy Systems* (pp. 1–15). Near Zero.) (accessed on 30 July 2020).

Jensen, T., Apen, A., Young, F., & Zeller, F., (2007). Detecting the attributes of wheat crop using digital imagery acquired from a low altitude platform. *Computers and Electronics in Agriculture, 59,* 66–77.

Kirschner, E. J., (1986). *Aerospace Balloons: From Montgolfiere to Space* (p. 21) TAB Books, Inc., Blue Ridge Summit, Pennsylvania, USA.

Klemas, V. V., (2010). Tracking oil slicks and predicting their trajectories using remote sensors and models. Case studies of the Sea princess and Deepwater Horizon oil spills. *Journal of Coastal Research, 26,*789–797.

Klemas, V. V., (2013). Airborne remote sensing of coastal features and processes. *Journal of Coastal Research, 29,*239–255.

Klemas, V. V., (2015). Coastal and environmental remote sensing from unmanned aerial vehicles: An overview. *Journal of Coastal Research, 31,*1260–1267 https://doi.org/10.2112/JCOASTRES-D-15–00005.1 (accessed on 30 July 2020).

Kostyrko, M., Simons, K., Smagur, E., & Hans, K., (2015). Low altitude aerial photography for archaeological studies-where is revolution leading us. *Archaeogis.,* 1–19 http://caaaconference.org/program (accessed on 30 July 2020).

Krausmann, F., Erb, K. H., Gingrich, S., Haberl, H., Bondeau, A., Gaube, V., Lauk, C., et al. (2016). Global human appropriation of net primary production doubled in the 20th century, *Proceedings of National Academy of Science. USA, 110,*10324–10329. doi: 10.1073/pnas.1211349110.

Krishna, K. R., (2012). *Precision Farming: Soil Fertility and Productivity Aspects* (p. 189). Apple Academic Press Inc., Waretown, New Jersey, USA.

Krishna, K. R., (2016). *Push Button Agriculture: Robotics, Satellite Guided Soil and Crop Management* (pp. 261–388). Apple Academic Press Inc., Waretown, New Jersey, USA.

Krishna, K. R., (2018). *Agricultural Drones: A Peaceful Purpose* (pp. 153–196). Apple Academic Press Inc., Waretown, New Jersey, USA.

Krishna, K. R., (2019). *Unmanned Aerial Vehicle Systems in Crop Production: A Compendium* (pp. 610–614). Apple Academic Press Inc., Waretown, New Jersey, USA.

Kushida, K., Yoshina, K., Nagano, K., & Ishida, T., (2009). Automated 3D forest surface model extraction from balloon stereo photographs. *Photogrammetric Engineering and Remote Sensing, 75,* 25–35.

Lambert C., (2018). *The Power of Aerostats* (pp. 1–8). Sky Sentry LLC https://aerospaceamerica. aiaa.org/departments/the-power-of-aerostats/ (accessed on 30 July 2020).

Lassey, K. R., Ulyatt, M. J., Martin, R. J., Walker, C. F., & Shelton, I. D., (1997). Methane emissions measured directly from grazing livestock in New Zealand. *Atmospheric Environment, 31,*2905–2914. https://doi.org/10.1016/S1352–2310(97)00123–4 (accessed on 30 July 2020).

LIDAR USA, (2019). *Topographic Mapping* (pp. 1–3). https://www.lidarusa.com/ surveyingmapping.html/ (accessed on 30 July 2020).

Lindstrand Technologies, (2019). *Highflyer Tethered Balloons* (pp. 1–5). Lind Strand Technologies USA Inc., South Boston, VA, USA.

Liu, X., Zhang, Y., Huey, L., Yokelson, R., Wang, Y., Jimenez, J., Campuzano-Jost, P., et al. (2016). Agricultural fires in the south eastern US: Emissions of trace gases and particulates and evolution of ozone, reactive nitrogen and organic aerosol. *Journal of Geophysical Research and Atmosphere, 121,* 7383–7414.

Livson, B., (2016). *Aerostats All Australia Mobile Coverage* (p. 65). Canberra, Australia, http://www.bal.com.au/AAA.pdf (accessed on 30 July 2020).

Lockheed Martin Inc., (2018). *Hybrid Airship* (pp. 1–5) https://www.lockheedmartin.com/ en-us/products/hybrid-airship.html (accessed on 30 July 2020).

Lowe, D. C., Brenninkmeijer, C. A. M., Brailsford, G. W., Lassey, K. R., & Gomez, A. J., (1994). Concentration and 13C records of atmospheric methane in New Zealand and Antarctica: Evidence for changes in methane sources. *Journal of Geophysical Research, 99,* 16913–16925.

Markets and Markets, (2018). *Aerostat Systems Market-Global Forecast to 2021* (pp. 1–74) https://www.marketsandmarkets.com/Market-Reports/aerostat-systems-market-193315635.html (accessed on 30 July 2020).

Marvel, K., Kravitz, B., & Caldeira, K., (2012). Geophysical limits to global wind power. *Nature Climate Change,* 22–23. doi: 10.1038/nclimate1683.

Matsuzaki, Y., & Fujita, I., (2017). *In situ* estimates of horizontal turbulent diffusivity at the sea surface for oil transport simulation. *Marine Pollution Bulletin, 117,* 34–40. doi: 10.1016/j.marpolbul.2016.10.026.

McGonegal, J., (2013). *Two Alums Dream of Tethered Aerostat Wind Farms.* Massachusetts Institute of Technology, Cambridge, Massachusetts, USA https://alum.mit.edu/slice/ two-alums-dream-tethered-aerostat-wind-farms (accessed on 30 July 2020).

Mijatovic, A., (2017). *Compact Aerial System: Specifications* (pp. 1, 2). Aero Drum Ltd. Technical Specifications. http://www.rc-zeppelin.com/compact-aerial-photography-systems.html (accessed on 30 July 2020).

Milkert, J., (2014). *Delft Aerospace Design Projects 2014: New Design in Aeronautics, Astronautics and Wind Energy* (pp. 326–328). TU Delft: Delft University of Technology, Delft, Netherlands.

Mines, F., (2016). *Draft history of Parachuting in Australia A draft history of Parachuting: Up to the Foundation of Sport Parachuting in 1958* (pp. 1–23). https://www.apf.com. au/APF-Zone/APF-Information/History-of-the-APF/Draft-History-of-Parachuting-in-Australia/default.aspx (accessed on 30 July 2020).

Missouri Alternative's Center, (2019). *Sustainable Agriculture Conference* (pp. 1–4). Missouri Alternatives Center's Electronic Newsletter. http://agebb.missouri.edu/mac/agopp/arc/agopp018.txt (accessed on 30 July 2020).

Missy, A., (2018). *Aerostats Ideal for Maritime Surveillance* (pp. 1, 2). TCOM, L. P. http://www.tcomlp.com/?attachment_id=rxjjdicjy/ (accessed on 30 July 2020).

Miyazo, Y., & Isobe, A., (2016). A combined balloon photography and buoy-tracking experiment for mapping surface currents in coastal waters. *J. Atmospheric and Oceanic Technology, 33,* 1237–1250. doi: 10.1175/JTECH-D-15–0113.1.

Myers, J. W., (1978). Balloon survey field season-1977. *Journal of Field Archaeology,* 5, 145–159.

Nandi, S., Hansda, T., Himangshu, H., & Pa,T., (2017). Geographic information in water engineering. *International Journal of Engineering Research 5,* 210–214

NASA Balloon Program Office, (2017). Scientific Balloons (pp. 1–7). National Aeronautics and Space Agency, Washington D.C. USA. https://sites.wff.nasa.gov/code820/ (accessed on 30 July 2020).

National Geographic, (2019). The American Prairie Reserve. A Garden of Conservation Technology. https://openexplorer.nationalgeographic.com/expedition/conservationtechnologygarden/view/6004729 (accessed on 30 July 2020).

National Instruments, (2019). *Delivering Services from the Sky with USRP and Lab VIEW Communications* (pp. 1, 2). http://sine.ni.com/cs/app/doc/p/id/cs-17702 (accessed on 30 July 2020).

Noli, D., (1985). Low altitude aerial photography from a tethered balloon. *Journal of Field Archaeology, 12,* 497–501.

Ozbay, G., Fan, C., & Yang, Z., (2017). *Relationship between Land Use and Water Quality and its Assessment using Hyper Spectral Remote Sensing in Mid-Atlantic Estuaries* (pp. 23–29). https://www.intechopen.com/books/water-quality/relationship-between-land-use-and-water-quality-and-its-assessment-using-hyperspectral-remote-sensin/ (accessed on 30 July 2020).

Pantuso, M., (2019). *Aeros-3200 Aerostat* (pp. 1–4). Aeroscraft Corporation, California, USA. www.aeroscraft.com/ (February 24th, 2019)

Parcak, S. H., (2016). *Satellite Imagery for Archaeology* (p. 237). https://archive.org/stream/SatelliteRemoteSensingForArchaeologyArch/0415448778archaeologyC_djvu.txt/ (accessed on 30 July 2020).

Perigrine Ltd., (2018). *Light Weight Helikites* (36 inch) (pp. 1–3). https://www.peregrinehawkkites.com/lightweight-helikite-instructions/ (February 18th, 2019)

Persistent Surveillance Services, (2019). *PS2 Surveillance Services 2019 GIS, Geomatics, Surveying and Inspection Helikites* (pp. 1–6) https://persistentsurveillanceservices.com/products/ (accessed on 30 July 2020).

Pocs, M., (2014). *A High-Altitude Balloon Platform for Exploring the Terrestrial Carbon Cycle* (Vol. 3, No. 2, pp. 1–16). DePaul Discoveries. http://via.library.depaul.edu/depaul-disc/vol3/iss1/2/ (accessed on 30 July 2020).

Poeplau, C., & Don, A., (2015). Carbon sequestration in agricultural soils via cultivation of cover crops-A meta-analysis. *Agriculture, Ecosystem and Environment, 200,* 33–41. doi: 10.1016/j.agee.2014.10.024. pp. 1–4.

Quets, J. J. El-Bana, M. I., Al-Rowaily, S. L., Assaeed, A. M., Temmerman, S., & Nijs, I., (2016). A mechanism of self-organization in a desert with phytogenic mounds. *Ecosphere* 7(11), e01494. 10.1002/ecs2.1494.

Raina, A. A., Gawale, A. C., & Rajkumar, S., (2017). *Design, Fabrication and Field Testing of Aerostat System* (pp. 1–248). National Seminar on Strategic application of lighter-than-air (LTA) vehicles at higher altitudes, Snow and Avalanche study establishment, Manali, Shimla, India.

Rajani, A., Pant, R. S., & Sudhakar, K., (2010). Dynamic stability analysis of a tethered aerostat. *Journal of Aircrafts, 47,* 1531–1538.

Ramos, J. J. G., Paiva, E. C., Maeta, S. M., Mirisola, L. G. B., Azinheira, J. R., Faria, B. G., Bueno, S. S., et al. (2000). Project Aurora: A status report. In: *Proceedings of the 3ʳᵈ International Airship Convention and Exhibition (IACE 2000)* (pp. 174–194). Friedrichshafen, Germany, The Airship Association: UK.

Rassinimalicki, (2019). *For Better Forestry and Agriculture* (pp. 1–12). Simosol Finland https://www.simosol.fi/?gclid=EAIaIQobChMIq5vSx42B4QIVkjUrCh0tDQnBEAAYBC AAEgLGcPD_BwE/ (accessed on 30 July 2020).

Raytheon, (2019). *JLENS: Joint Land attack Cruise Missile Defense Elevated Netted Sensor System* (pp. 1–6). Raytheon company, Waltham, Massachusetts, USA, https://www.raytheon.com/capabilities/products/jlens/ (accessed on 30 July 2020).

Rebecca, R., (2016). *On the Potential of Small UAS for Multispectral Remote Sensing in Large-Scale Agricultural and Archaeological Applications* (p. 327). University of Trier, Trier, Germany.

Redi, S., Aglietti, G. S., Tatnall, A. R., & Markvart, T., (2010). An evaluation of a high altitude solar radiation platform. *ASME-Journal of Solar Energy Engineering* (pp. 1–14) in print. https://www.researchgate.net/profile/Guglielmo_Aglietti/publication/221907407_ Aerostat_for_Solar_Power_Generation/links/0fcfd50c0580b4be07000000.pdf (accessed on 30 July 2020).

Research and Markets, (2018). *The Aerostat systems-Global Market Outlook 2017–2026* (pp. 1–68). ResearchandMarkets.com (accessed on 30 July 2020).

Ria-Novosti, (2019). *Russian Firm Builds Another Aerostat for China* (pp. 1–3). https://defense.pk/pdf/threads/china-buys-aerostat-from-russia.99699// (accessed on 30 July 2020).

Richardson, L., & Stommel, H., (1948). Note on eddy diffusion in the sea. *J. Meteorology, 5,* 238–240.

Riley, D. N., (1946). The technique of air-archaeology. *Archaeology Journal, 101,* 1–16.

Ritchie, G. L., Sullican, D. J., Perry, G. B., Rook, T. J., & Bednarz, C. W., (2008). Preparation of a low-cost digital camera system for remote sensing. *Applications of Engineering in Agriculture, 24,* 886–896.

Robinson, D. H., (1973). *Giants in the Sky: A History of Rigid Airships* (p. 400). University of Washington Press.

Rogers, M. E., (2001). *Helikite Elevated Platform (HEP) Final Report HEP phase 1* (pp. 1–39). Small business Innovative research (SBIR) topic A00–134. Carolina Unmanned Aerial Vehicle. Raleigh, North Carolina, USA. https://apps.dtic.mil/dtic/tr/fulltext/u2/a392293.pdf (accessed on 30 July 2020).

RT-Russian TV, (2019). *Prominent Russians: Dimitry Mendeleev* (pp. 1–15). Russiapedia. https://russiapedia.rt.com/prominent-russians/science-and-technology/dmitry-mendeleev/ accessed on 30 July 2020).

Santilli, F., Azara, S., Gorrei, L., & Perfetti, A., (2007). Evaluation of an aerial scaring device for birds' damage prevention to agricultural crops. *Proceedings of Conference: XIV ConvegnoItaliano Di Ornotologia* (pp. 14–146). doi: 10.4081/rio.2012.139.

Sara, G. B., (2013). Comparison of topographic surveying techniques in streams. Utah State University, Salt Lake City. *All Graduate Theses and Dissertations. 1516*, (p. 166) https://digitalcommons.usu.edu/etd/1516/ (accessed on 30 July 2020).

Seamans, T. W., Blackwell, B. F., & Gansowski, J. T., (2002). Evaluation of the allsopp helikite as bird scarer. *Proceedings of the Vertebrate Pest Conference* (pp. 23–27)

Shakatreh, H., Sawalmeh, A., Al-Fuquha, A., Dou, Z., Almaita, A., Khallel, A., Shamsaiha, O., et al. (2018). *Unmanned Aerial Vehicles: A Survey on Civil Application and Key Research Challenges* (p. 58). https://arxiv.org/pdf/1805.00881.pdf (accessed on 30 July 2020).

Sharma, P., (2017). *Precision Farming* (p. 256). Gene-Tech Books, New Delhi, India.

Shaw, J. A., Nugent, P. W., Kaufman, N. A., Pust, N. J., Mikes, D., & Knighton, W. B., (2012). Multi-spectral imaging systems on tethered balloons for optical remote sensing education and research. *Journal of Applied Remote Sensing, 6*, 1–17 doi: 10.1117/1. JRS.6.063613.

Shephard News Team, (2018). *Sky Star Aerostats for Central American Customer* (pp. 1, 2). https://www.shephardmedia.com/news/uv-online/rt-skystar-aerostats-central-american-customer/ (accessed on 30 July 2020).

Shirota, Y., Masaki, S., & Sanada, M., (1983). *Eye Spotted Balloons as a Device to Scare Grey Starlings* (p. 23). Japanese Ministry of Agriculture Contribution No 57–1228. doi: 10.1303/aez.18.545.

Shukla, K., (2018). *How Aerostats Enable Geo-Intelligence to Prevent Threats* (pp. 1–3). Geo-Spatial World. https://www.geospatialworld.net/ (accessed on 30 July 2020).

SkyDoc Systems INC., (2019). *Home of the Category-2 Hurricane Proof Aerostat* (p. 2). http://www.skydocballoon.com/ (accessed on 30 July 2020).

Spirig, C., Guenther, A., Greenberg, J. P., Calanca, P., & Tarvain, V., (2004). Tethered balloon measurements of volatile organic acids a boreal site. *Atmospheric Chemistry and Physics, 4*, 215–219.

Spring, A., Domingue, A., Mooney, A., Kerber, T., Lemmer, K., & Docherty, K., (2018). Development of a novel method for temporal analysis of airborne microbial communities of airborne microbial communities. *Midwest Ecology and Evolution Conference MEEC 2018 38th Annual Meeting* (pp. 1–17). Kellogg Biological Station Michigan State University. https://meec2018.files.wordpress.com/2018/04/program_meec2018–1.pdf (accessed on 30 July 2020).

Sputnik News, (2019). *Russian Aerostats Radar to be the Backbone of Indian Border Defense* (pp. 1–3). Sputnik News: Military and Intelligence. https://sputniknews.com/military/201703031051241800-russian-aerostats-radar-india-border-defense/ (accessed on 30 July 2020).

Staff Writers, (2019). *Airborne Response Supports Fire and Rescue Exercise with Drones and Aerostats* (pp. 1–5). http://www.spacedaily.com/reports/Airborne_Response_supports_National_Guard_and_Fire_Rescue_exercise_with_drones_and_aerostats_999.html/ (accessed on 30 July 2020).

Stafford, J. V., (2005). *Precision Agriculture 05* (p. 105). Precision Agriculture. Wageningen Academic Publishers, Wageningen. The Netherlands.

Steadler, T., (2018). *Soaring 'Super Towers' Aim to Bring Mobile Broadband to Rural Areas* (pp. 1–7). ITU News https://news.itu.int/soaring-supertowers-aim-to-bring-mobile-broadband-to-rural-areas/ (accessed on 30 July 2020).

Strand, T. G., Gullet, B., Urbanski, S., O'Neill, S., Potter, D., Aurell, J., Holder, A., et al. (2016). Grassland and forest under storey biomass emissions from prescribed fires in the

Southeastern United States. *International Journal of Wild land Fire, 25,*102–113 http://dx.doi.org/10.1071/WF14166/ (accessed on 30 July 2020).

Sweney, M., (2019). *EE Balloons and Drones to Help Fix Mobile Black Spots.* https://www.theguardian.com/business/2017/feb/21/ee-balloons-drones-blackspots-helikites-coverage (accessed on 30 July 2020).

Szpakowski, D., (2016). *Estimation of above Ground Biomass of Pasture Environments Using Structure from Motion* (pp. 1–60). Texas State Universities, College Station, Texas, USA, MS Thesis.

Tattaris, M., (2014). *Researcher Helps Remote Sensing Soar* (pp. 1–5). Lighter than Air Society. http://www.blimpinfo.com/airships/blimps/researcher-helps-remote-sensing-soar/ (accessed on 30 July 2020).

Tattaris, M., (2015). *Obregon Blimp Airborne and Eyeing Plots* (pp. 1–3). CIMMYT, Mexico, https://www.cimmyt.org/news/obregon-blimp-airborne-and-eyeing-plots/ (accessed on 30 July 2020).

TCOM, (2019). *12 M Tactical class Aerostat System* (pp. 1–4). https://www.youtube.com/watch?v=61bHs6rKFZ8/ (accessed on 30 July 2020).)

The Christian Science Monitor, (2018). *World's Largest Inflatable Vehicle- Bullet 58 Airship-Inflated and Ready for Action* (pp. 1–6). https://www.csmonitor.com/About/Contact (accessed on 30 July 2020).

The Editor, (2012). *Delivery of Small Tactical Multi-Payload Aerostat System (STMPAS)* (pp. 1–4). UAS Vision. https://www.uasvision.com/2012/07/18/delivery-of-small-tactical-multi-payload-aerostat-system-stmpas/ (accessed on 30 July 2020).

Trumble, S., Kasai, N., & Amanning, A., (2018). *The State of the Southern Border* (pp. 1–7). https://www.thirdway.org/memo/the-state-of-the-southern-border/ (accessed on 30 July 2020).

US Customs and Border Protection, (2013). *CBP Assumes Operational Control of Tethered Aerostat Radar Systems (TARS)* (pp. 1–3). https://www.cbp.gov/newsroom/national-media-release/cbp-assumes-operational-control-tethered-aerostat-radar-systems-tars# (accessed on 30 July 2020).

US Customs and Border Protection, (2014). *CBP Completes Upgrade of Aerostat Surveillance System* (pp. 1–4). https://www.cbp.gov/newsroom/national-media-release/cbp-completes-upgrade-aerostat-surveillance-system (accessed on 30 July 2020).

US Department of Homeland Security-Science and Technology Directorate, (2018). *Tethered Aerostat System Application Note* (pp. 1–18). https://www.dhs.gov/sites/default/files/publications/TetheredAerostat_AppN_0913–508.pdf (accessed on 30 July 2020).

US Embassy in Philippines, (2014). *Radar System to Philippines Navy* (pp. 1–3). https://ph.usembassy.gov/us-delivers-new-aerostat-radar-system-philippine-navy/ (accessed on 30 July 2020).

Valour Consultancy, (2019). *Our Fine Tethered Friends* (pp. 1–3). https://www.valourconsultancy.com/fine-tethered-friends/ (accessed on 30 July 2020).

Vasconcelos, Y., (2019). *Balloons for Internet Access* (Vol. 257, pp. 73–76.). Research Reports of Foundation for Research Support of the State of São Paulo (FAPESP).

Verhoeven, G. J. J., & Loenders, J., (2006). Looking through black-tinted glasses—a remotely controlled infra-red eye in the sky. In: *2ⁿᵈ International Conference, Remote Sensing. Archaeological, Proc. 2ⁿᵈ Int. Workshop—From Space to Place* (pp. 73–79) Rome, Italy.

Verhoeven, G. J. J., (2007). Becoming a NIR-sensitive aerial archaeologist. In: *Proc. SPIE—Remote Sensing for Agriculture, Ecosystems, and Hydrology IX* (Vol. 6742, pp. 67420Y-1–67420Y-13) Florence, Italy.

Verhoeven, G. J. J., (2012). Near Infrared aerial 'crop mark' archaeology: From its historical use to current digital implementation. *Journal of Archaeological Methods and Theory, 19,* 132–160.

Verhoeven, G. J. J., Loenders, J., Vermeulen, F., & Docter, R., (1994). Helikite aerial photography or HAP—a versatile means of unmanned, radio controlled low altitude aerial archaeology. *Archaeological Prospecting, 16,* 125–138.

Verhoeven, G. J., Smet, P. J., Poelman, D., & Vermeulen, F., (2009). Spectral characterization of a digital still cameras NIR modification to enhance archaeological observations. *IEEE Transactions on Geoscience and Remote Sensing, 47,* 3456–3468.

Vierling, L. A., Fersdahl, M., Chen, K., Li, Z., & Zimmerman, P., (2006). The Short-Wave Aerostat-Mounted Imager (SWAMI): A novel platform for acquiring remotely sensed data from a tethered balloon. *Remote Sensing and Environment, 103,* 255–264.

Vigilance, (2019). *Aerostat Systems* (pp. 1–3). Vigilance Group B. B. Netherlands. https://www.vigilance.nl/aerostat-systems.html (accessed on 30 July 2020).

Visiongain, (2019). *Military Aerostats Market Research Report-2016–2026* (pp. 165) https://www.visiongain.com/report/military-aerostats-market-report-2016–2026/ (accessed on 30 July 2020).

Wall, M. D., (2010). *Titan from Above: Blimp Survey Options to Study Saturn's Moon* (pp. 1–8). National Aeronautics and Space Agency. https://www.space.com/10544-titan-blimp-survey-options-study-saturn-moon.html (accessed on 30 July 2020).

Whatcom Co., (2008). *A Sampling of Agricultural Bird Scarer Tactics* (pp. 1–8). http://www.nwberryfoundation.org/raptors/raptor_pdf/bird-control-alternatives1.pdf (accessed on 30 July 2020).

White, J. W., Sanchez, P. A., Gore, M. A., & Bronson, K. F., (2012). Field based phenomics for plant genetics research. *Field Crops Research, 133,* 101–112.

White, S. M., & Madsen, E. A., (2016). Tracking tidal inundation in a coastal salt marsh with Helikiteairphotos: Influence of hydrology on ecological zonation at Crab Haul Creek, South Carolina. *Remote Sensing and Environment, 184,* 605–614. https://doi.org/10.1016/j.rse.2016.08.005 (accessed on 30 July 2020).

Whittelesey, J. H., (1970). Tethered balloon for archaeological photos. *Photogrammetric Engineering* (pp. 181–186). https://www.asprs.org/wp-content/uploads/pers/1970journal/feb/1970_feb_181–186.pdf (accessed on 30 July 2020).

Wikipedia, (2019). *High Altitude Wind Power* (pp. 1–14). https://en.wikipedia.org/wiki/High-altitude_wind_power (accessed on 30 July 2020).

Zhang, Q., (2015). *Precision Agriculture Technology for Crop Farming* (p. 360). CRC Press, Boca Raton, Florida, USA.

CHAPTER 5

Kites in Agrarian Regions

ABSTRACT

The emphasis of this chapter is on kites and their potential as providers of aerial photography of farms, crops and natural resources that are essential to farming companies and individual farmers. Reports dealing with usage of kites in farming sector are indeed feeble, if any. Firstly, historical aspects of kites and their usage by humans has been described. A table specifically summarizes and provides a timeline of events related to kites beginning with 200 B.C and up to 2017. History of aerial photography itself begins with first such effort using kites, by the Frenchman Arthur Batut in 1888. Several other historically important events like development and spread of different types of kites, such as the Rokkaku, Eddy, Box, Diamond, Flow farms, Spar-less parafoil kites, power kites, etc. have been mentioned. Also, use of kites in military conquests has been mentioned. Discussion encompass topics such as role of different types of kites in understanding the geo-morphological aspects. It includes collecting detailed spectral images of terrain, topography of land, soil types, natural vegetation and cropping patterns, water resources, irrigation systems, crops' nutrient (crop – N) and water status (crop's water status index). The potential of kites as providers of power through generation of wind power has been stressed as an aspect of good value to farming sector. Majority of reports about kites in farming, at present, agree that kites (including helikites) are of excellent value during bird scaring. Farms depend on gas bangers which are still not totally efficient in bird scaring. However, integrating helikites/kites with bird scaring sounds as bird scarers seems to offer better protection to crops, at heading and grain maturity. Several examples have been discussed where kite-aided bird scaring is effective. It is environmentally very safe because this method does not employ harmful pesticides, nor does it harm or kill bird species. Kites have played a role in offering aerial images relevant to ecological aspects and wide range of geographical regions, such as the arctic tundra, iceberg morphology, sandy low vegetation zones of arid deserts in Arabia, Sahelian agricultural belts, coastal vegetation and species diversity of plants, vegetational changes in response to development of infrastructure, etc.

5.1 INTRODUCTION

5.1.1 HISTORICAL ASPECTS OF KITES

Historical records available till data indicate that kites were invented in Asia. Chinese flew kites during 5th century B. C. Kites have been utilized during the military campaigns. For example, ancient records suggest that Korean generals had used kites with burning tails to scare away enemies. Chines had utilized kites with incendiaries during military campaigns in the medieval period. *Russian* records state that Prince Oleg of Novgorod used kites during the siege of Constantinople in the year 906 AD (American Kitefliers Association, 2019; Wikipedia, 2019b). Paintings from medieval period (1635) indicate use of kites as incendiaries. There are several paintings dating between 1700s to 1900s that depict kites flown for recreation, to measure distances and even for military purposes (Wikipedia, 2019b; Needham, 1965; see Table 5.1). Historically, kites were utilized for recreation and during traditional functions. Their usage for aerial photography and spectral analysis is comparatively recent.

The earliest of the kite aerial photographs were obtained by British meteorologist Archibald in 1887 and Arthur Batut in Labruguière (France), in 1888 (Benton, 2010; Archibald, 1887; KIPENED 10,2019; Marzolff, 2014). Arthur Batut took the first photographs above a place called Labraguire in France. The camera for aerial photography was held close to the kite. An altimeter was also mounted which recorded the altitude when the exposure was done. A fuse activated the shutter for photography (Kite Aerial Photography, 2019). During the same period, manned kite flying and aerial photography were proposed by Frenchman Marcel Maillot, Britisher Robert Baden-Powell, Americans Charles Lamson and William Abner Eddy, Australian Lawrence Hargrave (inventor of box kite in 1893 and kite train in, 1884) and French Captain Saconney. An Australian, Lawrence Hargrave invented "train kite' in 1884 and 'box kite' in 1893 (American Kitefliers Association, 2019). Within the realm of kite aerial photography, kite-aided aerial image of San Francisco obtained after the 1906 earthquake is famous.

TABLE 5.1 A Timeline of Events Related to Development and Usage of Kites

Year or Period	Historical events related to Kites	References
220 BC–206 BC	Kites were utilized by Chinese during the period of Han Dynasty	American Kitefliers Association, 2019
595 AD–673 AD	Kite flying spread to different regions in China, Korea and India. During Silla dynasty, it was used to carry fire balls into air	American Kitefliers Association, 2019
600 AD–700 AD	Evidences show that kites were introduced into Japan by the travelling Buddhist monks	American Kitefliers Association, 2019
1200 AD–1300 AD	Kites of different types were transported into various regions during Marco Polo's travels (13th century) surveillance, and bird scaring. Europeans were regularly flying the dragon or pennant shaped kites during medieval period. Romans used kites to project military banners	American Kitefliers Association, 2019
1400 AD–1500 AD	Malaysian historical records indicate that kites were used in plenty. Special type of tethers made of thread, gum and glass pieces were used during kite competitions in India and Malaysia	American Kitefliers Association, 2019
1483 AD–1530 AD	Paintings from Moghul period clearly show that diamond kites were in vogue during medieval period in Northern plains of India. Kites were also utilized to send messages to people inside the citadels and havelis	American Kitefliers Association, 2019
1500 AD–1700 AD	Dutch seafarers picked 'Diamond kite' from India and Malaysia and introduced it to Western Europe.	American Kitefliers Association, 2019
1603 AD–1868 AD	Japanese used kites as talismans. They were popular during the Edo period (i.e. 17th century AD) During 1712, kites were in vogue in Nagoya, Japan	American Kitefliers Association, 2019
1635 AD	A book titled 'Mysteries of Nature and Art' describes about Incendiary kites. Such kites are termed 'Fire Drakes'. It describes the methodology to prepare 'Fire drakes'.	Wikipedia 2019b

TABLE 5.1 (Continued)

Year or Period	Historical events related to Kites	References
1700 AD–1900 A.D	Kites were also used for scientific experimentation and related aspects.	American Kitefliers Association, 2019
	Benjamin Franklin and Alexander Wilson used kites to obtain greater understanding about static electricity. Several others such as Sir George Clayey, Samuel Langley, Lawrence Hargrave and Wright brothers designed new types of kites.	
1749 AD	Earliest recorded evidence about use of kites in experiments related to ambient weather were available from University of Glasgow, in Scotland. Two students, Alexander Wilson and Thomas Melville flew hexagonal kites and thermometer to detect, if the air above ground was cooler or warmer than on the ground surface.	Robinson, 2003;
1752 AD	Benjamin Franklin conducts an experiment using a kite and its tether, to prove the existence of Electricity in the lightning	Wikipedia, 2019b
1822–23 AD	Sir William Edward Parry and Reverend George Fisher studied ambient air temperature at different altitudes, using kites mounted with thermometers. They recorded air temperature at a place known as Igloolik situated in Northern American icy regions	Robinson, 2003
1822 AD	George Pocock who was a schoolteacher in Britain used multiple kites tied to carriage, to pull the van easily. It was a horseless vehicle. It marked the beginning of power kites.	American Kitefliers Association, 2019
1884 AD–1893 AD	Two different types of kites namely 'Box kite' (in 1884) and 'Train kite' (in 1893) were invented by an Australian named Lawrence Hargrave	Duffy and Anderson, 2016; Crouch, 2010; Robinson, 2002
1847 AD	Six-sided meteorological kites were used frequently by the meteorological office at Kew gardens in Kent, England	Robinson, 2003
1860 AD–1910 AD	This period is referred as 'Golden Age of Kites.' Kites were being used for meteorology, aeronautics, wireless communication and aerial photography. The interest in kite aerial photography and kite usage per se for scientific expedition reduced enormously after Wright Brothers invented airplane in 1903.	NASA, 2016
1870s AD	Samuel Franklin Cody designed first ever man-lifter Cody kite	Wikipedia, 2019b

Year or Period	Historical events related to Kites	References
1880s–1920 AD	Kites were used to lift cameras to conduct aerial survey at Blue Hill Observatory, Milton, Massachusetts, USA. Equipment related to recording weather parameters too were lifted by kites.	Digital Earth Watch, 2017; Marvin, 1897
	The data was used to prepare 24 hr charts of temperature, humidity altitude/air pressure in the atmosphere. Later, Eddy kites and Box kites with meteorographs became common.	
	Kites were also used study the phenomenon of static electricity in the air.	
1887 AD–1888 AD	Earliest kite aerial photography was conducted by a Frenchman Arthur Batut in 1888 and British meteorologist Douglas Archibald	Marzolff, 2014; KIPENED 10, 2019; Archibald, 1887; Kite Aerial Photography, 2019
1889 AD	Kite-aided sailing was initiated in English Channel	Wikipedia, 2019 a, b
1890 AD	William A Eddy designed the first diamond kites that are designated 'Eddy Kites'.	Parish, 2019
	He was inspired by the diamond kite models common in Java and Malaysia in Southeast Asia.	
1890s AD	The USA Weather Service initiated 17 'Kite Stations' in the New England zone of North America.	Robinson, 2003
	They were mandated to study the ambient atmosphere, particularly, its temperature and other features.	
	At least 3 different types of kites were utilized. They were high-wind kites (>30 mph wind speed),	
	Moderate-wind (12-30 mph wind speed)	
	kites and light-wind (8-10 mph wind speed) kites.	
1902 AD	First British Kite station was established at Crinan, in the west coast of Scotland	Robison, 2003

TABLE 5.1 *(Continued)*

Year or Period	Historical events related to Kites	References
1906 AD–1908 AD	The Cody kites named after its designer Samuel Cody were adopted by English War office.	NASA, 2016
	Cody kites have been used to draw boats and lift humans. Cody kites lifted a person to a record height of 1600 ft. In 1907, Alexander Graham Bell designed 'Tetrahedron Kites'. The first model called 'frost king' had 256 cells	
1914 AD–1945 AD	During World Wars I and II, military establishments of countries such as Great Britain, France,	American Kitefliers Association, 2019
	Germany, Russia used kites to make observations about enemy positions.	
	German Navy used box-type kite extensively to make observations on submarines.	
	The US Navy used Gibson-Girl Box kites and Garber's Target kite during World War II.	
1967 AD	Single kites were flown to an altitude of 22,500 ft. to 28,000 ft. from ground surface, by Prof. Phillip Kunz	Kitelines, 1980
1970 AD–1985 AD	Two-line foil-based power kites were designed by Ray Merry and Andrew Jones in 1970s in England.	Parish, 2017; Wikipedia, 2018
	It was first called two-line Flexifoil by the company that manufactured it. They are currently sold under brand names such as 'Stacker'. Later versions designed by Ted Daugherty are called 'Sparless Stunters.'	
1970s	Flow form kites were invented by Steve Sutton. They are commonly referred as *Sutton's Flowforms*	Sutton, 1999; Aber and Aber, 2016
1979	Train kites were flown to a height of 37,500 ft. from ground surface at Boonville, New York, USA	Kitelines, 1980
1993 AD	Helikites, a hybrid between helium balloon and kite was invented by Sandy Allsopp of England.	Helikite Hotspots Ltd. 2019; Allsopp Helikites Ltd. 2019; Sandy Allsopp (Personal communication)

Year or Period	Historical events related to Kites	References
1994 AD	Ralf Beutnagel a German kite flier developed a new model of kite known as 'Dopero.' It is meant for regions with low wind and is known to be best suited for aerial photography	Didakites, 2019; Beutnagel, 1995
2005 AD–till date	The Kite Aerial Photography was utilized to conduct geomorphological studies. They adopted high-resolution automatic cameras. Geographical features such as sand dunes in the deserts, tropical vegetation, rivers, lakes etc. were monitored, using KAP.	Lorenz and Scheidt, 2014; Zimbelman, and Scheidt, 2012
2006 AD	Ann Quemere makes a power kite-aided journey and crosses Atlantic Ocean solo, on August 14, 2006. It takes 55 hr for her to complete the trans-Atlantic journey	Best Breezes, 2009
2007AD	MS Beluga SkySails is the world's first container cargo ship powered by a kite. It was built in Germany. It is powered by a giant kite of 160 m^2 surface area (sail).	Wikipedia, 2019 a, b
2016 AD	Researchers at the University of Arkansas at Fayetteville, USA used kites to monitor Agricultural Experimental Stations. They screened soybean genotypes for drought tolerance, using kite aerial photography. They adopted kites with (infra-red) thermal sensors to obtain data about canopy temperature of soybean genotypes (i.e. Crop's Water Stress Index– CWSI).	University of Arkansas, 2014
2016 AD–2019 AD	Low and High-altitude wind energy generation using kite mounted light turbines at experimental scale became possible. World's first commercial kite driven power station was initiated in 2016 at a place called Stranraer, Scotland	Blackman, 2009; Mortenson, 2016; Johnston, 2016
2017 AD	Kites were used to ignite and burn agricultural farms in Gaza strip. Kites loaded with fuel or kites with ignited tails were released by Israelis across the border. These kites suppress crop production activities or destroy crops. Incidentally, these kites are not called agriculturally useful kites but they are better known as 'Incendiary kites' or 'Fire Drakes'. They may find applications in post-harvest burning of crop stubbles and crop residue.	Eglash, 2018; Altman, 2019

It was taken by an early pioneer in KAP, George Lawrence. He used a large panoramic camera and stabilizing rig that he designed (American Kitefliers Association, 2019). In 1906, Samuel Cody designed 'Cody box kites' that were utilized by the English War Office. The Cody kites drew attention of the military establishment after a demonstration that it could draw a boat in the English Channel. Alexander Graham developed Tetrahedron kites almost during the same period in USA, in 1907 (NASA 2016).

Kites were also examined as possible candidates to study the atmosphere. While dealing with various applications of kites, it seems, WH Dines proposed kites as useful platforms, to carry scientific equipment to assess atmosphere (Duffy and Anderson, 2016). Dines, it seems proposed kites as the possible low-cost method to obtain aerial photography. At present, kite-aided aerial photography is adopted to study a range of aspects relevant to geography, terrain, meteorology, agriculture, etc.

Now let us consider a few historical facts about sparless parafoil kites. The sparless foil kites are relatively recent introductions. Records show that in 1970s two-line power kites became popular. They were developed by Ray Merry and Andre Jones in England (Table 5.1). Such, 2 and 4-line flexifoil kites were used for traction. The concept was known as 'kite buggying.' An improved version of flexifoil kites was the 'Sparless Stunter.' It was designed by Ted Dougherty in mid-1980s. One of the largest Sparless Stunter kites of 460 ft^2 was flown in September 1988. A similar sparless kite was designed by Peter Lynn of New Zealand. It was demonstrated in 1991. A few versions of sparless kites produced in 1990s were 10 m^2 in size. Quadri-foils were utilized for traction. This was followed by evolution of 4-line buggy kites. Recent records show that, a quadri-foil kite with 4 lines was in vogue by mid-1990s. It was designed by Ted Dougherty. The quadri-foil kite was called 'Classic Kite.' It was rectangular in shape. Later, foil kite versions that were elliptical became more common. 'Quadritrac' is a later model used as tracking kite. In Europe, in the late 1990s, the kite buggy scene was becoming popular and there were many traction kite designs emanating from different companies (see Parish, 2007; Wikipedia, 2018, 2019a). A few kite developers were busy developing traction kites purely for speed. It seems these models were a bit difficult to fly and maneuver. 'Predator' a model designed by Peter Mirkovic is a good example. In the late 1990s, this was the most successful design in Great Britain. It was being used a lot in buggy races (Parish, 2007). Incidentally, in 2006, Ann Quemere completed a solo long journey from New York to Britanny, using boat with a power kite. She crossed the Atlantic Ocean, using a boat powered by a kite (Best Breezes, 2009).

Historically, kite flying has been a sport practiced in several regions, since a couple of millennia. It has been adopted by military establishments too. During recent decades, kites have fetched excellent aerial photography and useful digital data during scientific experimentation. Kite-aided aerial photography has been adopted for mapping the geographical aspects. Kites are legal to fly in many nations. However, there are regions in South Asia (e.g., Pakistan, India) where kites are banned in certain periods of the year and locations. Kites have caused detrimental effects to fliers due to excessive kite fighting and competition (American Kite fliers Association, 2019).

Historically, utilization of kites in agricultural farms is perhaps most recent. Kites are primarily adopted to obtain series of aerial images, using sensors that operate at different wavelength bands. They offer data about the terrain, soils, crops and their growth status. The sensors on kites could also depict details about crop's water status, disease/pest affliction, weed infestation, soil erosion, floods or drought affected patches in a field. At present, there are a few examples of kites being made specifically to suit farmers. Still, specialized kites known as 'Agricultural Kites' are yet to evolve and get popular across different agrarian regions. Low cost kites with sensors mounted on a rig were used by Bruno Gerard (ICRISAT, 2001). The aim was to assess landforms, the dunes and sandy soil of Sahelian West Africa. Major crops such as pearl millet, sorghum and cowpeas were monitored using kite aerial photography (ICRISAT, 2001; Tables 5.2). In Arkansas, soybean crop (genotypes) was recently assessed and ranked for drought tolerance, using kite-aided aerial photography (University of Arkansas, 2014; See Table 5.2).

We may note that, kite flying, and related activities are spread out all across the continents. Almost every town or a village enjoys kite flying. There are innumerable kite flier's associations, shops, and traders. Specialized agencies, associations, distributors, shops or even traders that deal exclusively with 'Agricultural Kites' are yet to appear on the scene. Perhaps agricultural kites will become popular with the advent of precision farming. No doubt, with introduction of precision farming, need for aerial photographs will become mandatory. Further, adoption of kites with sensors in farming will become a good reason to initiate 'Agricultural Kite Societies' with specific mandates. Efforts to proliferate agricultural kites will be a good idea. Perhaps, kites are the most efficient and least cost methods that offer aerial photographs. Kites offer digital data about land, water and crops to farmers.

TABLE 5.2 Utilization of Kite Aerial Photography to study Geomorphology, Natural Resources, Terrain, Natural Vegetation Agricultural Crops, and Fauna: A Few Recent Examples

Country/region	Salient features of the study conducted using kite aerial photography	References
Idaho, USA/ Bruneau dunes	Kite aerial photography has been adopted to study the Bruneau sand dunes region in Idaho.	Zimbelman and Scheidt, 2012
Vermont, USA/Swamps, Forest plantations and grasses	A delta kite was used to loft the rig with Power shot camera. The aerial imagery was obtained focussing swamps, grass patches, forest land and water ways at a place near Salisbury, Vermont, USA. The margins of the swamps and growth pattern of vegetation could be mapped.	Fastie, 2012
Kansas, USA/ Cheyenne Nature Preserve, Central Kansas	Low altitude (< 150 m altitude) kite aerial imagery was conducted over the Cheyenne National Preserve in Central Kansas. They aimed at assessing the landforms, its topography, marshes, water bodies, vegetation including its diversity and cropping zones. They adopted a Tetracam Agricultural Digital Camera mounted on a Delta Kite, initially. Later, the same spots were studied using large Rokkaku kite mounted with infra-red colour cameras	Aber et al. 2009
USA/ Wetlands and Riparian Regions of Colorado, USA	Salt Cedar is a rapidly spreading invasive species in the wetlands and riverine zones of Colorado. It was monitored, using Kite Aerial Photography. Further, biological control of Salt Cedar using a leaf eating beetle (*Diorhabda elongate deserticola*) was periodically examined, using kite aerial photography.	Aber et al. 2005
New Mexico, USA/ White sands region of New Mexico	Kite aided aerial photography was demonstrated by obtaining high resolution imagery using sensors. They aimed at studying the changes in sand dunes and their causes, in New Mexico, USA	Lorenz and Scheidt, 2014
California USA/ Salton Seacoast in Southern California	The KAP was utilized to monitor the sandy dunes of California.	Lorenz and Scheidt, 2014
California, USA/ Salinas Valley	The topography of agricultural fields, rivulets and other water resources also farm infrastructure could be photographed using KAP. Fields with cilantro at different stages from emergence, establishment of seedlings and till mature harvest were photographed. Such aerial images help in farm related decisions such as scheduling inputs and harvesting.	Benton, 2005

Country/region	Salient features of the study conducted using kite aerial photography	References
Peru/Andean Highlands	High-altitude terrain found in the Andean region of Peru was photographed in detail, using kite aerial photography. Digital elevation models were prepared. Areas covered included mountainous vegetation, meadows and wetlands.	Wigmore and Mark, 2017
Norway, Scandinavia/ Mountainous region	Mountainous terrain, vegetation and weather parameters were characterized using KAP. Spatial changes in geomorphology, and vegetation was recorded and mapped. Fine scale multi-temporal mapping of the terrain/vegetation helped in preparing ecosystem models.	Wundram and Loeffler, 2007
Poland/European Plains and Baltic coast	Kite aerial photography of the entire region of Northern plains and Baltic coast of Poland was conducted. Excellent relief maps depicting the terrain its topography, rivers, villages, general infrastructure and agricultural zones were prepared.	Aber and Galazka, 2000
Tunisia/Chott El Gharsa	Barchanoid sand dunes in the deserts of Tunisia were monitored using KAP. Details such as size, their shifting nature and sparse vegetation if any were photographed, using a Hero GO-PRO high resolution camera mounted on kites.	Lorenz and Scheidt, 2014
Tunisia/Kamech Catchment	Agricultural land prone to soil erosion and gully formation was photographed at decimetre resolution, using kite aerial photography.	Feurer et al. 2017; 2018
Niger/Sahelian West Africa	The KAP has been adopted to study the sandy tracts of a Sahelian location in West Africa. The topographical and soil fertility variations and cropping systems were mapped, using kite aerial photography	ICRISAT, 2001; Marzolff et al. 2003
Indonesia/Durai Island Anambas archipelago in South China Sea	The KAP was utilized to develop maps of coastal zones, its vegetation and influence of environmental parameters on underwater coral reefs and fauna	Currier, 2014
Srilanka/ Coastal belt in Ampara District	Coastal belt of Ampara district, particularly close to the harbour was studied, using Kite Aerial Photography. The shoreline and its changes were monitored, using. high-resolution KAP. Also, changes in vegetation, its diversity and growth (NDVI) were photographed, using multispectral sensors mounted on a rig.	Madurapperuma et al. 2018; 2019

There are few interesting facts to note. In the Middle east region, nations engaged in war have afflicted damage to agricultural crops by using kites. They have utilized kites with fuel-loaded tails that could be ignited via remote control. For example, kites with burning tails were landed in the farming zones of Gaza strip. It resulted in the burning of wheat and other crops and hay. It resulted in loss of food and feed stock. These kites too are employed in activities influencing agricultural crop production, but, in a detrimental way (Eglash, 2018; Eldar, 2018; Mraffko, 2018). They say, above the Gaza strip a burning kite is an indication of terrorist activity. Israelis have nicknamed these kites as 'Terror kites.' Some prefer to call them 'Incendiary kites.' Hopefully, kites that are deemed useful will not assume the status of a pest or a sky borne detrimental factor to agricultural farms.

5.1.2   TYPES OF KITES AND THEIR CLASSIFICATION

There are over 180 types of kites identified. Each type or class has several models of kites and their variations. Kites are named usually based on model, the material utilized to make a kite, size, shape, purpose, operating skills required, region where it is popular, etc. (Wikipedia, 2017, Murray et al., 2013). Kite types more frequently encountered during kite aerial photography are as follows: Diamond shaped Eddy, Delta, Rokkaku (hexagonal), Dopero, Box types, Train kites, Flowforms and Foil kites. They can be effectively utilized during aerial survey of terrain, natural resources, and agricultural cropping systems.

Aber and Aber (2016) while discussing about kites and their accessories state that there is no single type of kite or model that is best suited for 'kite aerial photography.' The bottom-line requirement is that the kite model should be able to lift the sensors required for photography. This suggestion holds true for kite-aided photography and spectral analysis conducted during agricultural crop production too. Further, they state that set of sensors and rig mostly weighs around 0.5 to 1.5 kg. As a thumb rule, for lighter winds (5–10 mph) we adopt large kites and smaller kites if wind (10–25 mph) is stronger.

Kites have been made using several different types of materials. A few commonly adopted materials are plastic sheets, cloth and paper. Textile material such as ripstop nylon, polyester, Dacron and silks are the frequently preferred materials, to make kites of different sizes.

Kites have different sized wings. Wings may be of different shapes. They could be mono-sail, flexible sail, or rigid wing made of toughened light plastic film. Wings could bi-plane or multiplane. Wings with narrow chord compared to the span have low aspect ratio. Kite's wings could be made of ram-air inflated cells (e.g., parafoil kites). It seems there are also kites with gyro wings. Multiple kites have several sub-kite units that are arranged as trains, chains coterie, single branching or multiple branching or any other convenient patterns. It seems over 200 kites grouped into one train has been flown efficiently into sky (e.g., parafoil kites).

We may also note that kites could be managed by a single person while conducting recreational flights or even kite aerial photography for scientific purposes. A single person may co-ordinate and control a few different kites simultaneously. Also, kite-fliers may control a series of kites (kite units) that operate as a single train kite and obtain aerial photography. Incidentally, kites could be a single unit or could be made of multiple units (e.g., train kites).

Aber and Aber (2016) broadly classify kites into 'Soft kites' and 'Rigid kites.' Both of the classes of kites have been adopted for general aerial photography. Particularly, during spectral analysis of agricultural terrain, natural vegetation and crops. Soft kites are easy to make. They have a low weight-to-surface ratio. Generally, soft kites are easy to launch. They can be folded into small suitcase and transported in times of need and lofted, quickly into breeze from any place. Softkites may have air cells that get inflated automatically with wind pressure (e.g., parafoil kites). Obviously, soft kites (e.g., Parafoil kites, Sutton Flowforms, etc.) have tendency to collapse when wind diminishes.

The other group of kites are classified broadly as 'rigid kites' by Aber and Aber (2016). Rigid kites have framework made of wood, bamboo or sticks that give the shape and form to the kite (e.g., Delta kites, diamond kites, Rokkaku hexagon kites, etc.). Sophisticated models have light weight graphite rods and nylon sheets. The rigid kites fly better in very light wind (4–7 mph) and gentle breeze (8–12 mph) conditions. Rigid kites do not have problem of getting deflated under low wind pressure. The frame offers rigidity and shape to the kite.

Let us consider a few types of kites that are relevant to our central theme, i.e., usage of kite in agriculture. These types of kites are utilized mostly for aerial photography, detecting and recording crop growth, disease/pest infestation if any, water needs of crops (i.e., Crop's Water Stress Index), bird scaring, wind power generation and even buggy or wagon pulling (i.e., power kites).

5.1.2.1 EDDY KITES

Eddy kites were invented by William A. Eddy of New Jersey, USA in 1890s. William Eddy was famous for his kite aerial photography of terrain and infrastructure. Also, for collecting meteorological data using probes mounted on rigs and attached to Kitelines. The original Eddy kite version did not possess tails. However, most diamond shaped Eddy kites at present are sold with tails. They have a straight or bow like spar to give rigidity and stability to the kite (Plate 5.1). The kite has a bowed spar which is attached at 19–20% distance lower down from top end of vertical spar. The sail is made of loose-fitting cotton or nylon or sometimes polythene or paper. The Eddy kite is amenable to make trains, particularly, if they are tail-less.

PLATE 5.1   Left: Eddy diamond kite; Right: A simple Eddy kite available commonly in any kite shop

Source: Dr. Jim Christianson, President, Skydog Kites LLC., 220 Westchester Rd., Colchester, CT 06415.

5.1.2.2   CODY KITES

The modern Cody kites are a product of over 100 years of modifications to its design and evolution. Cody kites are still popular and are used for variety of aerial photography related assignments. The double box types with wing flaps are still in vogue all over the world (Parish, 2017). It is a versatile kite design capable of lifting even a man or a wide range of payload items such as sensors, radars, etc. Modern Cody kites have light graphite spars and nylon sheets. Cody kites may have a top sail that adds to stability and power. Cody kites are made of different colors and designs. Cody kites are also flown purely for recreational purposes. Cody kites meant for recreation are easily recognizable in the sky. Cody kits are not costly. The price varies from a little

25 US$ to 500 US$ depending on the type and sophistication added. For aerial photography, Cody kites with facility for sensor and transmission of radio signals are utilized. They could cost more depending on sensor brands and digital processors. A few good examples of modern Cody kites are Lutz Treckzok's Cody and Dan Flintjer's 'Buffalo Cody Kites.'

5.1.2.3   DELTA KITES

There are several models of Delta kites produced by companies and individuals all across different continents. A few of them are easily recognized. They adopt definite dimensions and their ratios (Plate 5.2). Let us consider a couple of examples as described by Aber and Aber (2016). The Grand Delta kite is useful during aerial imagery of terrain and agricultural crops, in addition to general geographical aspects. This kite model has 11 ft. wingspan, a surface area of 30 ft². It weighs 540 g. Rigid rods made of light wood or aluminum or graphite support the frame of the kite. Such kites may have up to three tails to offer greater stability. Tails are usually 6–15 ft. in length. It is supposedly a relatively stable flyer (Plate 5.2). Therefore, it can offer sharp high-resolution imagery of crops, if adopted to assess crops.

The Delta-Conyne is actually combination of a delta and a triangular box kite (Aber and Aber, 2016). Delta-Conyne kites are preferable when wind is lighter. They may possess 10–15 ft. tails to offer stability. It weighs 1.4 kg. It is relatively a heavy kite. It can lift a rig of 1 kg under a wind speed of 8–10 mph.

Giant Delta kite has a larger wingspan of 19 ft. and a total surface area of 18 ft². It weighs 2.5 kg. It has a tail extending to 15–20 ft in length. It is a stable kite. It easily lifts a full complement of visual, multi-spectral, infrared and lidar on a rig. Yet, it is an easily foldable kite and fits a small suitcase that can be transported (Aber and Aber, 2019; Skydog Kites LLC. 2019; Plates 5.2 and 5.5).

5.1.2.3   DIAMOND KITES

Diamond kites are among most popular kite types worldwide, especially, for recreational purposes. They are easily maneuverable. They could be flown for hours at a stretch. They are also the popular types that are used during competitions and kite flighting events (Plate 5.3). Usually, fast types of diamond kites are flown by new commers to the sport. But, fighting kites and lines are provided to those kite fliers with certain degree of experience in flying techniques.

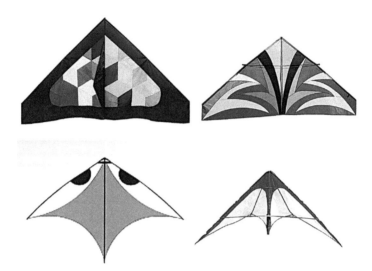

PLATE 5.2   Delta kites for recreation and Kite Aerial Photography.

Source: Dr. Jim Christianson, President, Skydog Kites LLC., 220 Westchester Rd., Colchester, CT 06415.

Note: Such delta kites of different sizes that are suitable for recreation and kite-aided aerial photography are available in kite shops all over the world. They are simple to make or manufacture in large numbers. They are low cost kites. The tethers could be attached with full complement of sensors, using a rig. Also, see https://www.youtube.com/watch?v=Fu4FrhRe_SY.

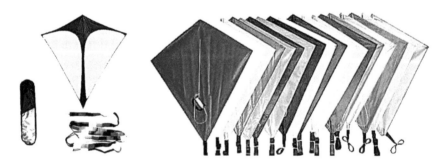

PLATE 5.3   Diamond Kites.

Note: Left: A single Diamond kite and its accessories; Right: A series of Diamond kites of different design sails.

Source: Jim Christianson, President, Skydog Kites LLC., 220 Westchester Rd., Colchester, CT 06415 and Dyna-kites Corporation Chicopee/Lancaster, Massachusetts, USA; http://www.dynakite.com/?gclid=EAIaIQobChMIsf_Xxeyv4wIVWiQrCh3OxwCqEAAYAiAAE-gLp9fD_BwE.

5.1.2.4 ROKKAKU KITES

Rokkaku kites are 6 sided. It is widely used in kite aerial photography. Rokkaku means 6-sided in the Japanese language. Rokkaku kites are built in different sizes and combinations of vertical and horizontal spreaders. The two most common ratios adopted while making a rokkaku kite are 3:4:5 and 4:5:6. The first number is the distance between two horizontal spreaders. Second is the span of the horizontal spreader and third is the overall height of the rokkaku kite (Shannon, 1996). We should keep the ratios of the kite's sides same while changing the overall size of the kite. The most popular ratio regarding size of the rokkaku kite is 3:4:5. Now, regarding the material utilized to manufacture a rokkaku kite. The most common fabric used is ripstop nylon. Heavier nylons are stiff. Cotton fabric too could be used to make rokkaku kites. Rokkaku kites are easy to bridle. Bowing strings on the kite is also important. Deeper bowing is needed if the kite is small in size. Bowing thread should be strong enough. Flight adjustments using bridles and bowing strings are common.

Let us consider a couple of examples of rokkaku kites. As stated earlier, they are kites with rigid framework. 'Duruma rokkaku' is said to be Japanese traditional model. It is an apt rigid kite under light breeze conditions of 5–10 mph (Aber and Aber, 2016). It weighs only 540 g. Therefore, it is a good choice for kite aerial photography. It is often provided with 5–6 ft. tail piece. The tail offers greater stability. Now, let us consider Giant rokkaku kites. They are usually 6–7 ft in width with a surface area of 36 ft^2. It often has a tail of 10–15 ft. in length. It is known to develop good pull even under light wind. It is not to be used if wind speed is greater than 10 mph. It is an apt kite for lifting cameras and rig for aerial photography.

5.1.2.5 BOX KITES

We may note that box kites are made in variety of dimensions, designs, sizes and weight. Material used to make the box kites too varies. For example, there are even box kites with 14 triangular faces (tetracaidekadeltahedral). They could be flown under 15–20 mph wind. They are heavy. Such box kites are purely recreational ones (Crouch, 2010; Spendlove, 1980; Plate 5.4).

The Tye-Dye box kite is 30" × 20" in size. It is made of ripstop nylon and fiber glass.

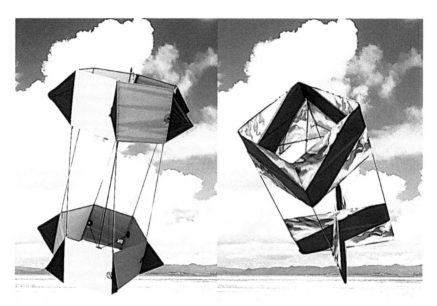

PLATE 5.4   Box kites. Left: Spinning Box kite; Right: Tye-Dye Box kite.

Source: Jim Christianson, Skydog Kites LLC.220 Westchester Rd., Colchester, CT 06415, USA.

Note: The dimensions of spinning box kite are 36" × 38". It is made of ripstop nylon and fiberglass.

5.1.2.6   DELTA-CONYNE KITES

The Delta-Conyne kites gained greater acceptance during 1970s and 1980s among the kite fliers (see Plate 5.5). As the name suggests, Delta-conyne kites are hybrid kites made by combining box kites (conyne portion) and delta kites. These Delta-conyne kites are relatively more stable in the air. They withstand wind turbulence better. A common Delta-Conyne kite could be made using 1'5 m wide fabric. We also require at least 5 m length of thin fiber glass. and PVC tubing to prepare a kite of this size. The fabric versions are usually heavier compared to plastic sheet. The spars on the box portion could be made of light weight wood or graphite rods. This kite could be provided with wings (or flaps). A bridle is usually passed through a punched hole. Sometimes, loops are sewn to connect the bridle. The kite is also provided with a cross-spar to give rigidity. The Delta-Conyne kites are easily amenable and lends itself to be flown as Train kite. It seems as many 30–40 kites could be attached one behind the other on a single kite line and flown. Usually, Kitelines are made of Dacron or twisted nylon (ripton nylon). Swivels are used for greater convenience (Batchelor, 2019). For aerial photography, we can use a rig with a set of sensors hanging from the kite line. We can select a range of different sized Delta-Conyne kites (Hunt, 2019; Plate 5.5).

PLATE 5.5   A Delta-Conyne kite.

Source: Jim Christianson, Skydog Kites LLC.220 Westchester Rd., Colchester, CT 06415, USA.

5.1.2.7   DOPEROS

The Dopero kites were first designed and tested by a German kite flier- Ralf Beutnagel in 1994. He developed it for areas with light wind. He initially faced certain difficulties in lofting it. So, he doubled the kite and allowed it to get lofted easily. They say, it is among the best suited kites for kite aerial photography. It is light in weight and can be easily transported and preserved (Didakites, 2019). It is an open vented version and does not need a tail to keep stability.

5.1.2.8   TRAIN KITES

Aber and Aber (2016) state that, multiple kites can enhance the lifting power. A few kites are aligned in series and flown into sky. Such train kites are suitable under light wind conditions. Train kites can hold a rig with set of sensors for kite aerial photography. Kite types such as delta, diamond, rokkaku, etc. are usually aligned together one behind another to form train kites. Soft kites too could be aligned to form a train kite. The Chinese Dragon kites are a good example of 'train kites.' Some of them could be 100 ft. in length with several of them attached in a series. The dragon kite's head is often made of bamboo covered with silk. Major applications of such train kites are in recreation and aerial photography.

5.1.2.9   FOIL KITES

Parafoil kites were first designed by Domina Jalbert in 1965. Lingard (1995) describes a parafoil kite as a tethered, ram-air inflated and sparless type. Foil kites do not possess a rigid frame. Parafoil kites are usually made of ripstop nylon. A parafoil kite has two layers of fabric, an upper and lower one connected with vertical cells tailored in between. The cells get filled with air and provide shape to the parafoil kite. The cells with air offer an aerodynamic shape to the kite. Parafoil kites are efficient. They are stable and are easy to maneuver. Parafoil kites are manufactured in wide range of sizes and shapes. The parafoil could be one large lifting parasol and a few smaller parasols clubbed into a bunch (Rose, 2014). Parafoils could be easily folded and stored until use (see Plate 5.6).

PLATE 5.6   A simple parafoil kite for recreation and kite aerial photography.

Source: Dr. Jim Christianson, Skydog Kite LLC. 220, Westchesteer Rd. Colchester, CT 06415, USA.

As stated earlier, foil kites are soft. They are devoid of spars (i.e., sparless). They consist of air cells for rigidity and shape. These cells open to the front to allow air to inflate the kite. Foil kites are used during kite surfing, kite land boarding, snowkiting, kite buggying, and kite energy systems and recreation. Foil kites could also be utilized in water surfing. Foil kites may have 2 or 3 or 4 lines while flying. The 2-line systems have rings, wrist bands or bar. The 4-line foil kites are controlled using bars or handles.

Parafoil kites have been utilized for wide range of activities related to recreation, aerial photography of natural resources, agricultural crop and urban infrastructure. They have been utilized to collect atmospheric data. Also, in dust profiling of ambient atmosphere above urban and rural areas (Reiche et al., 2012).

Parafoil kites sold for recreation or to pull beach buggies typically have following characteristics. They are all made of ripstop nylon. The wind velocity for usage ranges from 6 to 20 mph (or 10–32 kmph). The dimensions range from 13" × 21", 20" × 32", 28" × 38, 100" × 85". Every kite is supplied with at least 20–50 numbers of 500 ft. length of kite line. Most or all the common recreational kites are easily operated. They could be utilized by beginners or advanced well-trained kite fliers. Usually, kite control manuals meant for specific models of parafoils are made available. Literature about general principles of parafoil control are also made available (Coleman et al., 2017; Plate 5.7).

Parafoil kites serve as useful power kites. In fact, wide range of models and sizes of foldable foil kites are sold in different parts of the world (Plate 5.7). They are easy to package and store in smaller box. We have to note that, larger foil kites with better pull are required, if the intention is to drag a buggy over land or power a boat. Also, a larger foil kite is necessary, if, a full complement of sensors has to be lifted for kite aerial photography (Plate 5.8).

PLATE 5.7   Power foils. Left: Power foil-1.8; Right: Power foil-2.6.

Source: Dr. Jim Christianson, Skydog Kites LLC.220 Westchester Rd., Colchester, CT 06415, USA.

Note: Power foil 1.8 is a small traction foil for beginners. It is apt for conditions with stronger wind. Power foil 2.6 is meant for experienced kite fliers who are used to larger kites. Power foil 2.6 offers strong pull if flown in moderate to strong winds. Some characteristics of such a Power foil-2.6 are: Dimensions are 104" x 38". Material used to make the foil is ripstop nylon. It requires intermediate to advanced kite flying skill. It is suited for wind range of 8–25 mph. This kite is sold with a kite line of 35–70 kg.

PLATE 5.8   A larger sized power kite with accessories.

Source: Jim Christianson, Skydog Kites LLC.220 Westchester Rd., Colchester, CT 06415.

Note: Skydog Parafoil 9.3 is a big parafoil kite. It is nicknamed 'Big Dog' by the manufacturers. It is used for traction power. It can easily pull buggy and large boards with wheels. The characteristics of the power foil 9.3 are: Dimensions are 215" x 65". Material used for manufacture is ripstop nylon. It needs flier with advanced skills in kite flying. It is apt when wind is 2–8 mph. It is sold along with kite line.

We can identify at least three types of parafoil kites based on number of Kitelines attached to the foil.

1. *Single line:* These are relatively flat kites, often with multiple small keels to which the bridle lines are attached. These can look a bit like a flying inflatable mattress! Also, there are novelty kites that are technically parafoils, but come in all sorts of creature shapes and sizes (Plates 5.7 and 5.8).
2. *Dual line:* These are highly maneuverable stunt kites which pull hard and are flown from a stationary position. Although soft in construction, you still wouldn't want to be hit by one of these, in windy weather!
3. *Quad line:* Traction kites are designed for pulling buggies, mountain boards, or even a pair of in-line skates! Two lines are used for steering, while the other two are brake lines (Aber and Aber, 2016, 2019).

5.1.2.10   FLOWFORM KITES

Sutton's Flowform Kites were invented in 1970s by Steve Sutton (Sutton, 1999). The Flowform kite has soft air foil made of nylon. Air cells get filled through wind pressure and the air pressure is regulated using venting.

Sutton's Flowform kites are fairly stable and are good candidates for regular use above agricultural farms for aerial photography. Let us consider a couple of models of Sutton's Flowform kites utilized for general aerial photography. Aber and Aber (2016) describe 'Sutton Flowform-16' as wind inflated air foil kite. It is supposedly stable even when wind blows at 15–25 mph in a location. The Sutton's flow forms have long tails extending to 15 ft. The specific model described has 16 m² area. It weighs 310 g. This model can be packed easily since it has no rigid framework. It can be folded into small suitcase and transported. Skyfoil is yet another soft kite. It is 10 ft. wide and 8.5 ft. in length. It flies both with and without tails. It is useful during aerial photography. Sutton's Flowform 30 is a larger version in a series. It weighs about 650 g. It has a good weight to surface area ration. The lifting surface area is 2.8 m². This model, as well as Sutton's Flowform-16 is known to offer good lift and is apt to be used during aerial photography of terrain and crops in an agrarian region.

5.1.3   CHARACTERISTICS OF A KITE RELEVANT TO FLIGHT AND AERIAL PHOTOGRAPHY

Price (1996) states that it is important to judge the kite using characteristics most relevant to the purpose for which the kite is being flown. A few types of measurements such as size, weight, line pull, line angle, are easy. A few others that are commonly measured are the sail area, wind speed that kite encounters and most importantly the L/D ratio (i.e., lift to drag ratio). Further, it has been said that L/D ratio is among the most useful characteristic of a kite model that explains something about its efficiency. We can compare the kite models using lift to weight ratio.

Generally, wind should be measured near the kite itself. Wind speeds at the ground surface is often much lower than one experienced at altitudes of 100 to 400 ft. by the kite. It is preferable to attach a very light anemometer to measure the wind speed. An anemometer made of very light weight ping pong ball is suitable (Price, 1996). Usually, lighter wind will require use of larger kites. However, under stronger winds smaller kites are used

Wind speeds are classified as light air (1–5 kmph or 1–3 mph or 1–3 knots or *Beaufort 1*); Light breeze (6–11 kmph or 4–7 mph or 1–3 knots or *Beaufort 2*); Gentle breeze (12–19 kmph or 8–12 mph or 7–10 knots or *Beaufort 3*); Moderate breeze (20–28 kmph or 13–18 mph or 11–16 knots or *Beaufort 4*); Fresh breeze (29–38 kmph or 19–24 mph or 17–21 knots or *Beaufort 5*); Strong breeze (39–49 kmph or 25–31 mph or 22–27 knots or *Beaufort 6*); High Wind (50–61 km/h or 32–38 mph or 28–33 knots or *Beaufort 7*) (Parish, 2007).

The most important aspect of any kite used to lift a camera is stability. Generally, single lined kites are used as they allow very long line lengths and need less intervention from the flyer than steerable designs. Almost any stable kite design can be used to lift lightweight camera rigs (up to 500g or1 lb). As weight increases, specific designs are chosen for their additional flying characteristics, such as line pull, wind range, flying angle and ease of launch. Widely used designs are parafoil, rokkaku, delta (including variations such as delta conyne), dopero, and the new lighter-than-air helikite design. Of these, the parafoil is most popular as it generates a strong pull for its size, and it can be easily stored in a small space due to its sparless construction. Kites with spars such as the rokkaku or delta tend to fly at a higher angle than parafoils, which is of benefit when the space between the launch site and photographic subject is limited. A higher-flying angle permits the kite to lift more weight, as more of the generated force is acting vertically.

To carry a complex and relatively heavier camera rig a proportionately larger kite will be necessary. Otherwise we can make trains of rokkakus or diamond kites greater than 5,' to lift the cameras into air. We should note that larger kites will need stronger material and gluing method. Lines too will have to be tough and strong enough (Digital Earth Watch, 2017).

Tails are used on kites to stabilize them and reduce swaying as the wind blows. Usually, tails are larger depending on the size of the kites, say a diamond kite or delta kite. Fuzzy tails are used if one wishes to reduce the length of the tails and avoid entangling (Kite Aerial Photography, 2017; Fosset, 1994).

5.1.3.1   KITE AERIAL PHOTOGRAPHY (KAP): ACCESSORIES

Kite flying and capture of aerial photography involves utilization of a few accessories. A few are mandatory irrespective of the model employed for KAP. Other accessories depend on type of the kite and the purpose for which the kite is flown. In addition to the kite, we require Kitelines of apt size. Usually, kites are supplied with lines of 200 to 500 lb braided Dacron. Other material used for Kitelines are cotton thread, nylon thread and metallic wire. A simple spool usually stores 500 ft line. As a precaution, kite flyers often use a glove to handle the tether made of nylon or thin metallic wire. Gloves are made of thick leather in order to protect fingers, hands/palm. Larger and thicker cattle gloves too are preferable to be used while flying kites in stronger wind. Larger spools called 'strato-spools' hold about 1000 ft. of kite line. Kite handle and anchor straps are supplied. So that, kites packaged in boxes or bags could be transported, using a personal car. Kite transportation is also done using a trailer. On rough terrain, cargo dolly's are utilized to transport a kite, cameras and spools of Kitelines.

Recently, Aber and Babb (2018) have reviewed our knowledge about various methods, computer software packages, and their utility in processing, specifically, the kite-derived aerial images. According to them, it is still a 'low-tech' proposition. Historically, KAP is a technique that predates the modern methods of aerial robots (UAVS), aerostats, aircrafts, or satellites. It involves simple 'tethered platforms (kites). Further, they state that KAP existed since 1880s and almost coinciding with general photography itself. The KAP was a subdued methodology during most of the 1900s. Now, the KAP seems to have regained a certain level of interest, since low-cost high-resolution aerial photography is sought frequently. At present, reports dealing with KAP include studies on geomorphology, monitoring natural resources, vegetational and faunal changes, monitoring agricultural crops and studying their growth and productivity patterns, etc.

Most important accessory for aerial photography is the sensors (photo-graphic cameras) that operate at different wavelength bands. Indeed, there are large number of brands of sensors that can serve the requirements of clients dealing with aerial photography (Krishna, 2019). Most commonly, visual cameras (red, green, blue), multi-spectral cameras, near infra-red cameras, cameras, red-edge cameras or LIDAR are mounted on to the rig. These sensors are needed to obtain digital data and images relevant to crop growth. Especially, to measure parameters such as NDVI, GI, Leaf area index (LAI), Chlorophyll index, Crop's water stress index (CWSI). The rig and entire set of sensors need to be light in weight, so that, the set could be lifted by the kite line.

Whatever is the purpose of kite aerial photography and its utility in scientific studies, processing the image and transmitting it to users is a basic requirement. Aber and Babb (2018) have evaluated at least three different image processing software that are frequently adopted during KAP. To obtain high quality KAP, farmers may have to adopt software packages that are most suitable ones, such as PIX4D mapper, Agisoft's Photoscan or Microsoft ICE etc. There are indeed several software packages that could get evaluated prior to final selection for KAP. They say, selection of a software package depends on the algorithms, cost, and purpose it serves.

Photogrammetric methods that help in 3D and 2D image processing could be useful. Aber and Babb (2018) have compared 3 software packages namely, 'Photomodeler Scanner,' 'Photoscan Professional' and 'Pix4D Mapper.' Of course, by using kites as low-cost platforms. It is said that, for scientific and educational purposes software such as Photoscan is useful. It costs about 1.5 to 4000 US$ to install the software and process kite-derived images. Pix4D Mapper that is frequently used on autonomous drone aircrafts is a costlier proposition for kite enthusiasts. Normally, it is recommended to use cameras that provide GPS tags to the photos. Overall, software packages used on images derived from sensors on different platforms could also be adopted for KAP. A few software packages offer free image compositing and ortho-rectification. A few may be better for photogrammetric analysis and 3D image processing. A software to compare a set of wide range of other software that are utilized during KAP is required. It could be able to compare the algorithms required, the cost (i.e., subscription) per unit time, per unit images, number of ortho-images it can process, resolution, quality of final image and purposes it serves, etc.

5.2   UTILIZATION OF KITES: A GENERALIZED VIEW

There are indeed several types of kites and each has a range of applications. There are also kites meant for specific tasks. For example, Eddy and Delta kites could be used for aerial photography. Dragon kites, Train kites and Diamond kites are useful in recreation and advertisement of events or items, A quad-line parafoil kite is useful in traction and as a power source during surface water surfing, A simple list for various applications of kites includes; aerial photography, teaching aerodynamics, transport (power and buggy kites), decoration, fishing, sports and recreation, military, scientific experimentation, industrial uses, wind energy generation, acting as light beacon, advertising and most importantly in the present context in aiding agricultural procedures (Wikipedia, 2019a; Aber and Aber, 2003). According to Alevizos (2012), there are a few beneficial traits of kites that make them suitable during aerial photography. They are, in fact, a new research tool for those interested in geomorphology, archaeology, studying insects, birds and microbial spores in the atmosphere, etc. Klemas (2015) states that, kite aerial photography has been used for military surveillance, scientific surveys, meteorological observations and several other tasks related to natural sciences. The identification and study of landforms have also been approached using KAP, including Quaternary landscapes in Poland (Aber and Galazka, 2000). Also, researchers have used KAP to help map ice and periglacial landforms in Alaska (Boike and Yoshikawa, 2003).

For the present, KAP has been utilized frequently to study the geomorphology and natural resources. The KAP has been put to use in different continents to depict wide range of geographical features. To quote a few examples, Boike and Yashikawa (2003) have made some interesting observations of permafrost and periglacial regions using kite-aided aerial photography. The kite aerial photography (KAP) has offered excellent images of ice wedges, undulations and surface features of glaciers, icebergs and their surface features. Similarly, KAP has helped in studying the tropical geographical features and diversity of vegetation and crops. KAP has also been examined as a method for aerial photography of vegetational diversity and land features of tropical, semi-arid, riverine and mountainous tracts. For example, in the West African Sahel, kites have been explored as a possible low-cost method to study the dunes, their features, and crops such as pearl millet, sorghum grasses and cowpea. The KAP has been shown as a useful method to study the shoreline features and vegetation. For example, there are reports that suggest usage of KAP to conduct ecological investigation of the coastal regions in Srilanka and Indonesia. Overall, KAP is a relevant method to study the geomorphology and ecological aspects of different regions on earth.

The acquisition of remotely sensed data from a range of cheap, lightweight platforms such as tethered kites and blimps(e.g., Aber et al., 2002; Marzolff et al., 2003; Boike and Yoshikawa, 2003; Smith et al., 2009; Vericat et al., 2008; Seang and Mund, 2006) is becoming common. Kites serve as an ideal tool for basic mapping purposes. The aerial view provided from the kite has proven useful in describing the geographic features of the location, in Alaska (Boike and Yoshikawa, 2003). Further geomorphological study has been conducted on sand dunes by Lorenz and Scheidt (2014) and surface features relevant to Quaternary studies by Aber and Gałazka (2000).

Kite aerial photography (KAP) is proposed as a creative tool for geography field teaching (Sander, 2014). The KAP can be integrated as field experiment, surveying technique or group activity. The aerial images acquired can be instantaneously integrated in geographic information systems and used for discussion. The use of KAP in field teaching aims at explaining spatial variation. The KAP improves observational skills in real environments through sharp images. It enables students to interpret remotely sensed data from a nadir perspective. The KAP allows a new vantage point to the fieldwork.

Kite aerial photography consulting agencies that offer aerial photography and digital data are yet to become popular with farmers. Farmers may then resort to such agencies which will fetch them imagery showing the status of their crop. Such KAP agencies may also help farmers in using the digital data in the precision equipment such as GPS-guided seed planters, fertilizer applicators and pesticide sprayers. There are, at present, a couple of kite aerial photography-based sales and consulting firms. They offer aerial images of urban settings, natural resources, water ways, crops, etc. (Aber and Aber, 2019). Such services have to be extended to agricultural operations. Kite aerial photography based agricultural agencies could supply images periodically, from the time land preparation begins through all the crop growth stages until harvest.

5.2.1   GEOMORPHOLOGY AND MAPPING USING KITE AERIAL PHOTOGRAPHY

Studies on geomorphology often need a clear and detailed depiction of the site, obtained using aerial photography. Several different types of platforms such as satellites, airplanes, small drone aircrafts, microlights, blimps, parafoils, aerostats, balloons and kites have been adopted. Usually, high resolution sensors such as visual, (red, green, blue), near infra-red, infra-red, red-edge and LIDAR are used to obtain images from vantage points. Each type of platform has its advantages and lacunae. In the present context we are concerned with geomorphological studies of agricultural terrain, using kites as platforms. There are not many studies made about agricultural regions, landforms, soil type, water resources and crops, using kites. Therefore, to obtain a better perspective, several examples that deal with use of KAP to study geo-morphological features, in general, has been included in this chapter.

Some of the advantages associated with using KAP to study geomorphology are listed by Currier (2014). They are: (a) The rules and regulations on using of KAP is much less compared to other aerial vehicles. (b) Geographical surveys conducted using KAP requires relatively lower levels of investment. (b) It involves lower level of technological skills than using other means such as autonomous drone aircrafts, airplanes,' or satellites. (c) KAP needs only few hours of training in kite flying and aerial photography. (d) It is easy to transport the equipment and kites from one location to other.

Lorenz and Scheidt (2014) have described a simple parafoil kite attached with digital camera (still and video) as a useful apparatus to conduct geomorphological studies. It seems the kite plus sensors set up costs less than US$ 400 (2012 price level). It occupies only 2 l in volume when collapsed. Therefore, it is compact and easy to maneuver, and transport whenever needed. They have stated that the kite-based platform is efficient in obtaining aerial photography. For example, kite plus sensors were used to study changes in dunes in coastal region. Further, they state that the kite plus sensor system offered excellent digital elevation models (DEMs) of terrain. The digital surface models (DSMs) are highly useful while studying the geographical aspects of a location. It has been pointed out that best results (aerial photos) were obtained with parafoil 10' (Product no, 5012). It withstands disturbance created by winds of 6mph. If one anticipates stronger breeze, then smaller, compact kite (Parafoil kite 5' Product No, 5272) could be utilized (Lorenz and Scheidt, 2014; Lorenz et al., 2010).

Boike and Yoshikawa (2003) used a camera suspended from a kiteline to obtain high resolution pictures in Alaska. The idea was to perform geometric analyzes of ice-wedge networks and to map water and vegetation. Rigid Delta kites were used for light to moderate winds. But, parafoil kites were flown in moderate to strong winds. The payload ranges from 0.5 to 2kg and the size of the kite depends on wind conditions. The photographs provide an important data set for monitoring changes in permafrost patterns, periglacial processes and coastal vegetation overtime. Home-made kites attached with camera rigs have also been used in archaeology. In some developing countries, the KAP has been adopted to map local vegetation, obtain tree population counts in villages, and assess complex humanitarian emergencies (Klemas, 2015; Sklaver et al., 2006; Murray et al., 2013; Oklahoma State University, 2019).

Delta kites with rigs that support visual, near-infra-red and infra-red cameras have been used to collect imagery of swamps, vegetation in its surroundings and open expanses in the Salisbury area of Vermont, USA. The rigs obtained an image every 10 seconds, using AuRiCo KAP controller. Later, the ortho-images were stitched and processed to arrive at visual and infra-red images (Fastie, 2012). Microsoft ICE was utilized for processing aerial images (ortho-rectification). Another computer program for processing ortho-mosaics that was utilized to develop the images of swamps, forest plantation and vegetation is called 'MapKnitter.'

Aerial photography using cameras attached to tethers of kites requires some standardization. It depends immensely on wind conditions, the gimbal that holds the sensors and height to which the kite has been flown. For example, trials conducted by Anderson (2002, a, (b) in Dallas, Texas, USA shows that kites flown to 100 to 200 ft suffered due to erratic wind that sometime died out and conditions became still. That makes it difficult to hold the kites at a height. The kites flown at 300–400 ft altitude provided sharp images of terrain. Therefore, for geomorphological studies such a setup of simple camera and kite flown at 400 ft could serve the purpose. The images could be relayed and stored in digital format (Anderson 2002, a,b).

Aber and Galazka (2000) have used kite aerial photography very effectively for studying geographical aspects of a region. For example, in Poland, they used KAP to obtain ortho-mosaics and process them to get details of the terrain. The area they photographed and mapped included Baltic coast in North to mountainous tracts in southern Poland. The kite aerial photography expeditions also included the east of Warsaw and till Gdansk in northwest. To conduct geographical study, they transported the equipment necessary for KAP such as the kites, lines, cameras and computer-based image processors, using a pickup van. It means that all accessories for KAP was easily transportable. They have reported that kite aerial photography was conducted in weather pattern that was near perfect for the purpose. They adopted kite models such as Sutton's *Flowforms* and Large Delta-Conyne types of kites. They were attached with remote controlled and automatic camera rigs. The expedition to map Polish geomorphology extended from 1998 October till September 2000. It provided details of mountains, vegetation, coastal zone and its vegetation, villages and their infrastructure, agricultural zones surrounding village inhabitations, etc. From this study it is clear that, kite aerial photography is indeed suitable for making useful imagery of land and its features. The digital data obtained from vantage heights could be most useful to clients. The KAP involves low altitude aerial imagery, say, at 150 ft. above ground surface. Therefore, it offers excellent images of villages, and agricultural belts. KAP is cost effective.

Reports from Sahelian zone has clearly shown that kite aerial photography could be adopted to study the sandy semi-arid zone. Particularly, to study the region, its topography, formation of sandy undulated surface of fields and crop's growth status. In fact, photography generated using kite aerial photography has been compared with data from traditional survey methods. The KAP was found to fetch satisfactory imagery. The imagery could be utilized to plan seeding and other agronomic procedures (ICRISAT, 2001; Marzolff et al., 2003).

Currier (2014) states that, KAP offers low-cost aerial imagery of terrain and its features. For example, kites with sensors were used to obtain ortho-mosaics. Appropriate image developing computer programs such as Agisoft's Photoscan are used. This way, ortho-rectified images of the coastal zone of Indonesian archipelago could be obtained. Over 350 ortho-mosaics were pooled and tailored to get excellent images of Durai islands, (Indonesia), its coastal region, the vegetation and areas that supports coral reefs. Currier (2014) states that, among various methods of aerial imagery, KAP is a low-cost method that requires much less sophistication than flying an aerial robot like small fixed winged drone aircrafts or multi-copters. KAP seems simple and straight forward. The KAP was utilized to develop a 1: 2000 map of Durai Islands. It depicted coral reef zones, tortoise inhab-iting zones and vegetation. Such maps were primarily meant to understand ecological aspects of islands in the Indonesian. KAP could also help us in getting digital elevation models (DEMs) of archipelago. Particularly, its terrain with undulations, vegetation, streams, coastal areas, etc. Digital elevation models help in judging the soil erosion, impact of weather, floods, soil erosion, and drought on geomorphology (Aber et al., 2010; Marzolff and Poesen, 2009; Smith et al., 2009; Dandois and Ellis, 2010; Aber et al., 1999).

As stated earlier, kite aerial photography has been effectively utilized to monitor coastal geomorphology and its ecological implications. For example, at Oluvil in Ampara District of Srilanka, researchers adopted kite aerial photography. The aim was to study the terrain near the harbor and neighboring areas on the coast. Aspects such as coastal topography, vegeta-tion, its diversity and biomass (NDVI) were all monitored. The influence of sea level changes and construction of harbor on coastal geomorphology was studied. Kite aerial photography was useful in obtaining data such as NDVI, GI, Digital Elevation Model (DEM) of vegetation, high resolution maps of coast, etc. (Madurapperuma and Dyllesse, 2018; Madurapperuma et al., 2019). Dyllesse et al. (2012) have compared different platforms used for aerial imagery. They believe that for developing countries such as Srilanka the kite aerial photography is affordable and could be practiced. Kite aerial photography is a low-cost technique. At the same time, it offers high resolu-tion images.

The KAP has also been utilized to study the inter-tidal landscapes. This is in addition to aerial imagery of coastal land, its vegetation and growth (NDVI). Detailed digital data of landscapes and tides were obtained using infrared sensors (Mathew, 2000).

5.3   UTILIZATION OF KITES IN AGRICULTURAL CROP PRODUCTION

Kite aerial photography (KAP) is obviously a type of imagery adopting kites as the platforms for electro-optical sensors. The process involves using a kite with rigs to lift a visual or a multi-spectral camera. The cameras could be operated using remote control or could be pre-programmed (i.e., automatic shutter control). Such a set up helps to obtain aerial imagery of terrain, natural vegetation, crops, etc. There are indeed a wide range of sensors that could be carried on a rig attached to a kite. It ranges from simple visual camera, near infrared, infra-red, red-edge, thermal cameras and lidar. Currently, digital cameras with facility for rapid transfer of data and images are most popular. Most camera brands with 3 megapixel and above range are suitable for simple aerial photography. Kites flying at 150 to 200 ft. altitude are utilized (Kite Aerial Photography, 2017; Aber et al., 2010). Video cameras too could be lifted using appropriate kites. Previous experiences suggest that kite aerial photography (video) done at 150 ft. to 200 ft. above the terrain offers excellent imagery.

Several different types of kites are suitable for lifting the remote-controlled camera rigs. Usually, the rig has a point and shoot camera that weighs 14–16 ounces. The kites meant for aerial photography will have to lift a pay load (cameras) that weigh 1.5 lbs to 2.0 lbs (650 to 900 g). The camera rigs can be simple, consisting of a trigger mechanism with a disposable camera. It could also be made of complex apparatus using radio control and digital cameras. The camera can also be attached directly to the kite. However, it is usually secured to an adjustable rig. The camera rig is usually suspended from the kite line, at a distance from the kite. This distance reduces excessive movement being transmitted from the kite to the camera. It allows the kite to be flown into higher, stable air before the camera is attached. If possible, the camera is set to a high shutter speed to reduce motion blur. Cameras using internal image stabilization features can increase the number of sharp photos. In order to take photographs that are oriented correctly with the horizon, a suspension method is used to allow the rig to automatically level itself under the kite line. Basically, kite aerial photography involves attaching the cameras (sensors) to the kite itself or suspending it from the lines. Whatever be the case, cameras have to be tightly held at an altitude and with least disturbance. This is to ensure that aerial photography is sharp and without haze or blur. A simple method involves suspending a rigid material from the kite's line and attaching the cameras to it. Gravity helps in keeping the rig at a level irrespective of the angle of the kite's line. The rig and pendulum oscillate but cameras are held steady. Often specialists in kite aerial photography adopt the 'Picavet cross.' The Picavet suspension is named after the French inventor Pierre Picavet. The Picavet cross has a rigid wooden or aluminum cross upon which the cameras stay steady. The dimensions and shape of the Picavet cross may depend on the purpose. Overall, the Picavet suspension provides better stability to cameras (Kite Aerial Photography, 2019; Aber and Aber, 2019; Aber et al., 2010).

Shutter control for aerial imagery is an important aspect of kite aerial photography. Shutter control actually depends on the model of the cameras and the requirements. The camera's shutters could be operated using radio-controlled servos or using infra-red signals or sometimes wired connection that runs along the kite line. Usually, cameras are operated after the rig has taken the correct altitude envisaged and the lenses are focused at correct angles. Regular gimbals too could be utilized to obtain correct angles and aerial photographs of desired location, along with GPS tags. There are radio-controlled camera packages that are apt for kite aerial photography. The camera rig has pan and shutter release servos, battery pack and electronic time. Many of the brands are offered with set of accessories, so that, the cameras could be directly mounted on to the rig (Kite Aerial Photography, 2017; Aber and Aber, 2019, 2003).

A few common precautions listed during training a person in kite aerial photography are as follows:

a)   Adopt the correct brand and size of the kite;
b)   Match the kite model with rig and camera set up to be adopted for kite aerial photography;
c)   Use the correct recommended line;
d)   Use a reel while bringing down the kite to wrap up the line. This helps in avoiding knotting of the line;
e)   Select an appropriately sized spool or reel;
f)   Mark every 100 ft. length on the kite line. It helps in judging the distance that kite has travelled away;
g)   Check the wind speed. Note that wind is more turbulent at lower levels near the ground surface and calmer at higher altitudes;
h)   Do not fly the kite near airports, trees, tall buildings or places that harbor hazards (Scothaefner, 2019; Aber and Aber, 2019, 2003).

5.3.1 KITES TO STUDY AGRO-ENVIRONMENT, LAND RESOURCES, AND CROPS

Kites and their utility in meteorological measurements has been reviewed earlier (Robinson, 2003; Aber et al, 2019). Meteorological kites are in vogue since mid-18th century (i.e., 1749 AD.). Meteorological kites were adopted for the first time to measure temperature of ambient atmosphere at Glasgow University, Glasgow, Scotland. Weather readings like temperature, relative humidity, wind velocity, etc. were recorded, using a 'meteorograph' mounted to kite line. In practical field locations, kites with 'meteorographs' have been lofted till 1000 ft. from ground surface. However, a weather kite could be flown till 10,000 ft. altitude to get temperature data (Robinson, 2003). During the period until mid-20th century, kites that were stable in the atmosphere were utilized for generating weather data. During this period kite flying for meteorological data collection was vigorously pursued. However, with the advent of helium balloons, airplanes and aerostats the use of kites has receded. Several types of kites are popular among meteorologists. For example, larger versions of hexagonal Rokkakus, Box kites (double box), Eddy kites, Diamond and delta kites have all been useful in obtaining aerial imagery of a location and simultaneously in collection of useful meteoro-logical data. During 1900s, kite designs were made simpler by the French weather experts. They made lighter kites that were effective. Such kites possessed meteorographs with ability for self-recording of metrological data. Similarly, at the Blue Hill Observatory, Boston, USA, kites with camera rigs and meteorographs with facility to record and store data were produced. They collected data about, ambient air temperature, wind velocity, rela-tive humidity, altitude and barometric pressure were in vogue. Since, early 1900s, several nations/regions have adopted kites, to secure weather data. For example, nations such as France, Great Britain, Egypt, USA, Denmark, Sweden and India had all used kites, to secure weather data periodically (Robinson, 2003). At present, kites could be easily mounted with set of instruments that record a range of data and provide simultaneously the aerial imagery of the location, the clouds, atmospheric conditions, terrain and its features.

Reiche et al. (2012) believe that compared to several other methods of remote sensing using sensors and air-borne probes, kite-aided aerial survey and analysis of atmosphere might be easier to perform. They state that kite-aided systems are inexpensive, highly flexible and are suited well to assess remote areas. Kite-based aerial probes have been adopted in conducting atmospheric profiling. They have been regularly used to assess parameters such as temperature, relative humidity, wind speed and dust. Reiche et al. (2012) have actually assessed atmospheric parameters in an agrarian region of Inner Mongolia. Further, they aimed at measuring particulate matter in the atmosphere up to 100 m altitude from crop's canopy. They used parafoil kites to obtain atmospheric samples. The parafoil kite was large enough (4 m x 2 m parafoils). It could carry a payload of 6 kg instrumentation. The payload included environmental dust monitor, anemometer and GPS receiver. They also tried to standardize a kite-based profiling of particulate -matter concentration in the atmosphere, up to 100 m altitude from crop's canopy. No doubt, optimal wind velocity during kite-based sample collection is a requirement.

An interesting suggestion to use kite-aided atmospheric dust profiling emanates from the fact that, particulate dust is caused by windstorms, high speed wind and erosion of topsoil. Sahelian zone is often affected by dust storms that can affect seed emergence and seedling establishment. Even normal growth of seedlings may get affected due to interference with photosynthetic activity. Farmers and agricultural agencies could perhaps adopt kites, to assess atmospheric conditions. Then, arrive at appropriate predictions and adopt remedial measure. Kites, no doubt, are a cost-efficient proposition to Sahelian farmers and agricultural agencies (Reiche, 2012; Marzolff and Poessen, 2009).

One of the reports emanating from Great Plains of USA suggests that kite aerial photography of agricultural terrain and crops has been a routine since a decade. For students dealing with agricultural technology at Beatrice, in Nebraska, USA, KAP is taught with an aim to decipher the details of crops, the growth variations if any in cereal fields, insects, disease and weed infestation. They say traditional scouting is tedious compared to KAP. A set of kite, kite-line, sensors and radio view finder and shutter operation instrumentation costs US$ 1000 (2002 price levels) (Douglass, 2003). Most of the photographs obtained using visual, NIR, IR and LIDAR could be easily archived. It could be repeatedly viewed and decisions regarding crop inputs could be made accordingly. The kite utilized is 15–60 ft^2 in size. It had a kite-line of 300 lb. The kite was usually flown up to 1000 ft for aerial photography. At this altitude each photograph covers an area of 5–10 ha. It shows up the details of crop and its variations (Douglass, 2003). It is clear that KAP could be routinely adopted on cereal crops in the Great Plains region. It has already been done for the past decade in Nebraska, USA. Regular training about KAP and its usage in farming seems to be the requirement.

Benton (2005) states that, by using KAP, we can swiftly obtain images of large stretches of crops, the terrain, mountains, rivulets, farm vehicles, etc. For example, in the Salinas valley of California, cilantro crop was photographed along with land topography of the location, water resources and farm infrastructure. Such aerial photos help farmers to judge the crop, its nutritional and water status. It helps to manage the crops by channeling inputs appropriately. Fields with cilantro crop just planted or those with seedlings that are still immature and those ready for harvest could be easily detected by using KAP.

Low cost rokkaku kites have been adopted to obtain aerial photos of land, a small water body such as pond, hills, forest stands, agricultural crops, etc. (Khan et al., 2015). Kite aerial photos were shot with Canon A810 (visible light) and NIR images with model Pentax WG-10 (near-infrared and multi-spectral) placed on a Rokkaku kite. KAP was utilized to obtain images of wheat fields at various stages of growth, different moisture and nutritional stages (Janez and Sasa, 2017, 2019). Healthier plants reflected more of infrared photons. In addition to farm structures and wheat crop, the NIR and visible band width photographs were obtained from a Roman archaeological site. The historical site was imaged using NIR and multi-spectral sensor placed on the rokkaku kites. The site included a gothic church and burial grounds of Roman era at a place near Ljublijana, Slovenia. Images obtained using LIDAR gave greater details about the river, the wheat fields and their growth variations.

As stated earlier, kite aerial photography has been adopted to study the natural resources, vegetation and agricultural crops encountered in the Sahelian West Africa. Sahelian zone supports large number of farming units that experience resource constraints. Such farmers may find use of kites to obtain aerial photography of their fields much easier. It is cheaper than to invest in high cost drone aircraft derived images or satellite images. Kites may be ideal to study land resource, its topography and even fertility variations in the remote locations. Further, it has been stated that, kite aerial photography could be utilized to characterize spatial variations in the shrub vegetation, study the biodiversity of botanical species and study the undulations in sandy terrain. Researchers at the ICRISAT's Sahelian Center (Niamey, Niger) say that kite aerial photography may help in monitoring the major crops such as millet, sorghum cowpea and groundnuts grown by Sahelian farmers (ICRISAT, 2001).

5.3.2   KITES IN ASSESSING CROP GROWTH THROUGH AERIAL PHOTOGRAPHY

Basically, kite aerial photography (KAP) of crops is similar to that performed using small UAVs (air crafts), autonomous parafoils, blimps, or aerostats. It involves use of a series of sensors such as visual bandwidth (R, G and B), near-infra-red, infrared, and LIDAR. The sensors are placed at vantage points above the crop canopy by hanging the camera rig from the kite line. Since the kite line is not very high in altitude, the sensors get focused accurately and offer high resolution images of crop canopy, disease/pest attack if any, crops' water status, drought or flood effects, etc. The spectral data recorded in digital format can often be directly utilized in the autonomous precision farming vehicles. So, KAP is useful to those who opt to use precision farming methods. A kite can be lofted repeatedly and in quick intervals. Therefore, we can make a detailed study of the phenology of the crop at various stages of the crop, by imaging the crop in short intervals. Verhoeven(2012) state that, generally, a UAS such as small unmanned autonomous aircraft is preferred to study the phenomics of crops such as wheat, sorghum or other cereals. However, kites too have been utilized to study the phenology of crops in detail (Aber et al., 2002; Boike and Yashikawa, 2003; Fosset, 1994)

The crop growth and its productivity depend immensely on the dynamics of soil-N. Particularly, its availability to roots, its absorption and yield formation. Hence, several different methods have been designed to study the crop's N status and adopt remedial measures accordingly. Recent trends involve observation of crops using sensors placed on platforms such as small UAV aircrafts, helikites and kites. Kite aerial photography using a set of visual, infra-red and lidar is possible. The green leaf index (chlorophyll index) is directly correlated with crop's N status. Hence, it is a good indicator of crop's-N status. There are reports suggesting that kite aerial photography using light weight camera rigs is efficient in assessing sugarcane growth (NDVI) and N status (Leaf chlorophyll index). Let us consider this example. Sugarcane is an important crop in southern Louisiana. The crop has stringent requirements for soil-N. To study the soil-N variation and sugarcane crop's N status and its needs, researchers have evaluated the use of kite aerial photography. Particularly, to ascertain leaf-N status. McCollam (2016) reports that kite aerial photography as a method depends much on the wind conditions and stability of the rig and cameras. A stable camera offers high resolution images of sugarcane crop. Kite aerial photography is an ideal tool to study sugarcane growth and N status, if wind conditions permit. Major advantages of using KAP is that it is easy, requires minimum skills in kite flying and could offer high resolution spectral data about sugarcane. KAP is less costly for farmers to adopt. KAP could be a good bet to study the in-season fertilizer-N requirements of sugarcane crop.

5.3.3   KITES TO ASSESS CROP'S WATER STRESS INDEX (CWSI)

Kite aerial photography involves use of appropriate kite model, its correct size and apt cameras. When we intend to estimate water status of crops it is essential that KAP has facility for infrared photography. Also, timing and stage of the crop should be appropriate so that soil temperature corrections can be made easily. The kite infrared photography has been attempted successfully by Aber et al. (2009). The timing of infrared kite aerial photography is very important. Usually, it is preferable to time it when soil and crop temperatures could be deciphered as distinctly separate. Further, it has been reported that, kite aerial photography as standardized by Aber et al. (2001) involved two sets of rigs and cameras. One set was modified to hold a camera with facility for visual color photography. A second rig had a camera with facility for infra-red color photography. Such data derived from the infrared photography, particularly, crop canopy temperature is essential during assessment of crop's water status. Field maps showing variations in crop's water stress index (CWSI) helps farmers to use them (digital data) in the precision farming vehicles meant for irrigation.

Researchers at the University of Arkansas have evaluated a series genotypes of soybean, using kite aerial imagery. They have obtained infrared aerial imagery and thermal maps of soybean canopy, using thermal sensors (University of Arkansas, 2014). The basic understanding is that crop's canopy is cool when it has sufficient water in its tissue. The plant transpires greater amount of water and so it is cool. The canopy temperature is higher if water is scarce in the soybean tissue. Crop's water stress index and its variation (maps) could be easily obtained using kite aerial imagery. Researchers have found that some genotypes do not wilt quickly despite scarcity of water. The influence of drought is delayed because they conserve water in the stem and leaves. A few others have deeper roots. Therefore, they draw moisture from lower strata of soil. Under both situations, the kite aerial imagery will tell us that canopy is cooler and water status of soybean genotype is congenial. Further, it has been pointed out that data drawn about crop's canopy is less accurate, if infrared imagery is done at morning time from 9–11 hrs. It could get sufficiently confounded because soil and crop temperature readings may overlap. Correction for background soil reflection is needed. It is preferable to make kite-aided aerial infrared photography around 1400 hrs when soil and crop canopy temperature differ significantly. Researchers at the University of Arkansas have suggested that, kite aerial photography, which is not costly, but at the same rapid could be utilized, Particularly, to assess soybean genotype for drought tolerance (University of Arkansas, 2014).

Aber and Nagasako (2018) have reported the use of a color infra-red photography that is improvised compared to traditional infra-red sensors. For the past two decades they have adopted color-infra-red photography, using a digital camera such as SONY α 6000. The images obtained using blue, green, near-infra red band width appear blue, green and red in false color format, respectively. The vegetation appears orange in color. Crop's response to soil moisture variation could be deciphered and mapped accurately.

5.3.4   KITES IN PRECISION FARMING

Aerial imagery required for conducting Precision Agriculture (PA) procedures are derived, using different autonomous and semi-autonomous platforms. A few platforms experimentally evaluated for obtaining aerial photography at visual and infra-red bandwidth and digital data are kites (Aber et al., 2002; Aber and Aber, 2019) balloons (Amoroso and Arrowsmith, 2000); parafoils and unmanned aerial aircrafts (Krishna, 2015, 2018, 2019). Such data could be used during precision farming.

At present, literature about role of kites in precision farming is feeble, if any. That is because, there are very few attempts made to study crops using KAP. This technique has not been and utilized to obtain useful digital data required for precision farming. Also, attempts to obtain aerial imagery of land, its topography, soils, soil moisture levels and water resources and cropping systems are feeble. The KAP has not been used to any extent to demarcate 'management blocks' that are required during precision farming. Yet, the forecast is that soon as the KAP gets popular, there will be many agencies distributing aerial imagery and digital data obtained via KAP, to farmers and crop production companies.

Recent trends in precision farming suggest that with appropriate modifications of farm vehicles, inputs could be efficiently utilized. Precision farming principles could be applied to regions that support small or large farming units. Prompt use of aerial robots such as fixed -winged or copter drones may be costly. However, low cost methods such as kites could be adopted during precision farming. Kites provide aerial photography that pinpoint areas with topographical features, soil type, soil-N and water dearth. The KAP offers the crucial digital data depicting nutrient (leaf N) and waters status (CWSI) *in situ*. The ground station iPad with suitable programs can then offer accurate prescriptions in digital format and they could be used in the variable-rate applicators (Cook et al.,2013; Stafford, 2000).

5.3.5   KITES TO STUDY PESTS AND DISEASES AFFECTING CROP PLANTS

There are certain basic requirements to detect pest/disease affected patches of crop, using KAP. Ability to distinguish and decipher healthy and diseased/ insect attacked patches as different, using their spectral signatures is essential. Sometimes, it is easy to do so. Since, the affected regions of the crop's canopy clearly show up the crop loss. Sensors with visuals band width lenses are good enough to offer clear images of affected and healthy regions of the crop. However, we may also utilize multi-spectral cameras, infrared and red-edge band width cameras, to detect disease/insect attacked patches. The advantage of mapping disease/pest attacked zone using KAP is yet to be realized clearly by farmers. Prior to it, we need to be aware of the spectral signatures of different types of diseases/insect damage that affect crops grown in a region. Hence, there are not many examples where in KAP has been effectively adopted. This is a situation similar to other platforms and sensor combinations. There is need for data banks with spectral data about a healthy crop canopy or that affected by disease/pest. Yet, there are a couple of examples wherein KAP has been adopted to detect crop disease progression.

Kite aerial photography as a method to detect the spread of virus disease was examined some 15 years ago at the International Crops Research Institute for the Semi-arid Tropics-Sahelian Center in Niamey, Niger. Researchers adopted KAP to obtain digital data about the spread of soil-borne diseases of groundnut. They conducted the image analysis. It revealed suppression of crop's growth due to virus attack that could be deciphered as 'yellow patches.' Areas with virus tolerant good growth appeared as blue colored region. This way spatial variation of groundnut soil borne virus disease could be studied. Remedial measures could be adopted using the maps showing distribution of soil-borne virus disease (ICRISAT, 2001).

Standing crops in the field are damaged by several different detrimental factors such as natural disasters (e.g., soil erosion), insect pests, diseases, weeds that out compete crops, even animals that destroy crops by eating and trampling. Animals may often trespass into crop fields by damaging the fence or penetrating it. Then, they randomly eat and/or trample standing crops. For example, wild boar (*Sus scrofa*) affects several different crops. In this case, Diam et al. (2012) state that, kite aerial photography serves well by showing up spots that are affected by wild boar attacks. They used a shock proof Go Pro HD Hero 2 camera attached to kite's line. The images were georeferenced then processed, using Agisoft's Photoscan. They have concluded that kite aerial photography is a low-cost technique to detect damaged areas of crops in large fields.

5.3.6 MANAGEMENT OF BIRD DAMAGE IN AGRICULTURAL FARMS USING KITES

Bird damage to cereal grain crops grown in vast expanses or to horticultural fruit crops grown in plantations are common. It can be severe depending on the geographic location, the vastness of cropping belts, season, bird species and their population. Crop protection measures that reduce bird damage are available. They have to be adopted in a timely fashion and accurately. There are several methods of bird control that could be adopted singly or in a series.

A report emanating from Pennsylvania State Extension Services states that crop loss in horticultural zones due to birds is increasing cost of production to growers. Estimates of damage reported are 30% to 35% of small berry production, 7% of wine and table grapes, 13% of apples and pears, 16% of stone fruits, and 22% of the nut crops. This includes whole fruits being consumed by birds, fruit knocked off bushes or canes, and fruit made unsalable due to pecks, holes, slashes, etc. (Brittingham and Falker, 2010; Fraser et al., 1998; Ontario Ministry of Agriculture, 2006; Stetson, 2010). It is useful to know the species of birds that destroy the crop's produce. Also, we should obtain accurate data regarding the percentage loss that birds can afflict to the particular crop's produce. Such data helps in channeling the remedial measures appropriately. We can then try to improve the kite-aided bird control measures. We can adopt kite-aided bird control and integrate the procedure along with other methods (Plate 5.9).

Further, it has been pointed out that farmers often limit their bird control procedure to a single method, like a banger or propane gun, and leave it operating without checking its effectiveness. We may note that, birds exposed to a frequently repeated stimulus will get habituated quickly. Therefore, the bird control tool will lose its effectiveness regarding scaring them away.

Hence, the suggestion is to adopt a few different methods. Therefore, kite-aided bird control could be considered as yet another approach that needs to be effectively integrated to achieve bird control.

Fazluluhaque and Broom (1985) have evaluated use of traditional gas bangers to reduce damage to spring cabbage crop. They have compared the efficacy of gas bangers with kites flown at 300 ft. above the cabbage fields. They report that Wood pigeon somehow avoided flying into fields with kites flown above them. The protection continued for three months without the Wood pigeons getting used to kites. Generally, cabbage fields with kites suffered less damage compared to traditional gas bangers. Introduction of flying kites above the cabbage fields reduced the damage from 14.7% to 0.6%. It was suggested that, kites should be flown early in the morning to achieve good control over birds entering the fields.

Sorghum is a cereal grain crop cultivated worldwide for grains, fodder, alcohol and biofuel. In the tropics and semi-arid tropics, it is a predominant cereal. Bird damage to sorghum panicles at ripening stage is frequently observed. In some regions, birds are considered serious pests on sorghum. They say, bird damage on sorghum depends on the field location, its size, season, planting density and genotype. Bird scaring using gas bangers are common. However, recently, kites and helikites flown above sorghum fields seem to reduce damage to sorghum grains. Bird infestation is generally low in fields where kites are flown into the sky and scaring acoustics are used. A few suggestions indicate that innate resistance of panicles and grains to bird damage, say due to high phenolics, color, etc. plus acoustics could reduce grain damage perceptibly (Mofokeng and Shargie, 2016; also see Plate 5.9). In other words, it is preferable to use kites/helikites to scare birds. Plus, use other methods too to control the avian pests.

Tubaro (1999) has analyzed the effects of bioacoustics on bird species. It has been stated that a particular acoustic noise may attract a certain species of birds and scare a few others. Acoustics has to be adopted after careful trials in the fields with grain crops. Acoustic stimulus could be a repellent or an attractant. A certain degree of caution about the net effectiveness with regard to bird control is required. Plus, trial and error may be needed while adopting kite-aided bird control in crop land. There are a few other points to note. Bajoria et al. (2017) state that, in general, a frightening stimulus scares off Wood pigeons. However, it may also scare away or create an alarm for other non-target bird species.

In summary, kite-aided bird control is a feasible method. It is a low-cost agro-technique. It integrates with other methods of bird control. It could be most effective yet environmentally safe. Because, other methods do adopt harmful chemicals. *During kite-aided bird scaring, the birds are not sacrificed.* One more suggestion is that, we have to experimentally evaluate different methods of bird control. Integrated approach to achieve bird control should include kite-aided bird control (Plate 5.9).

PLATE 5.9   A Bird scarer kite.

Source: Robert Sutton, President Sutton Agricultural Enterprises Inc., Salinas, California, USA.

Note: Top: A small bird scarer kite attached to a pole in the field; Bottom: Large bird scarer kite used in farms. See https://www.youtube.com/watch?v=Rcu3g8BdzYI.

5.4   KITES TO STUDY AGRO-ECOLOGICAL ASPECTS

Kites could be utilized to study ecological aspects of natural and experimental vegetations. Let us consider a few examples where kite-aided aerial photography and spectral data has been useful in understanding ecological aspects of geographic location. Kite aerial photography has been suggested as a method to study the environmental changes in the permafrost areas. The kite-aided aerial photography reveals both spatial and temporal changes of permafrost and its influence on the sparse vegetation. The ecological implications of changes in permafrost on arctic fauna and surface vegetation (e.g., algae) could be studied, using kite-aided aerial photography (Boike and Yashikawa, 2003).

Next, studying geomorphological and eco-systematic aspects of mountainous terrain is not easy. Often it might be very difficult to negotiate the sharp gradient, plus, thick vegetation if it happens to be a highly vegetated zone. Obtaining aerial imagery that depicts sharp changes regarding vegetation, moisture, temperature and relative humidity based on altitude and ground features is also difficult. However, kite aerial photography conducted periodically all through the seasons might help us to study the mountainous zone, in greater detail. Aerial photography might offer a detailed yet overall perspective of the hilly tracts and their eco-systematic aspects. For example, Wundram and Loffler (2007) have reported about one such study of mountainous terrain in Norway and its geomorphology. The basic reasoning is that mountains and their ecosystems may have direct influence on global weather changes. They have adopted low cost kite aerial photography to assess terrain, its vegetation (NDVI), spectral properties of vegetation, water bodies and landforms. The low budget aerial photography using kites and sensors was efficient. The digital data from kites helped in making digital surface models (DSMs) of ecosystem.

Further, Wundram and Loffler (2007) state that, kite aerial photography conducted from 150 ft. above ground surface offers sharp and high-resolution images of mountainous vegetation, in Norway. The region could also be picturized using satellites such as IKONOS, Landsat, etc. Of course, the same zone could be studied using satellites, drone aircrafts, parafoils, zeppelins, aerostats, helikites. No doubt, kites too could be adopted to make low-altitude high-resolution imagery of the same terrain and study the eco-systematic changes in detail. The equipment used during kite aerial photography is both simple and cost efficient. Wundram and Loffler (2007) have adopted two different types of kites to act as platforms for aerial photography. They are *Flowform* without spars and 2 m^2 in size. Secondly, Dopero kites of 6 m^2

size were used under low wind conditions (Beutnagel, 1995). The camera adopted during the study was a Canon Powershot 570 with automatic timer for shutter activity. The kite aerial photography could depict changes in the central mountainous region of Norway. It could specifically show up changes that occur in vegetation based on altitude (See Wundram and Loffler, 2007; Beutnagel, 1995).

Primarily, kites are cost effective aerial platforms. Kites are easy for those who wish to study ecological aspects of a location. Kites are good platforms for light weight photographic equipment. Now, let us quote an example wherein KAP has been used consistently, to study the changes in the habitat. The Chesapeake area in Massachusetts (USA) is home to salt marshes. The salt marshes support a diverse flora and fauna that are spatially and temporally dynamic. Aerial photography using an unmanned aerial vehicle is possible. As stated above, kite-aided aerial photography has been used effectively to study marshes around Chesapeake Bay (CBEC, 2019; Plate 5.10). Reports suggest that kites with sensors are also useful to monitor atmospheric conditions. The kite aerial photography has helped in monitoring the behavior of invasive plant species in marshes. Kites could be lofted repeatedly and quickly to study the invasive grass species such as *Phragmites*. They have found that *Phragmites* grass covers, crowds and deters other species from proliferating in the marshes. At the same time, kites carrying atmospheric probes could record data above each spot from where ecological data has been accrued using kite-aided photography (CBEC, 2019; see Plate 5.10). We may note that, National Aeronautics and Space Agency (NASA) too utilizes tethered kites to assess terrain and atmospheric conditions (NASA 2019; Padgett, 2019; Plate 5.10). The sensors are grouped in an 'Aeropod' (sensors in a gimble) attached to kite's tether.

Now let us consider an example, where in kite aerial photography has been adopted to track the ecological consequences of releasing a biological control agent – a leaf eating beetle to reduce the population of a weed. In western United States of America, kite aerial photography has been utilized to monitor the ecological consequences of an invasive plant species such as Salt cedar (*Tamarix ramosissima*). This shrub is supposedly robust in growth. It spreads rapidly on the terrain. To suppress this shrub, researchers have spread a leaf eating beetle known as *Diorhabda elongate* (Aber et al., 2005; Eberts et al., 2003; DeLoach and Gould, 1998). The high-resolution imagery from sensors on kites was useful in marking areas where biological control of Salt cedar was effective. The digital data obtained was suitable for quantitative analysis of decrease in invasive shrub population (Aber et al., 2010).

PLATE 5.10 Kite with probe to conduct ecological investigations.

Left: A kite with atmospheric probes; Right: An aeropod invented at Goddard Space Flight Center facility. It can be fitted with sensors for aerial photography. It is aerodynamically stable so that images are sharp and accurate. This particular aeropod hangs from the tether that is used to fly the kite.

Note: The electro-optical and electro-chemical probes collect data relevant to assess atmospheric conditions such as spectral images, temperature, relative humidity, particulate matter, gaseous composition, emissions, etc.

Source: Chesapeake Bay Environmental Center, Massachusetts, USA, NASA AeroKAT Educational Program Washington D.C.

Aber et al. (2009) have made an extended study of marshes, agricultural zones and general vegetation that is traced in the nature conservancy area in Central Kansas, USA. The idea was to assess the influence of environmental changes on marsh species, water bodies and agricultural cropping zones. They have adopted both Delta (8.2 m^2) and Rokkaku types (3.3 m^2) of kites. They were mounted with a radio-controlled rig having Tetracam ADC cameras. The cameras provided excellent imagery. The water bodies were darker patches. Whereas pink-red and magenta depicted crops and vegetation. Gravely areas were brighter in appearance. The color infra-red camera offered images and digital data. Such data was suitable to make accurate analysis of effects of weather on vegetation.

Report by Currier (2014) makes it clear that KAP could be used to assess the influence of weather parameters, streams and coastal belt on the vegetation of small islands. For example, KAP offered a series of images of vegetation of Durai islands in the Indonesian archipelago. The spatial differences in geomorphology and their impact on native population of tortoises (*Chelonia mydas and Eretmochelys imbricata*) and forest vegetation could be studied, using KAP. The KAP offered imagery that could supplement studies aimed at understanding the changes in habitats of forest tree species, tortoises and coral reefs.

As stated earlier, kite aerial photography is a low-cost technology. It is a better proposition when terrain and atmospheric conditions are not congenial, such as in Antarctica. It seems kite aerial photography is efficient in obtaining imagery of coasts and intertidal zones, where in, penguin populations are traced. KAP was in fact utilized to monitor penguin resting grounds, their population changes and migratory trends (Aber and Babb, 2018; Bryson et al., 2013; Fraser et al., 1999). It shows that KAP could be adopted even in locations with harsh weather parameters. Of course, KAP needs a constant less turbulent wind breeze, to keep the kite without disturbance. Overall, kite aerial photography is a versatile method. It is suitable to study ecological aspects of diverse geographical locations. It offers detailed data both in visual photographic and digital format.

Andersen et al. (2014) have stated that KAP has potentially greater role to play in studying and understanding wetland ecosystem. They list the following accessories as the minimum requirement, if one wants to use KAP on a wetland location to obtain aerial photography and digital data. They have also mentioned a few good examples in support of KAP. They say that, kite-based aerial photography offered to the wetland scientific community is an inexpensive, user-friendly remote sensing technique. It has numerous applications in wetlands research. Using KAP, they have enumerated the wetland plant species in the Rio Bosque Wetland Park (Texas, USA). This wetland zone has 372 acres of land area that needs remedial treatment. It has invasive species that over crowds the native plant species and affect various other eco-systematic functions. They have classified the botanical species, both native ones and invasive types. Not only wetlands, they also quote a few other examples of KAP, where in, it has been evaluated in the Arctic tundra region. Specifically, KAP has been utilized to study the ponds in the tundra region. Particularly, the changes in the water level, the vegetation, margins of the pond, etc. Excellent digital elevation models (DEMs) of the tundra ponds and surroundings have been obtained using such inexpensive kite aerial photography.

5.5 GENERATION OF WIND ENERGY USING TURBINES ON KITES: YET ANOTHER SOURCE OF ENERGY TO FARMS

It has been stated that, wind power generated using traditional tall concrete towers and large turbines are generally costly to operate. Frequency of brake-downs and difficulty in tracing a constant breeze are problems. Instead, kites plus turbines flown at higher altitudes and large enough to generate sizable wind power seems a better alternative. Aerospace researchers at the Delft University (Netherlands) state that, kite power is efficient in terms of capital required and maintenance costs. Similarly, 'Kite Power Systems' – a wind power generating company situated in United Kingdom says sooner we may find kite-aided wind power generation as an efficient method in Scottish area. There are plans to demonstrate kite-aided wind energy systems that produce 3 megawatt and 500 kw larger versions (Arnolds, 2019).

A couple of companies in Oregon, USA have been working on tethered rigid kite that spins an electrical generator. It usually deploys at 300 to 500 ft. above farmers' field. Therefore, it flies at altitude well within the Federal Aviation Agency (USA) stipulation. The energy generated by the turbine could be utilized *in situ* by the farmer or channeled to grid and sold for others to use. A single kite flown turbine offers electrical energy enough to serve 5 mid-sized homes. Prototypes of the kite flown turbines are being tested in Oregon and other locations in USA (Mortenson, 2016).

High altitude wind turbines located on tethered aerostats has been discussed in the previous chapter. They are gaining in importance, since they generate electricity from a renewable source. The energy generating system is environmentally safe and cost efficient (see Altaeros Energies Inc., 2019; Glass, 2018; Sharma, 2019). All of the above arguments also hold good for turbines located on tethered kites flown to higher altitudes. Kites with light weight wind turbines can generate an extra amount of electricity in farms and other locations. Farmers with kite turbines may actually benefit significantly. They may satisfy their own electricity requirements to a certain extent, if they flew kite turbines. Blackman (2009) states that, it is possible to generate electricity by flying the kites with turbines to an altitude of 30,000 ft. above ground surface. Then, transmit the electrical energy through electric wires entwined with tethers. Researchers at Stanford Environmental Science group have stated that, at high altitudes (30,000 ft. from ground), energy that could be generated using high altitude kite turbines is equivalent to over 100 times the actual requirements of a medium sized farm. The high-altitude wind currents are a good repository of clean energy for kite turbines to harness. A few European weather research centers too have come to the same conclusion about utility of kite turbines. This is after analyzing data about wind current above the European plain's region (Blackman, 2009; Glass, 2018; Altaeros Energies Inc., 2019). Further, it has been pointed out that high altitude wind jet streams occur in many regions of the world. For example, above North America, Australia, Fareast including

South Korea, Japan, etc. Clearly, kites can be flown to harness this renewable energy for farmers to utilize. Wind currents are well distributed at high altitudes. So, one of the suggestions is to loft a few kites from a region and make a grid, so that, electricity is generated stored for regular use at any time.

Netherlands is famous for its efforts in harnessing wind energy for agricultural and domestic household purposes. Reports from Delft University of Technology (Netherlands) state that kites are being actively explored, tested modified and assessed for generating wind power. Kites and turbines of different dimensions are being evaluated and standardized as a source of renewable energy. They have adopted a kite 25 m^2 size and a light wind turbine to generate electricity. They used 20 KW hr generators that were connected via tethers. The flow of energy is dependent on generally uniform wind that is possible at high altitudes. It seems a 25m^2 kite with wind turbine could generate electricity enough for 40 households or 15–20 farms. Some of the key aspects that need greater attention and improvisation are automatic control using sophisticated electronics, synchronization with power grid and channeling the electrical power generated to a power grid for use by several farmers on the ground. Overall, researchers at Delft Technology University say that, the kite-aided high-altitude wind power generation has drawn good attention from rural and urban people. They may get popular in due course. The cost of generation of electricity using kite turbines is now 8 cents per unit. It could reduce to 2 cents, if the instruments for the technique are produced in high number (Owens, 2013).

Coleman et al. (2017) has pointed out that there is greater interest in cost effective parafoil kites owing to possibility of generating wind energy. A Parafoil kite could serve as a useful platform for a wind turbine that operates at 1000 m altitude above ground. It has also induced research that aims at modifying and improving the efficiency of parafoil kites, particularly, in terms of airborne wind energy generation (Archer, 2013; Archer and Caldiera, 2009 and Coleman et al., 2017).

One of the estimates states that, wind breeze available at altitudes above 500 m (i.e., 1640 ft.) are generally consistent and stronger. They are most suitable for wind power generation using kites that loft the suitable turbine. It seems wind power generated at that altitude can alone suffice the electrical energy requirements of the globe. In fact, it offers several folds more energy compared to yearly requirement of the world at present levels. However, at present only 4% of energy generated is derived using high latitude wind power. Reports suggest that wind power is gaining acceptance and turbines lofted to high altitudes are getting bigger in size. It needs a kite with thick tether to loft a wind turbine and conduct the energy generated to ground station. Since kites are low cost and need minimal sophistication and technical assistance, they are being preferred in many locations in Europe. A complete set of kite-aided wind turbine costs only 1 to 10% of the usually constructed tall concrete towers and wind turbine.

5.6   KITES IN FARMING: FUTURE COURSE

Kite aerial photography is a low-cost method that offers spectral data about the agricultural terrain and crops. It is a low technology method that could be easily adopted in any agrarian region of the world. It can be practiced by farmers with perhaps shortest duration of training. The essentials of obtaining the digital data and high-resolution visual photography of the land resources, its topography and crop's canopy are just similar to any method. For instance, it involves use of platforms such as satellites, small UAV (aircrafts), autonomous parafoils, microlights, blimps or aerostats. So, the agricultural agencies that process the imagery obtained using other methods can process KAP and digital data. KAP is amenable for repeated use. It can offer data that depicts both spatial and temporal changes of a crop's canopy. These above facts should make the KAP most sought after in different agrarian belts. However, kites cannot be used simultaneously, say, to spray farm chemicals, etc. like the sprayer drones (Copter UAVs). The digital data from KAP has to be introduced into precision sprayer drones or other precision vehicles (land vehicles) by the farmer. Yet, it is not a major constraint. Kites are easy to store and transport. It could be a major advantage, if compared with some large drone aircrafts, autonomous parafoils, microlights or blimps. Kites are sold almost everywhere all over the world and there are innumerable companies/distributors who deal with kites *per se*. Most immediate requirement is to manufacture kites meant exclusively for agricultural uses. Also, train as many farmers to use them effectively. Future should be exceedingly good for KAP in the farm world. KAP has the potential to replace other platforms and dominate the agricultural sky.

KEYWORDS

- **crop water stress index**
- **digital elevation models**
- **digital surface models**
- **kite aerial photography**

- **leaf area index**
- **normalized difference *vegetation* index**
- **precision agriculture**
- **unmanned aerial vehicle**

REFERENCES

Aber, J. S., & Aber, S. W., (2003). Applications of kite aerial photography: Property survey. *Transactions of the Kansas Academy of Science, 106,*107–119 https://www.researchgate.net/publication/232674573_Applications_of_kite_aerial_photography_Property_survey (accessed on 30 July 2020).

Aber, J. S., & Babb, T. I., (2018). The challenges of processing kite aerial photography imagery with modern photogrammetry techniques. *International Journal of Aviation, Aeronautics and Aerospace, 5*(2), 1–32. https://doi.org/10.15394/ijaaa.2018.1210 (accessed on 30 July 2020).

Aber, J. S., & Galazka, D., (2000). Potential of kite aerial photography for quaternary investigations in Poland. *Geological Quarterly, 44,* 33–38.

Aber, J. S., & Nagasako, T., (2018). Color-infrared kite aerial photography with a mirror less digital SLR camera. *Transactions of the Kansas Academy of Science, 121,*319–329. https://doi.org/10.1660/062.121.0401 (accessed on 30 July 2020).

Aber, J. S., Aaviksoo, K., Karofeld, E., & Aber, S. W., (2002). Patterns in Estonian bogs as depicted in color kite aerial photographs. *Suo (Mires and Peat), 53*(1), 1–15.

Aber, J. S., Aber, S. W., & Leffler, (2001). Challenge of Infra-red kite aerial photography. *Transactions of the Kansas Academy of Science, 104,* 18–27. https://www.jstor.org/stable/3628086?seq=1 (accessed on 30 July 2020).

Aber, J. S., Aber, S. W., & Pavri, F., (2002). Unmanned small format aerial photography from kites for acquiring large scale high resolution, multi-view angle imaging. Pecora 15/Land satellite information IV/ISPRS commission I/FIEOS2002. Conference Proceedings. https://pdfs.semanticscholar.org/eb56/c6119f51f57a9342422e4f6c67faa898541b.pdf (accessed on 30 July 2020).

Aber, J. S., Aber, S. W., Buster, L., Jensen, W. E., & Sleezer, R. L., (2009). Challenge of infra-red kite aerial photography: A digital update. *Transactions of the Kansas Academy of Science, 112,* 31–39.

Aber, J. S., Eberts, D., & Aber, S. W., (2005). Applications of kite aerial photography: Biocontrol of Salt Cedar (*Tamerix*) in the western United States. *Transactions of Kansas Academy of Science, 108,* 63–66 https://www.jstor.org/stable/3628207?seq=1 (accessed on 30 July 2020).

Aber, J. S., Marzolff, I., & Ries, J. B., (2010). *Small Format Aerial Photography: Principles, Techniques and Geoscience and Applications* (pp. 213–228). Elsevier, Amsterdam, The Netherlands.

Aber, J. S., Sobieski, R., Distler, D. A., & Nowak, M. C., (1999). Kite aerial photography for environmental site investigations in Kansas. *Transactions of the Kansas Academy of Science, 102,* 57–67.

Aber, S., & Aber, S. W., (2016). *Kite Aerial Photography Equipment* (pp. 1–15). http://www. geospectra.net/kite/equip/equip.htm/ (accessed on August 9th, 2020).

Aber, S., & Aber, S. W., (2019). *Kite Aerial Photography Consulting and Sales* (pp. 1–4). http://www.geospectra.net/kite/consult/consult.htm/ (accessed on August 9th, 2020).

Agrawal, A. K., (2015). *Wind Energy Kites* (pp. 1, 2). Down to Earth. https://www.downtoearth. org.in/news/wind-energy-kites-46982 (accessed on 30 July 2020).

Alevizos, E., (2012). *Low-Altitude Coastal Aerial Photogrammetry for High Resolution Seabed Imaging and Habitat Mapping of Shallow Areas* (pp. 1, 2). EDU General Assembly, Vienna, Austria. https://ui.adsabs.harvard.edu/abs/2012EGUGA..14..561A/abstract/ (accessed on August 9th, 2020).

Allsopp Helikites Ltd., (2019). *Aerial Photography* (pp. 1–7). Allsopp Helikite Ltd, Hampshire, England, United Kingdom, http://www.allsopphelikites.com/index. php?mod=page&id_pag=33 (accessed on 30 July 2020).

Altaeros Energies Inc., (2019). *Clean Energy* (pp. 1–3). Altaeros Energies, Somerville, Massachusetts, USA.

Altman, J., (2019). *Terror by Fire Heating up Middle East* (pp. 1–3). The Medialine. https:// themedialine.org/by-region/terror-by-fire-heating-up-middle-east/ (accessed on 30 July 2020).

American Kitefliers' Association, (2019). *History of Kites: Early Adventures* (pp. 1–7). AKA Cedar Ridge, California, USA. http://kite.org/education/history-of-kites/ (accessed on 30 July 2020).

Amoroso, L., & Arrowsmith, R., (2000). *Balloon Photography of Brush Fire Scars East of Carefree* (pp. 1–24). Arizona. Tempe, Arizona: Arizona State University, Department of Geological Sciences. Available at: http://activetectonics.asu.edu/Fires_and_ Floods/10_24_00_photos (accessed on 30 July 2020).

Andersen, C. G., Zesati, S. A. V., Lougheed, V., & Tweedie, C., (2014). *Kite-based Arial Photography (KAP) a Low Cost, Effective Tool for Wetland Research* (pp. 28–32). Wetland Science and Practice. https://www.researchgate.net/publication/269336116_Kite-based_Aerial_Photography_KAP_A_Low_Cost_Effective_Tool_for_Wetland_Research/ (accessed on 30 July 2020).

Anderson, W., (2002a). *Tale of Two Extremes* (pp. 1–3). Kite Aerial Photography E-Resources KAPER News http://www.davidhunt.me/SiteMap.html/ (accessed on August 9th, 2020).

Anderson, W., (2002b). *My First Experience with Kite Aerial* (pp. 1–3). Kite Aerial Photography E-Resources. http://www.davidhunt.me/articles/ARTICLE_020304_anderson.html KAPER (accessed on 30 July 2020).

Archer, C., & Caldeira, K., (2009). Global assessment of high-altitude wind power. Energies, 2,307–319.

Archer, C., (2013). An introduction to meteorology for airborne wind energy. In: Ahrens, U., Diehl, M., & Schmehl, R., (eds.), Airborne Wind Energy (pp. 81–94). Springer. Berlin, Germany.

Archibald, D., (1887). The story of the earth's atmosphere. Quoted In: *Kite Aerial Photography* (pp. 1–7,174). (Wikipedia, 2019). https://ipfs.io/ipfs/ QmXoypizjW3WknFiJnKLwHCnL72vedxjQkDDP1mXWo6uco/wiki/Kite_aerial_ photography.html (accessed on 30 July 2020).

Arnolds, R., (2019). *Will Kite Power Fly High as the Next Renewable Energy Solution* (pp. 1–3). Earth.com. https://www.earth.com/news/kite-power-renewable-energy/ (accessed on 30 July 2020).

Bajoria, A., Mahto, N. N., Bopanna, C. K., & Pant, R., (2017). Design of a tethered Aerostat System for Animal and Bird Hazard mitigation. *2017 First International Conference on Recent Advances in Aerospace Engineering (ICRAAE), 66*, 93–99. doi: 10.1109/icraae.2017.8297244.

Batchelor, P., (2019). *Delta Conyne* (pp. 1, 2). http://www.batchelor.nethttps://www.batchelors.net/kites-in-the-classroom/61-delta-conyne.html (accessed on 30 July 2020).

Benton, C. C., (2005). *Cilantro Fields, Salinas Valley California. Kite Aerial Photography* (pp. 1–3). http://kap.ced.berkeley.edu/gallery/gal090.html (accessed on 30 July 2020).

Benton, C. C., (2010). *Kite Aerial Photography-The First Kite Photographs* (pp. 1–3). http://kap.ced.berkeley.edu/background/history1.html (accessed on 30 July 2020).)

Best Breezes, (2009). *Ann Quemere Completes Atlantic Powered by Kites* (pp. 1–6) http://best-breezes.squarespace.com/journal/2006/8/23/anne-quemere-completes-atlantic-crossing-powered-by-kite.html (accessed on 30 July 2020).

Beutnagel, R., (1995). *Dopero* (pp. 1–12). https://www.kiteplans.org/planos/dopero/dopero_2.html/ (accessed on 30 July 2020).

Blackman, C., (2009). *Kites Flying in High Altitude Winds Could Provide Clean Electricity* (pp. 1–6). https://phys.org/news/2009–06-kites-high-altitude-electricity.html (accessed on 30 July 2020).

Boike, J., & Yashikawa, K., (2003). Mapping of periglacial geomorphology using kite/balloon aerial photography. *Permafrost and Periglacial Processes, 14*, 81–85. doi: 10.1002/ pp. 437.

Brittingham, M. C., & Falker, S. T., (2010). *Controlling Birds on Fruit Crops* (pp. 1–4). College of Agricultural Sciences Agricultural Research and Cooperative Extension, Penn State Extension. https://extension.psu.edu/controlling-birds-on-fruit-crops/ (accessed on August 9th, 2020).

Bryson, M., Johnson-Roberson, M., Murphy, R. J., & Bongiorno, D., (2013). Kite aerial photography for low-cost, ultra-high spatial resolution multi-spectral mapping of intertidal landscapes. *PLOS One, 8*(9). https://www.ncbi.nlm.nih.gov/pmc/articles/PMC3777947// (accessed on August 9th, 2020).

CBEC, (2019). *CBEC AeroKATS Program.* Chesapeake Bay Environmental Center, Massachusetts, USA. https://www.bayrestoration.org/aerokats/ (accessed on 30 July 2020).

Coleman, J., Ahmad, H., & Toal, D., (2017). Development of testing of a control system for the automatic flight of tethered parafoil. *Journal of Field Robotics, 34*, 519–538.

Cook, S. E., O'Brien, B. O., Corner, B. J., & Oberthur, T., (2013). Is precision agriculture irrelevant to developing nations? In: Stafford, J., & Werner, A., (eds.), *Precision Agriculture.* (pp. 432–447). Wageningen Academic Publishers, Wageningen, The Netherlands.

Crouch, T. D., (2010). *Hargrave Box Kites* (pp. 1, 2). Encyclopedia Britannica. https://www.britannica.com/topic/Hargrave-box-kite (accessed on 30 July 2020).

Currier, K., (2014). Mapping with strings attached: Kite aerial photography of Durai Island, Anambas Islands, Indonesia. *Journal of Maps, X*, x-X, dx. doi.org/10.1080/17445647.2014.925839.

Daim, A., Hauke, L., & Keuling, O., (2012). Field mapping of economic damages in agricultural crops caused by wild boar (*Sus scrofa*) with kite aerial photography (KAP) and GIS support. *9th International Symposium on Wild Boar and other Suids* (pp. 1–3) at Hannover, Germany. https://www.researchgate.net/publication/266326315_Field_mapping_of_economic_damages_in_agricultural_crops_caused_by_wild_boar_Sus_scrofa_with_kite_aerial_ photography_KAP_and_GIS_support doi: 10.13140/2.1.1853.3124 (accessed on 30 July 2020).

Dandois, J. P., & Ellis, E. L., (2010). High spatial resolution three-dimensional mapping of vegetation spectral dynamics using computer vision. *Remote Sensing of Environment, 136,*259–276.

Dellysse, J. E., Madurapperuma, P. D., & Kuruppuarachi, K. A. J. M., (2012). *Preliminary Study on Biomass Mapping Along the Coastal Zone of Hambantota Region* (pp. 497–500) Srilanka. The Open University of Srilanka, Colombo.

DeLoach, C. J., & Gould, J., (1998). Biological control of exotic, invading salt cedar (*Tamarix spp*) by the introduction of *Tamarix*-specific control insects from Eurasia. In: Aber, J. S., Eberts, D., & Aber, S. W., (eds.), *Applications of kite aerial photography: Biocontrol of Salt Cedar (Tamarix) in the Western United States. Transactions of Kansas Academy of Science, 108,* 63–66 http://issg.org/database/species/references.asp?si=72&fr=1&sts=&lang=TC/ (accessed on August 9th, 2020).

Didakites, (2019). *Dopero Kites.* http://www.didak.com/shop/en/store/kites/one-line-kites/ dopero (accessed on 30 July 2020).

Digital Earth Watch, (2017). *Getting images: Kites and Balloon Aerial Photography* (pp. 1–14). The Lawrence Hall of Science, Berkeley, USA.

Douglass, R., (2003). *Crop Scouting* (pp. 1–6). Southern Community college, Beatrice, Nebraska, USA. http://www.davidhunt.me/articles/ARTICLE_030222_cropscout_ douglass.html (accessed on 30 July 2020).

Duffy, J., & Anderson, K., (2016). *A 21ˢᵗ Century Renaissance of Kites as Platforms for Proximal Sensing* (pp. 1–8). https://www.researchgate.net/publication/300098890_A_21st-century_ renaissance_of_kites_as_platforms_for_proximal_sensing/ (accessed on August 9th, 2020).

Eberts, D., White, S., Broderick, S. M., Nelson, S., Wynn, S., & Wydoski, R., (2003). *Biological Control of Salt Cedar* (pp. 1–18) Peublo, Colorado. Technical Memorandum No: 8220–03–06. United States Bureau of Reclamation, Denver, Colorado, USA.

Eglash, R., (2018). *Gazan's Challenge Israelis High-Tech Defenses in the Flaming Kites* (pp. 1–39). Washington Post. https://www.washingtonpost.com/world/gazans-challenge-israels-high-tech-defenses-with-flaming-kites/2018/06/17/4c27eb16–6b2b-11e8-a335-c4503d041eaf_story.html?noredirect=on&utm_term=.324b851a6286/ (accessed on 30 July 2020).

Eldar, S., (2018). *Will Gaza Kites lead To War Israel* (pp. 1–3). Al-Monitor. https://www. al-monitor.com/pulse/originals/2018/06/israel-palestinians-gaza-strip-hamas-kites-idf-intifada.html/ (accessed on 30 July 2020).

Fastie, C., (2012). *Swamp NDVI* (pp. 1–3). http://fastie.net/salisbury-swamp/ (accessed on 30 July 2020).

Fazlulhaque, A., & Broom, D. M., (1985). Experiments comparing the use of kites and Gas Bangers to protect crops from Wood pigeon damage. *Agroecosystems and Environment, 12,*219–228 doi: 10.1016/0167–8809(85)90113–6.

Feurer, D., Planchon, O., El-Maaoui, M. A., & Simane, A. O., (2018). Using kite for 3D mapping of gullies at decimeter-resolution over several kilometers: A case study on the Kamech Catchment, Tunisia. *Natural Hazards and Earth System Science, 18,*167–1582.

Feurer, D., Planchon, O., El-Maaoui, M. A., Booussema, R., (2017). *Potential of Kite-Borne Photogrammetry for Decametric and Kilometric Square 3D Mapping: An Application for Automatic Detection.* doi: 5194/nhess-2017–60.

Fosset, R. C., (1994). *Aerial Photography by Kite* (p. 16). http://www.kapshop.com/p35/ Aerial-Photography-By-Kite/product_info.html (accessed on 30 July 2020).

Fraser, H. W., Fisher, H. K., & Frensch, I., (1998). *Bird Control on Grape and Tender Fruit Farms, Ontario Ministry of Agriculture, Food and Rural Affairs, Factsheet,* 98–105.

Fraser, W. R., Carlson, J. C., Duley, P. A., Holm, E. J., & Patterson, D. L., (1999). Using kite-based aerial photography for conducting a detailed penguin censuses in Antarctica. *Water Birds: The International Journal of Water bird Biology, 22*(3), 435–440. http://doi. org/10.2307/1522120 (accessed on 30 July 2020).

Glass, B., (2018). In: Steadler, T. (ed.), *Soaring 'Super Towers' Aim to Bring Mobile Broadband to Rural Areas* (pp. 1–7). ITU News https://news.itu.int/soaring-supertowers-aim-to-bring-mobile-broadband-to-rural-areas/ (accessed on 30 July 2020).

Helikite Hotspots Ltd., (2019). *Airborne Wifi* (pp. 1–3). https://www.helikite-hotspot.com/ about/ (accessed on 30 July 2020).

Hunt, D., (2019). *Delta-Conyne Kites* (pp. 1–3). KAPER Kite Aerial Photography E-Resource. http://www.davidhunt.me/basics/dc_R.html/ (accessed on August 9th, 2020).

ICRISAT, (2001). It's a bird! It is a plane. No, it's a super scientist. *ICRISAT Newsletter SATrends, 13*, 1, 2. http://www.icrisat.org/what-we-do/satrends/01dec/1.htm (accessed on 30 July 2020).

Janez, M., & Sasa, S., (2017). Seeing the Invisible (pp. 1–12). KAPJASA https://kapjasa. wixsite.com/kap-jasa/single-post/2019/06/17/Seeing-the-Invisible (accessed on 30 July 2020).

Janez, M., & Sasa, S., (2019). *The World Viewed from a Kite* (pp. 1–6). KAPJASA https:// kapjasa.wixsite.com/kap-jasa (accessed on 30 July 2020).

Johnston, I., (2016). *One of the World's First Kite Driven Power Stations Set to Open in Scotland* (pp. 1–4). https://www.independent.co.uk/environment/kite-power-station-scotland-wind-turbine-plant-electricity-a7348576.html (accessed on 30 July 2020).

Khan, A. F., Khurshid, K., Saleh, N., & Yousef, A. A., (2015). A low cost rokkaku kite set up for aerial photography. International archives of photogrammetry. *Remote Sensing and Spatial Information Sciences, XL-3/W2,* 103–108.

Kipened 10, (2019). Arthur Batut 1888. In: Earth from above. *International Conference on Kite Aerial Photography* (pp. 1–3). https://wokipi.com/Kapined/art-va.html/ (accessed on August 9th, 2020).

Kite Aerial Photography, (2017). *Cobra Kites* (pp. 1–4). http://www.cobrakite.com/kap.html/ (accessed on August 9th, 2020).

Kite Aerial Photography, (2019). *A Bit of History: Arthur Batut of Labraguire* (pp. 1, 2) France. http://kap.ced.berkeley.edu/background/history1.html (accessed on 30 July 2020).

Kitelines, (1980). World records in kiting: Questions, answers and challenges. *Kitelines, 3,* 31–35.

Klemas, V. V., (2015). Coastal and environmental remote sensing from unmanned aerial vehicles: An overview. *Journal of Coastal Research, 31,* 1260–1267. https://doi.org/10.2112/ JCOASTRES-D-15-00005.1 (accessed on 30 July 2020).

Krishna, K. R., (2015). *Push Button Agriculture: Robotics, Drones and Satellite guided Soil Fertility and Crop Management* (p. 499). Apple Academic Press Inc., Waretown, New Jersey, USA.

Krishna, K. R., (2018). *Agricultural Drones: A Peaceful Pursuit* (p. 456). Apple Academic Press Inc., Waretown, New Jersey, USA.

Krishna, K. R., (2019). *Unmanned Aerial Vehicle Systems in Crop Production: A Compendium* (p .710). Apple Academic Press Inc., Palm Bay, Florida, USA.

Lingard, S. J., (1995). Ram-air parachute design. *13ᵗʰ AIAA Aerodynamic Decelerator Systems Technology Conference* (pp. 1–51). Clearwater Beach FL. USA https://pdfs.semanticscholar.org/745e/11fceda2f5544b2e9559c0ee7d6e450d6b85.pdf/ (accessed on 30 July 2020).

Lorenz, R. D., & Scheidt, S. P., (2014). Compact and inexpensive kite apparatus for geomorphological field aerial photography with some remarks on apparatus. *Geographical Research Journal 3, 4,* 1–8. https://doi.org/10.1016/j.grj.2014.06.001 (accessed on 30 July 2020).

Lorenz, R. D., Jackson, B., & Barnes, J., (2010). Inexpensive time-lapse digital cameras for studying transient meteorological phenomena: Dust devils and playa floods. *Journal of Atmosphere and Ocean Technology, 27,*246–256.

Madurapperuma, B. D., Dellysse, J. E., Zahir, I. L. M., & Iyooba, A. L., (2019). Mapping shoreline vulnerabilities using kite aerial photographs at Oluvil harbor in Ampara. In: *Proceedings of the 7ᵗʰ International Symposium SEUSL* (pp. 197–204). http://ir.lib.seu.ac.lk/handle/123456789/3010 (accessed on 30 July 2020).

Madurapperuma, D. D., & Dyllesse, J. E., (2018). Coastal fringe habitat monitoring using kite aerial photography: A remote sensing-based case study. *Journal of Tropical Forestry and Environment, 8,* 25–35.

Marvin, C. F., (1897). *A Monograph on the Mechanics and Equilibrium of Kites* (p. 71). The University of Michigan and Weather Bureau W. B. No.122, Department of Agriculture, Washington, DC; USA.

Marzolff, I., & Poesen, J., (2009). The potential of 3D gully monitoring with GIS using high-resolution aerial photography and a digital photogrammetry system. *Geomorphology, 111,* 48–60.

Marzolff, I., (2014). The sky is the limit? 20 years of small aerial photography taken from UAS for monitoring geomorphological processes. Proceedings of European Geological Union General Assembly, Vienna, Austria. *Geophysical Research Abstracts, 16,*7005.

Marzolff, I., Ries, J. B., & Albert, K. D., (2003). Kite Aerial Photography for gully monitoring in Sahelian landscapes. *Proceedings of Second workshop of the EARSeL, Special Interest group on Remote Sensing for Developing Countries.* (pp. 1–8) Bonn, Germany, https://www.researchgate.net/publication/258407297_Kite_Aerial_Photography_for_Gully_Monitoring_in_Sahelian_Landscapes/ (accessed on 30 July 2020).

Mathew, S. A., (2000). *KAP for Multispectral Mapping of Intertidal Landscapes* (pp. 1–3). https://publiclab.org/notes/mathew/05–20–2014/plos-one-kap-for-multi-spectral-mapping-of-intertidal-landscapes/ (accessed on 30 July 2020).

McCollam, G., (2016). *Correlating Nitrogen Application Rates in Sugarcane with Low-Cost Normalized Difference Vegetation Index (NDVI)* (pp. 1–31). Ellendale Farms LLC. Louisiana https://projects.sare.org/sare_project/fs14–282// (accessed on 30 July 2020).

Mofokeng, M. A., & Shargie, N. G., (2016). Bird damage and control strategies in grain sorghum production. *International Journal of Agriculture and Environmental Research IJAAER, 2,*264–269.

Mortenson, E., (2016). *Oregon Firm Developing Airborne Wind Energy System* (pp. 1–5). https://www.capitalpress.com/state/oregon/oregon-firm-developing-airborne-wind-energy-system/article_4e1227e1–4a47–5261–907a-76008e16d9fd.html (accessed on 30 July 2020).

Mraffko, C., (2018). *Fire Kite from Gaza a Burning Issue in Israel* (pp. 1–3). https://news.yahoo.com/fire-kites-gaza-burning-issue-israel-005810119.html; _ylt=Awrx5R1uFCJdXGsAZYC7HAx.; _ylu=X3oDMTBybjhmNXEwBGNvbG8Dc2czB HBvcwM1BHZ0aWQDBHNlYwNzcg--/ (accessed on 30 July 2020).

Murray, J. C., Neal, M. J., & Labrosse, F., (2013). Development and deployment of an intelligent Kite Aerial Photography Platform (iKAPP) for site surveying and image acquisition. *J. Field Robot., 30,* 288–307.

NASA, (2016). *Kites: Principles of Flight: A Background* (pp. 1–65). Aeronautics Research Mission Directorate-NASA. National Aeronautics and Space Agency, Washington DC., USA http://www.nasa.gov/sites/default/files/atoms/files/kites_t4.pdf (accessed on 30 July 2020).

NASA, (2019). *NASA Kite Invasion Spurs Ever-Growing Education Program* (pp. 1–8). NASA Technology Transfer program, Washington D.C. https://spinoff.nasa.gov/Spinoff2018/ ee_8.html (accessed on 30 July 2020).

Needham, J., (1965). *Science and Civilization in China* (pp. 576–577). 4 Cambridge University Press, Cambridge, United Kingdom.

Oklahoma State University, (2019). *Kite Aerial Photography* (pp. 1–4). http://4h.okstate.edu/ projects/science-and-technology/kite-aerial-photography/rs/KiteAerialPhotography.pdf/ (accessed on 30 July 2020).

Ontario Ministry of Agriculture, (2006). *Food and Rural Affairs, Publication 360, Guide to Fruit Production: Bird Control.* Grape Growers of Ontario, http://www.grapegrowersofontario. com/1159/.

Owens, N., (2013). *Delft Professor Puts Kites High on List for Renewable Energy* (pp. 1–7). Phys.org. https://phys.org/news/2013-07-delft-professor-kites-high-renewable.html/ (accessed on August 9th, 2020).

Padgett, D., (2019). Globe atmospheric protocols and kite remote sensing: Innovations in exposing HBSU students to climate sciences. *Proceeding of American Meteorological Society-2019* (pp. 1, 2). https://ams.confex.com/ams/2019Annual/webprogram/ Paper354620.html (accessed on 30 July 2020).

Parish, T., (2007). *Power Kite: A Brief History* (pp. 1, 2). Ezine articles. http://EzineArticles. com/712347 (accessed on 30 July 2020).

Parish, T., (2017). *Modern Cody Kites* (pp. 1–5). Mybestkite.com http://www.my-best-kite. com/cody-kites.html (accessed on 30 July 2020).

Parish, T., (2019). *Some History and Personal Experience* (pp. 1–7). My Best Kites. https:// www.my-best-kite.com/eddy-kite.html (accessed on 30 July 2020).

Price, R., (1996). Measuring the kite characteristic. *American Kitefliers' Association Journal, the Aerial Eye, 2,* 4–22.

Reiche, M., Funk, R., Zhang, Z., Hoffmann, C., Li, Y., & Sommer, M., (2012). Using a parafoil kite for measurement of variations in particulate matter—a kite-based dust profiling approach. *Atmospheric and Climate Sciences* 2(01), 41–51. doi: 10.4236/acs.2012.210006.

Robinson, M., (2002). *Hargrave: The Noble Inventor* (pp. 1, 2). http://kitehistory.com/ Miscellaneous/Hargrave.htm/ (accessed on August 9th, 2020).

Robinson, M., (2003). *Meteorological Kites: Scientific Kites of the Industrial Resolution* (pp. 1–5). http://www.kitehistory.com/Miscellaneous/meteorological_kites.htm/ (accessed on August 9th, 2020).

Rose, M., (2014). *What is Parafoil Kite* (pp. 1–3). Great Canadian Kite Company. https:// www.greatcanadian company.com (accessed on 30 July 2020).

Sander, L., (2014). Kite aerial photography (KAP) as a tool for field teaching. *Journal of Geography in Higher Education, 38,* 425–430. https://www.tandfonline.com/doi/abs/10.10 80/03098265.2014.919443/ (accessed on August 9th, 2020).

Scothaefner, C., (2019). *Kite Aerial Photography: An Introduction* (pp. 1–3). http://scotthaefner.com/publications/archives/AsianPhotography.pdf (accessed on 30 July 2020).

Seang, T. P., & Mund, J. P., (2006). Geo-referenced balloon digital aerial photo technique: A low-cost high-resolution option for developing countries. In Proc. Map Asia 2006, 5th annual conf. on geographic information technology and application. *J. Agric. Eng. Res., 76*(3), 267–275.

Shannon, K., (1996). Rokkaku tips and techniques. American Kitefliers Association, New Mexico, USA. *The Aerial Eye, 2,* 12, 13.

Sharma, M., (2019). *Wind Power* (pp. 1, 2). Power News https://steelguru.com/power/could-high-flying-drones-power-your-home-one-day/539782 (accessed on 30 July 2020).

Sklaver, B. A., Manangan, A., Bullard, S., Svanberg, A., & Handzel, T., (2006). Rapid imagery through kite aerial photography in a complex humanitarian emergency. *International Journal of Remote Sensing, 22*(21, 22), 4709–4714.

Skydog Kites LLC, (2019). *Delta Kites* (pp. 1–5). https://www.gamesofberkeley.com/brands/skydog-kites-llc/ (accessed on August 9th, 2020).

Smith, M. J., Chandler, J. H., & Rose, J., (2009). High spatial resolution data acquisition for the geosciences: Kite aerial photography. *Earth Surface Processes and Landforms, 34,* 155–161. http://doi.org/10.1002/esp.1702 (accessed on 30 July 2020).

Spend Love, J., (1980). *Innovations: 14-D Box Kite* (Vol. 3, pp. 13–15). The Kitelines.

Stafford, J. V., (2000). *Implementing Precision Agriculture in the 21st Century* (pp. 733–737). Wageningen Academic Publishers, Wageningen, the Netherlands.

Stetson, D. I., (2010). *Birds on Tree Fruits and Vines, Pest Notes* (pp. 1–4) University of California State-wide Integrated Pest Management Program, Agriculture and Natural Resources. http://ipm.ucanr.edu/PMG/PESTNOTES/pn74152.html/ (accessed on August 9th, 2020).

Sutton, K., (1999). *From Chute to Kite: How the Classic Flow form Came to Be* (Vol. 13, pp. 39–41). Kitelines.

Tubaro, P. L., (1999). Bioacoustics applied to systematics, conservation and management of natural populations of birds. *Sociedad Espaniola de Etologia Etalogia, 7,* 19–32. https://www.sciencebase.gov/catalog/item/505772c9e4b01ad7e027c1aa/ (accessed on 30 July 2020).

University of Arkansas, (2014). *Kites, Balloons Collect Aerial Data for Soybean Drought Tolerance Research* (pp. 1, 2). Plant management network, University of Arkansas, Fayetteville, Arkansas, USA http://www.plantmanagementnetwork.org/pub/crop/news/2014/AerialData/ (accessed on 30 July 2020).

Verhoeven, G. J. J., (2012). Near Infrared aerial 'crop mark' archaeology: From its historical use to current digital implementation. *Journal of Archaeological Methods and Theory, 19,* 132–160.

Vericat, D., Brasington, J., Weaton, J., & Cowie, M., (2008). *Accuracy Assessment of Aerial Photographs Acquired Using Lighter-Than-air Blimps: Low-Cost Tools for Mapping River Corridors* (pp. 985–1000). https://doi.org/10.1002/rra.1198 (accessed on 30 July 2020).

Wigmore, O., & Mark, B. G., (2017). High-altitude kite mapping: Elevation of the kite aerial photography (KAP) and structure from motion digital elevation model in the Peruvian Andes. *International Journal of Remote Sensing 39,* 1–21 doi: 10.1080/01431161.2017.1387312.

Wikipedia (2018). *Foil Kites* (pp. 1–8). https://en.wikipedia.org/wiki/Foil_kite (accessed on 30 July 2020).

Wikipedia, (2017). *Kite Types*. https://en.wikipedia.org/wiki/Kite_types (accessed on 30 July 2020).

Wikipedia, (2019a). *Kite Applications* (pp. 1–14). https://en.wikipedia.org/wiki/Kite_applications (accessed on 30 July 2020).

Wikipedia, (2019b). *Kites* (pp. 1–17). https://en.wikipedia.org/wiki/Kite. (accessed on 30 July 2020).

Wundram, D., & Loffler, J., (2007). Kite Aerial Photography in high mountain ecosystem research. *Writings of Geography and Spatial Research 43*, 15–22.

Zimbelman, J. R., & Scheidt, S. P., (2012). *Investigation of Reversing Sand Dunes at the Bruneau Dune, Idaho as Analogues for Features on Mars* (pp. 1, 2). Paper presented at Fall meeting of AGU American Geophysical Union, Fall Meeting 2012, San Francisco, California. Abstract ID: P21D-1870, http://adsabs.harvard.edu/abs/2012AGUFM.P21D1870Z/ (accessed on 30 July 2020).

Index

9 781774 637623